The International Law on Climate Change

Global climate change is a topic of continuously growing interest. As more international treaties come into force, media coverage has increased and many universities are now starting to conduct courses specifically on climate change laws and policies. This textbook provides a survey of the international law on climate change, explaining how significant international agreements have sought to promote compliance with general norms of international law. Benoit Mayer provides an account of the rules agreed upon through lengthy negotiations under the United Nations Framework Convention on Climate Change (UNFCCC) and multiple other forums on mitigation, geoengineering, adaptation, loss and damage, and international support. *The International Law on Climate Change* is suitable for undergraduate and graduate students studying climate, environmental or international law. It is supported by a suite of online resources, available at www.internationalclimatelaw.com, featuring regularly updated lists of complementary materials, weblinks and regular updates for each chapter.

Benoit Mayer is an assistant professor in the Faculty of Law at The Chinese University of Hong Kong, where he teaches international law and climate law. His research on the international law on climate change has been published in leading journals including the *European Journal of International Law*, the *Asian Journal of International Law*, *Transnational Environmental Law*, *Climate Law* and *Climatic Change*.

The International Law on Climate Change

Benoit Mayer
The Chinese University of Hong Kong

CAMBRIDGE
UNIVERSITY PRESS

CAMBRIDGE
UNIVERSITY PRESS

University Printing House, Cambridge CB2 8BS, United Kingdom

One Liberty Plaza, 20th Floor, New York, NY 10006, USA

477 Williamstown Road, Port Melbourne, VIC 3207, Australia

314–321, 3rd Floor, Plot 3, Splendor Forum, Jasola District Centre, New Delhi – 110025, India

79 Anson Road, #06-04/06, Singapore 079906

Cambridge University Press is part of the University of Cambridge.

It furthers the University's mission by disseminating knowledge in the pursuit of education, learning, and research at the highest international levels of excellence.

www.cambridge.org
Information on this title: www.cambridge.org/9781108419871
DOI: 10.1017/9781108304368

© Cambridge University Press 2018

First published 2018

A catalogue record for this publication is available from the British Library.

Library of Congress Cataloging-in-Publication Data
Names: Mayer, Benoit, author.
Title: The international law on climate change / Benoit Mayer.
Description: New York: Cambridge University Press, 2018.
Identifiers: LCCN 2018000611 | ISBN 9781108419871 (hardback)
Subjects: LCSH: Climatic changes – Law and legislation. | Global warming – Law and legislation. | Environmental law, International. |
International law. | BISAC: LAW / International.
Classification: LCC K3585.5.M39 2018 | DDC 344.04/6342–dc23
LC record available at https://lccn.loc.gov/2018000611

ISBN 978-1-108-41987-1 Hardback
ISBN 978-1-108-41229-2 Paperback

Contents

Figures

Tables

Preface

Climate change is one of the greatest concerns of our time. For more than a quarter of a century, efforts have been made to mobilize international law as a tool to tackle climate change. Through the outcomes of protracted international negotiations and extensive doctrinal research, a new field of study has gradually emerged in international law. The international law on climate change is a system of State obligations to tackle climate change. It seeks to protect not only the sovereign rights of every State, but also the effective enjoyment of human rights, the interests of future generations and humankind as a whole, as well as other forms of life on Earth. The task is a formidable one: an attempt at altering the way we are changing our world. Some of the most complex negotiations ever undertaken have only touched the surface of the problem. The challenges are daunting, but the stakes are high and failure is not an option. It is hardly an overstatement that the fate of humankind depends on the international law on climate change.

As a field of study, the international law on climate change has largely focused on current developments, at the expense of laying the foundations of a new discipline. Toward the end of each year, the representatives of virtually every State meet for a session of the Conference of the Parties to the 1992 UN Framework Convention on Climate Change (UNFCCC). Agreements are regularly concluded, like, most recently, the Paris Agreement of 2015. This ritual event, heavily reported by the international media, set the cadence in the field. The need to report on the latest outcomes of international negotiations and to analyze them has impeded the conduct of a more systematic and comprehensive inquiry into the international law on climate change.

Rules agreed by States through dedicated negotiations parts of a joint effort to address climate change – the UNFCCC regime – are an important component of the international law on climate change, but the latter is not confined to the former. Another component of the international law on climate change relates to relevant rules adopted in diverse other international regimes, for instance, under the Convention on International Civil Aviation of 1944 or the Montreal Protocol on Substances that Deplete the Ozone Layer of 1987. Overall, the general and abstract norms and principles that States have accepted as general international law, such as the no-harm principle and the law on State responsibility, form the last component of the international law on climate change.

These components differ in terms of their origin and significance. The rules dedicated to tackle climate change, adopted through international negotiations, are likely to reflect the interests of powerful nations, often through a complacent approach to their excessive levels of greenhouse gas emissions. By contrast, the norms reflected in the general practice of States that they have historically accepted as law impose fairer principles and, typically, far more demanding obligations on States, including on the most powerful ones. As explained throughout this book, States have never agreed, either explicitly or tacitly, that international climate agreements would set aside the principles of general international law. Instead, the UNFCCC regime should be approached as a regime which seeks to promote compliance with general international law. It does so, in particular, by defining collective objectives and national commitments over specified periods.

This book seeks to provide what could be the first comprehensive account of the international law on climate change as a discipline. As such, although this book contains a detailed presentation of the rules agreed upon by States within the UNFCCC regime and beyond, it does not stop there. There is little doubt that, if an international court or tribunal were to determine the obligations of States in relation to climate change, it would recognize obligations rooted in general international law. Therefore, this book also contains a detailed analysis of the relevant norms of general international law. Thus, it seeks to bring all the components of the international law on climate change together in a unique, comprehensive and coherent account of what international law has to say about climate change.

As a full-fledged discipline rather than a mere field of study, the international law on climate change would not only analyze ongoing developments, but would also contribute more actively to these developments. It would situate international climate agreements in the broader context of its two other components, with due consideration in particular to the relevant norms of general international law. A more consistent understanding of the international law on climate change could facilitate international negotiations by determining a benchmark for reasonable expectations of a fair and equitable outcome. It would help municipal courts decide cases based not just on the commitments specifically agreed upon by national governments, but also, beyond, on customary international law. One day, an international court or tribunal could engage with this discipline in a contentious or advisory case. Beyond climate change, this would help fulfill the promise of international law of promoting justice in international relations.

This book has been designed to be readily accessible by undergraduate and postgraduate students in law with or without any background in international or environmental law. If used as a teaching material, it should offer a number of opportunities for students to engage in debates. Thus, while this book could help train professionals and researchers in the field, it could also serve as a meaningful case study in a field of law to foster a critical intellectual engagement with the law more generally. In addition, it is hoped that this first attempt at a comprehensive presentation of the international law on climate change as a whole may also be useful to legal scholars or anyone else interested in how international law addresses one of the defining problems of our time.

Benoit Mayer

Acknowledgments

This book benefited from the insights of many in the field and beyond. I had the chance to test some of my ideas through diverse interactions with, among many others, Antony Anghie, Eyal Benevisci, Ben Boer, Anatole Boute, Chen Yifeng, Simon Chesterman, Marie-Claire Cordonier Segger, Christel Cournil, James Crawford, François Crépeau, Myriam Feinberg, Stephen Gardiner, Markus Gehring, Gregory Gordon, Hsu Yao-Ming, Richard Janda, Koh Kheng Lian, Douglas Kysar, Jolene Lin, Lye Lin Heng, Frédéric Mégret, Shinya Murase, Jarna Petman, Song Ying, M. Sornarajah, Usha Tandon, Qin Tianbao, Mikko Rajavuori, Surabhi Ranganathan, Christina Voigt, Shan Wenhua, Yee Sienho, Prabhakar Singh, Wang Canfa, Zhang Hao, Alexander Zahar and Zhao Yuhong. I also benefited immensely from comments received from several anonymous peer-reviewers invited by Cambridge University Press and from peer-reviewers commenting on previous publications.

This book builds on my lecture notes developed while teaching a course on "International Climate Change Law and Politics" in the International Law Institute at the University of Wuhan, China. Support was provided first by the University of Wuhan, then by the Chinese University of Hong Kong (CUHK). Yao Junqian in Wuhan, then Winnie Cheung, Athena Kong and Sean O'Rilley in the Chinese University of Hong Kong provided precious assistance. Financial assistance was received from the Faculty Funding Support for Teaching Development and Research-Related Activities in CUHK.

The book could not come to reality without the help of commissioning editor Joe Ng, content manager Thomas Haynes, project manager Nicola Howcroft and copy-editor Jon Lloyd, to whom I would like to express my profound gratitude.

Abbreviations and Acronyms

AAU	Assigned Amount Unit
AOSIS	Alliance of Small Island States
ATS	Australian Treaty Series
AWG-DP	Ad Hoc Working Group on the Durban Platform for Enhanced Action
AWG-KP	Ad Hoc Working Group on Further Commitments for Annex I Parties under the Kyoto Protocol
AWG-LCA	Ad Hoc Working Group on Long-Term Cooperative Action under the Convention
AWG-PA	Ad Hoc Working Group on the Paris Agreement
BECCS	Bioenergy with Carbon Capture and Storage
BINGOs	Business and Industry NGOs
C2ES	Center for Climate and Energy Solutions
CAIT	Climate Analysis Indicators Tool
CBD	United Nations Convention on Biological Diversity
CCS	Carbon Capture and Storage
CDM	Clean Development Mechanism
CDP	Carbon Disclosure Project
CER	Certified Emission Reduction
CESCR	Committee on Economic, Social and Cultural Rights
CFCs	Chlorofluorocarbons
CIFOR	Center for International Forestry Research
CMA	Conference of the Parties Serving as the Meeting of the Parties to the Paris Agreement
CMP	Conference of the Parties Serving as the Meeting of the Parties to the Kyoto Protocol
COP	Conference of the Parties

CORSIA	Carbon Offsetting and Reduction Scheme for International Aviation
CPI	Climate Policy Initiative
DSB	Dispute Settlement Body
ECJ	European Court of Justice
ECtHR	European Court of Human Rights
EECC	Eritreat-Ethiopia Claims Commission
EEDI	Energy Efficiency Design Index
EIA	Environmental Impact Assessment
ENGOs	Environmental NGOs
EPA	Environmental Protection Agency
ERT	Expert Review Team
ERU	Emission Reduction Unit
ETS	Emission Trading Scheme
EU	European Union
GATT	General Agreement on Tariffs and Trade
GCF	Green Climate Fund
GDP	Gross Domestic Product
GEF	Global Environment Facility
GhG	Greenhouse Gas
HBFCs	Hydrobromofluorocarbons
HCFCs	Halomethane
HFCs	Hydrofluorocarbons
HRC	Human Rights Council
IACrHR	Inter-American Court of Human Rights
IAR	International Assessment and Review
ICA	International Consultation and Analysis
ICAO	International Civil Aviation Organization
ICJ	International Court of Justice
IEA	International Energy Agency
IGES	Institute for Global Environmental Strategies
IISD	International Institute for Sustainable Development
ILA	International Law Association
ILC	International Law Commission
ILM	International Legal Materials
IMO	International Maritime Organization

INC/FCCC Intergovernmental Negotiating Committee for a Framework Convention on Climate Change

INDC Intended Nationally Determined Contribution

IPCC Intergovernmental Panel on Climate Change

IPOs Indigenous Peoples Organizations

IRENA International Renewable Energy Agency

ISO International Standardization Organization

ITLOS International Tribunal for the Law of the Sea

JI/KP Joint Implementation under the Kyoto Protocol

JI/UNFCCC Joint Implementation under the United Nations Framework Convention on Climate Change

LGMA Local Government and Municipal Authorities

LULUCF Land Use, Land-Use Change and Forestry

MARPOL International Convention for the Prevention of Pollution from Ships

MEF Major Economies Forum on Energy and Climate

MRV Measurement, reporting and verification

NAZCA Non-State Actor Zone for Climate Action

NDC Nationally Determined Contribution

NGO Non-Governmental Organization

OCHA Office for the Coordination of Humanitarian Affairs

OECD Organization for Economic Co-operation and Development

OHCHR Office of the High Commissioner for Human Rights

OSPAR Convention Convention for the Protection of the Marine Environment of the North-East Atlantic

PCIJ Permanent Court of International Justice

PFCs Perfluorocarbons

QELRC Quantified Emission Limitation or Reduction Commitment

RCPs Representative Concentration Pathways

REDD+ Reducing Emissions from Deforestation and Forest Degradation in Developing Countries

REN21 Renewable Energy Policy Network for the 21st Century

RINGOs Research and Independent NGOs

RMU Removal Unit

SBI Subsidiary Body for Implementation

SBTA Subsidiary Body for Scientific and Technological Advice

SEEMP Ship Energy Efficiency Management Plan

SEI Stockholm Environment Institute

SIDS Small Island Developing States

SRFC Sub-Regional Fisheries Commission

SUV Sports Utility Vehicle

tCO2eq Tonne of carbon dioxide equivalent

TUNGOs Trade Union NGOs

UNCED United Nations Conference on Environment and Development

UNCHD United Nations Conference on the Human Environment

UNCLOS United Nations Convention on the Law of the Sea

UNDP United Nations Development Programme

UNEP United Nations Environment Programme

UNESCO United Nations Educational, Scientific and Cultural Organization

UNFCCC United Nations Framework Convention on Climate Change

UNHCR United Nations High Commissioner for Refugees

UNRIAA United Nations Reports of International Arbitral Awards

UNTS United Nations Treaty Series

UNU-EHS United Nations University Institute for Environment and Human Security

USSR Union of Soviet Socialist Republics

WCED World Commission on Environment and Development

WIM Warsaw International Mechanism for loss and damage associated with climate change impacts

WMO World Meteorological Organization

WRI World Resources Institute

WTO World Trade Organization

WWF World Wide Fund for Nature

YOUNGOs Youth NGOs

Table of Authorities

TREATIES

1940s

Convention on International Civil Aviation, December 7, 1944, 15 *UNTS* 295.
Charter of the United Nations, June 26, 1945, 1 *UNTS* XVI.
Statute of the International Court of Justice, June 26, 1945, 3 *Bevans* 1179.
General Agreement on Tariffs and Trade, October 30, 1947, 55 *UNTS* 194.
Convention on the International Maritime Organization, March 6, 1948, 289 *UNTS* 48.

1950s

Convention for the Protection of Human Rights and Fundamental Freedoms, November 9, 1950, 213 *UNTS* 222.
Convention Relating to the Status of Refugees, July 28, 1951, 189 *UNTS* 150, as modified by the Protocol relating to the Status of Refugees, January 31, 1967, 606 *UNTS* 267.
Agreement between the United Nations and the World Meteorological Organization, Approved by the General Assembly of the United Nations, December 20, 1951, 123 *UNTS* II-415.

1960s

International Convention on the Elimination of All Forms of Racial Discrimination, December 21, 1965, 660 *UNTS* 195.
International Covenant on Economic, Social and Cultural Rights, December 16, 1966, 993 *UNTS* 3.
International Covenant on Civil and Political Rights, December 16, 1966, 999 *UNTS* 171.
Protocol Relating to the Status of Refugees, January 31, 1967: see Convention Relating to the Status of Refugees, July 28, 1951.
Vienna Convention on the Law of Treaties, May 23, 1969, 1155 *UNTS* 331.
American Convention on human rights ("Pact of San Jose"), November 22, 1969, 1144 *UNTS* 123.

1970s

Convention on Wetlands of International Importance Especially as Waterfowl Habitat, February 2, 1971, 996 *UNTS* 246.
Convention for the Protection of the World Cultural and Natural Heritage, November 16, 1972, 1037 *UNTS* 151.
London Convention on the Prevention of Marine Pollution by Dumping of Wastes and Other Matter, December 29, 1972, 1046 *UNTS* 120.
Protocol Relating to the 1973 International Convention for the Prevention of Pollution from Ships: see International Convention for the Prevention of Pollution from Ships.

Convention on the Prohibition of Military or Any Other Hostile Use of Environmental Modification Techniques, December 10, 1976, 1108 *UNTS* 151.

International Convention for the Prevention of Pollution from Ships, November 2, 1973, 1340 *UNTS* 184, revised by the 1978 Protocol Relating to the 1973 International Convention for the Prevention of Pollution from Ships, February 17, 1978, 1340 *UNTS* 61.

Agreement on Technical Barriers to Trade, April 12, 1979, 1868 *UNTS* 120.

Convention on the Conservation of Migratory Species of Wild Animals, June 23, 1979, 1651 *UNTS* 333.

Convention on the Elimination of All Forms of Discrimination against Women, December 18, 1979, 1249 *UNTS* 13.

1980s

African Charter on Human and People's Rights ("Banjul Charter"), June 27, 1981, 1520 *UNTS* 217.

United Nations Convention on the Law of the Sea, December 10, 1982, 1833 *UNTS* 3.

Convention against Torture and Other Cruel, Inhuman or Degrading Treatment or Punishment, December 10, 1984, 1465 *UNTS* 85.

Vienna Convention for the Protection of the Ozone Layer, March 22, 1985, 1513 *UNTS* 293.

Montreal Protocol on Substances that Deplete the Ozone Layer, September 16, 1987, 1522 *UNTS* 3.

International Convention on the Rights of the Child, November 20, 1989, 1577 *UNTS* 3.

1990s

International Convention on the Rights of all Migrant Workers and Members of their Families, December 18, 1990, 2220 *UNTS* 3.

United Nations Framework Convention on Climate Change, May 9, 1992, 1771 *UNTS* 107.

United Nations Convention on Biological Diversity, June 5, 1992, 1760 *UNTS* 79.

Convention for the Protection of the Marine Environment of the North-East Atlantic, September 22, 1992, 2354 *UNTS* 67.

Instrument for the Establishment of the Restructured Global Environmental Facility, March 14–16, 1994, 33 *ILM* 1283.

General Agreement on Trade-Related Aspects of Intellectual Property, April 15, 1994, 1869 *UNTS* 299.

United Nations Convention to Combat Desertification in those Countries Experiencing Serious Drought and/or Desertification, Particularly in Africa, October 14, 1994, 1954 *UNTS* 3.

Protocol to the Convention on the Prevention of Marine Pollution by Dumping of Wastes and Other Matter, November 7, 1996, 1046 *UNTS* 138.

Kyoto Protocol to the United Nations Framework Convention on Climate Change, December 11, 1997, 2303 *UNTS* 162.

Rome Status of the International Criminal Court, July 17, 1998, 2187 *UNTS* 90.

2000s

Stockholm Convention on Persistent Organic Pollutants, May 22, 2001, 2256 *UNTS* 119.

Convention on the Protection of the Underwater Cultural Heritage, November 2, 2001, 41 *ILM* 37.

International Convention on the Rights of Persons with Disabilities, December 13, 2006, 2515 *UNTS* 3.

International Convention for the Protection of All Persons from Enforced Disappearance, December 20, 2006, 2716 *UNTS* 3.

2010s

Doha Amendment to the Kyoto Protocol, December 8, 2012, in the annex of decision 1/CMP.8.

Minamata Convention on Mercury, October 10, 2013, 55 *ILM* 582.

Paris Agreement, December 12, 2015, in the annex of decision 1/CP.21.

Amendment to the Montreal Protocol on Substances that Deplete the Ozone Layer, October 15, 2016, (2017) 56 *ILM* 196.

DECISIONS ADOPTED BY THE PARTIES TO INTERNATIONAL CONVENTIONS

UN General Assembly (UN Charter of June 26, 1945)

Resolution 217 A, "Universal Declaration of Human Rights" (December 10, 1948).

Resolution 1721 (XVI) C, "International co-operation in the peaceful uses of outer space" (December 20, 1961).

Resolution 2626 (XXV), "International Development Strategy for the Second United Nations Development Decade" (October 24, 1970).

Resolution 3202 (S-VI), "Declaration on the Establishment of a New International Economic Order" (May 1, 1974).

Resolution 41/128, "Declaration on the Right to Development" (December 4, 1986).

Resolution 44/207, "Protection of global climate for present and future generations of mankind" (December 22, 1989).

Resolution 44/228, "United Nations Conference on Environment and Development" (December 22, 1989).

Resolution 45/212, "Protection of global climate for present and future generations of mankind" (December 21, 1990).

Resolution 55/2, "United Nations Millennium Declaration" (September 8, 2000).

Resolution 60/1, "2005 World Summit Outcome" (October 24, 2005).

Resolution 66/288, "The future we want" (July 27, 2012).

Resolution 69/283, "Sendai Framework for Disaster Risk Reduction 2015–2030" (June 3, 2015).

Resolution 70/1, "Transforming our world: the 2030 Agenda for Sustainable Development" (September 25, 2015).

Conference of the Parties to the UN Framework Convention on Climate Change of May 9, 1992

Decision 1/CP.1, "The Berlin Mandate: review of the adequacy of Article 4, paragraphs 2 (a) and (b), of the Convention, including proposals related to a protocol and decisions on follow-up (April 7, 1995).

Decision 2/CP.1, "Review of first communications from the parties included in Annex I to the Convention" (April 7, 1995).

Decision 3/CP.1, "Preparation and submission of national communications from the parties included in Annex I to the Convention" (April 7, 1995).

Decision 4/CP.1, "Methodological issues" (April 7, 1995).

Decision 5/CP.1, "Activities implemented jointly under the pilot phase" (April 7, 1995).

Decision 7/CP.1, "The report on implementation" (April 7, 1995).

Decision 11/CP.1, "Initial guidance on policies, programme priorities and eligibility criteria to the operating entity or entities of the financial mechanism" (April 7, 1995).

Decision 14/CP.1, "Institutional linkage of the Convention secretariat to the United Nations" (April 7, 1995).

Decision 16/CP.1, "Physical location of the Convention secretariat" (April 7, 1995).

Decision 3/CP.2, "Secretariat activities relating to technical and financial support to Parties" (July 19, 1996).

Decision 5/CP.2, "Linkage between the Ad Hoc Group on Article 13 and the Ad Hoc Group on the Berlin Mandate" (July 19, 1996).

Decision 9/CP.2, "Communication from parties included in Annex I to the Convention: Guidelines, schedule and process for consideration" (July 19, 1996).

Decision 11/CP.2, "Guidance to the Global Environment Facility" (July 19, 1996).

Decision 3/CP.3, "Implementation of Article 4, paragraphs 8 and 9, of the Convention" (December 11, 1997).

Decision 4/CP.3, "Amendments to the list in Annex I to the Convention under Article 4.2(f) of the Convention" (December 11, 1997).

Conference of the Parties to the UN Framework Convention on Climate Change serving as the Meeting of the Parties to the Kyoto Protocol of December 11, 1997

Conference of the Parties to the UN Framework Convention on Climate Change Serving as the Meeting of the Parties to the Paris Agreement of December 12, 2015

Conference of the Parties to the Convention on Biological Diversity of June 5, 1992

Resolution MEPC.229(65), "Promotion of technical cooperation and transfer of technology relating to the improvement of energy efficiency of ships" (May 17, 2013), in IMO document MEPC 65/22.

Meetings of the Parties to the Montreal Protocol of September 16, 1987

Decision I/12E, "Clarification of terms and definitions: developing countries" (May 2–5, 1989).
Decision II/10, "Data of developing countries" (June 27–29, 1990).
Decision III/3, "Implementation Committee" (June 19–21, 1991).
Decision III/5, "Definition of developing countries" (June 19–21, 1991).
Decision IV/7, "Definition of developing countries" (November 23–25, 1992).
Decision XXVII/1, "Dubai Pathway on Hydrofluorocarbons" (November 1–5, 2015).

Meeting of States Parties to the Convention on the Protection of the Underwater Cultural Heritage of November 2, 2001

Resolution 5 / MSP 5 (April 28–29, 2015).

OSPAR Commission (OSPAR Convention of September 22, 1992)

OSPAR Convention decision 2007/1 to Prohibit the Storage of Carbon Dioxide Streams in the Water Column or on the Sea-Bed (June 25–29, 2007).
OSPAR Convention decision 2007/2 on the Storage of Carbon Dioxide Streams in Geological Formations (June 25–29, 2007).

Conference of the Contracting Parties to the Ramsar Convention of February 2, 1971

Resolution VIII.3, "Climate change and wetlands: impacts, adaptation, and mitigation" (Valencia, November 18–26, 2002).
Resolution X.24, "Climate change and wetlands" (Changwon, October 28–November 4, 2008).
Resolution XI.14, "Climate change and wetlands: implications for the Ramsar Convention on Wetlands" (Bucharest, July 6–13, 2012).
Resolution XII.11, "Peatlands, climate change and wise use: implications for the Ramsar Convention" (Punta del Este, June 1–9, 2015).

UNESCO World Heritage Committee (World Heritage Convention of November 23, 1972)

Decision 29 COM 7B.a (Durban, July 10–17, 2005).
Decision 32 COM 7A.32 (Québec City, July 2–10, 2008).

INTERNATIONAL LAW COMMISSION

ILC, *Draft Articles on Responsibility of States for Internationally Wrongful Acts, with commentaries*, in (2001) *Yearbook of the International Law Commission*, vol. II, part two.
ILC, *Draft Articles on Diplomatic Protection, with Commentaries*, in (2006) *Yearbook of the International Law Commission*, vol. II, part two.
ILC, *Draft Principles on the Allocation of Loss in the Case of Transboundary Harm Arising out of Hazardous Activities*, in (2006) *Yearbook of the International Law Commission*, vol. II, part two.
ILC, *Fragmentation of International Law: Difficulties Arising from the Diversification and Expansion of International law* (April 13, 2006), doc. A/CN.4/L.682.
ILC, *Guiding Principles Applicable to Unilateral Declarations of States Capable of Creating Legal Obligations, with Commentaries Thereto*, in (2006) *Yearbook of the International Law Commission*, vol. II, part two.

DECISIONS OF INTERNATIONAL COURTS AND TRIBUNALS

Permanent Court of International Justice

PCIJ, *Status of Eastern Carelia*, Advisory Opinion of July 23, 1923, PCIJ Ser. B, No. 5.

PCIJ, *Mavrommatis Palestine Concessions (Greece v. United Kingdom)*, Judgment on jurisdiction of August 30, 1924, in Series A, No. 2.

PCIJ, *Factory at Chorzów (Germany v. Poland)*:
- Judgment on jurisdiction of July 26, 1927, in Series A, No. 9.
- Judgment on the merits of the claim for indemnity of September 13, 1928, in Series A, No. 17.

International Court of Justice

ICJ, *Monetary Gold Removed from Rome in 1943 (Italy v. France, United Kingdom and United States)*, Judgment of 15 June 1954.

ICJ, *Corfu Channel (United Kingdom v. Albania)*, Judgment of April 9, 1949.

ICJ, *Northern Cameroons (Cameroon v. United Kingdom)*, Judgment of June 15, 1954.

ICJ, *Legal Consequences for States of the Continued Presence of South Africa in Namibia (South West Africa) Notwithstanding Security Council Resolution 276*, Advisory Opinion of June 21, 1971.

ICJ, *Military and Paramilitary Activities in and against Nicaragua (Nicaragua v. United States)*, Judgment of June 27, 1986.

ICJ, *Certain Phosphate Lands in Nauru (Nauru v. Australia)*, Judgment of June 26, 1992.

ICJ, *East Timor (Portugal v. Australia)*, Judgment of June 30, 1995.

ICJ, *Legality of the Use by a State of Nuclear Weapons in Armed Conflict*, Advisory Opinion of July 8, 1996.

ICJ, *Legality of the Threat or Use of Nuclear Weapon*, Advisory Opinion of July 8, 1996.

ICJ, *Gabčíkovo-Nagymaros Project (Hungary v. Slovakia)*, Judgment of September 25, 1997.

ICJ, *Pulp Mills on the River Uruguay (Argentina v. Uruguay)*, Judgment of April 20, 2010.

ICJ, *Accordance with International Law of the Unilateral Declaration of Independence in Respect of Kosovo*, Advisory Opinion of July 22, 2010.

ICJ, *Whaling in the Antarctic (Australia v. Japan: New Zealand Intervening)*, Judgment of March 31, 2014.

ICJ, *Obligations Concerning Negotiations Relating to Cessation of the Nuclear Arms Race and to Nuclear Disarmament (Marshall Islands v. United Kingdom)*, Judgment of October 5, 2016.

International Tribunal for the Law of the Sea

ITLOS, *The MOX Plant Case (Ireland v. United Kingdom)*, Order for Provisional Measures of December 3, 2001.

ITLOS, *Case Concerning Land Reclamation by Singapore in and around the Straits of Johor (Malaysia v. Singapore)*, Order for Provisional Measures of October 8, 2003.

ITLOS Seabed Dispute Chamber, *Responsibilities and Obligations of States Sponsoring Persons and Entities with Respect to Activities in the Area*, Advisory Opinion of February 1, 2011.

ITLOS, *Request for an Advisory Opinion Submitted by the Sub-Regional Fisheries Commission (SRFC)*, Advisory Opinion of April 2, 2015.

Arbitral Tribunals

United States-Germany Mixed Claims Commission, Administrative Decision No. II, November 1, 1923, VII *UNRIAA*, 23.

Responsabilité de l'Allemagne à raison des dommages causés dans les colonies portugaises du sud de l'Afrique (Portugal v. Germany), Arbitral Award of July 31, 1928, II *UNRIAA* 1011.

Trail Smelter (U.S. v. Canada), Arbitral Award of March 11, 1941, (1949) III *UNRIAA* 1938, at 1965.

Case Concerning the Audit of Accounts in Application of the Protocol of 25 September 1991 Additional to the Convention for the Protection of the Rhine from Pollution by Chlorides of 3 December 1976 (Netherlands v. France), Arbitral Award of March 12, 2004, (2004) XXV *UNRIAA* 267.

Iron Rhine Railway (Belgium v. *Netherlands)*, Arbitral award of May 24, 2005 (2005) XXVII *UNRIAA* 35.

Eritrea-Ethiopia Claims Commission (EECC), decision of August 17, 2009, Final Award on Ethiopia's Damages Claims, in (2009) XXVI *UNRIAA* 631.

Regional Human Rights Courts

IACrHR, *Velásquez Rodríguez* v. *Honduras*, Judgment of July 29, 1988, Ser. C, No. 4 (1988).

ECtHR Grand Chamber, *Al Skeini* v. *United Kingdom*, Judgment of July 7, 2011.

WTO Dispute Settlement Body

DS135: *European Communities – Measures Affecting Asbestos and Asbestos Containing Products.*

DS58: *United States – Import Prohibition of Certain Shrimp and Shrimp Products.*

MUNICIPAL AND EU LAWS AND POLICIES

China

Measures for the Operation and Management of CDM Projects (2005).

Ecuador

Constitution (September 28, 2008).

European Union

Council Decision 2002/358/EC concerning the approval, on behalf of the European Community, of the Kyoto Protocol to the United Nations Framework Convention on Climate Change and the joint fulfilment of commitments thereunder (April 25, 2002), doc. 32002D0358.

Directive 2003/87/EC establishing a scheme for GhG emission allowance trading within the Community (October 13, 2003), doc. 32003L0087.

Directive 2008/101/EC of 19 November 2008 amending Directive 2003/87/EC so as to include aviation activities in the scheme for GhG emission allowance trading within the Community (January 13, 2009), doc. 32008L0101.

Directive 2008/101/EC of 19 November 2008 amending Directive 2003/87/EC so as to include aviation activities in the scheme for greenhouse gas emission allowance trading within the Community (January 13, 2009), doc. 32008L0101.

Directive 2009/31/EC on the geological storage of carbon dioxide (April 23, 2009), doc. 02009L0031.

Directive 2010/30/EU on the indication by labelling and standard product information of the consumption of energy and other resources by energy-related products (May 19, 2010), doc. 32010L0030.

Directive 2011/92/EU on the assessment of the effects of certain public and private projects on the environment (December 13, 2011), doc. 02011L0092.

Fiji

Constitution (August 22, 2013).

France

Civil Code (Georges Rouhette and Anne Rouhette-Berton trans.).

India

Constitution (November 26, 1949).

Energy Conservation Building Code (May 2007).

Tunisia

Constitution (June 26, 2014).

United States

Clean Air Act, 42 U.S.C.A. § 7521(a)(1).
Energy Independence and Security Act, December 19, 2007, Pub. L. 110–140.
Energy Policy Act, August 8, 2005, Pub. L. 109–158.
Senate resolution 98, 105th Cong., 143 Cong. Rec. S8138-39 (July 25, 1997).
US Code Chapter 85 (1963), Title 42, *Air Pollution Prevention and Control*, Subchapter II, "Emission Standards for Moving Sources."

DECISIONS BY MUNICIPAL AND EU COURTS

Austria

Bundesverwaltungsgericht (Administrative Court), case W109 2000179-1/291E, judgment of February 2, 2017.
Verfassungsgerichtshof (Constitutional Court), case E 875/2017, judgment of June 29, 2017.

European Union

Case C-366/10, *ATA* v. *Secretary of State for Energy*, judgment of December 21, 2011, doc. 62010CJ0366.

India

Supreme Court, *Narmada Bachao Andolan* v. *Union of India*, judgment of March 15, 2005.

Japan

Tokyo High Court, decision of December 26, 2012 on the Kotopajang dam.

The Netherlands

District Court of the Hague, *Urgenda Foundation* v. *The State of the Netherlands*, judgment of June 24, 2015.

New Zealand

Supreme Court, *Teitota* v. *Chief Executive of the Ministry of Business, Innovation and Employment*, judgment of July 20, 2015, [2015] *NZSC* 107.

Pakistan

High Court of Lahore, *Ashgar Leghari* v. *Federation of Pakistan*,
*   Order of September 4, 2015.
*   Order of September 14, 2015.

South Africa

High Court (Western Cape Division), *Earthlife Africa Johannesburg* v. *Minister of Energy*, judgment of April 26, 2017, [2017] *ZAWCHC* 50.

United States

District Court, N.D. California, *Native Village of Kivalina* v. *ExxonMobil*, decision of September 30, 2009, 663 F.Supp.2d 863 (N.D.Cal. 2009).
District Court, N.D. Texas, Fort Worth Division, *ExxonMobil* v. *Healey*, decision of October 13, 2016, Civil Action No. 4:16-CV-469-K.

Court of Appeals for the Ninth Circuit, *Native Village of Kivalina* v. *ExxonMobil*, decision of September 21, 2012, 696 F.3d 849 (9th Cir. 2012).

Supreme Court, *Massachusetts* v. *Environmental Protection Agency*, judgment of April 2, 2007, 549 US 497.

Supreme Court, *American Electric Power* v. *Connecticut*, June 20, 2011, 131 *S.Ct.* 2527.

Supreme Court, *Native Village of Kivalina* v. *ExxonMobil*, (2013) 133 S.Ct. 2390.

Supreme Court, *Utility Air Regulatory Group* v. *EPA*, June 23, 2014, (2014) 573 *US* 573.

Supreme Court, *West Virginia et al.* v. *Environmental Protection Agency*, No. 15A773.

1

Introduction

Climate change is one of the greatest challenges of our time – an age where our impacts, as human societies, have extended beyond our immediate environment and are now affecting entire planetary equilibria. The stakes of addressing climate change are higher than traditional questions in international law about peace among nations, individual welfare or economic development. Climate change challenges just about everything we know and care about, from individual rights and welfare to social harmony and civilization, environmental protection and ecological balance. Cataclysmic scenarios of runaway climate change raise chilling prospects where even our survival as a species could potentially be threatened.

Yet, tackling climate change is a daunting task. Profound transformations are needed in our economies, societies, political systems and individual ways of life. Our model of development needs to be fundamentally altered in order to take planetary limits into consideration. Law needs to play a central role in these changes, but existing governance institutions are ill-fitted to the task. States remain the main center for decisions, but national interests, as they are generally understood, are often inconsistent with a rational utilization of global commons. By seeking to satisfy populations, democratic institutions are often inclined to take decisions patently contrary to the interests of future generations. After more than a quarter of a century of some of the most complex transnational political debates, international efforts to address climate change have achieved little success.

This introduction lays the foundations for the following chapters. Firstly, it takes stock of what science has to tell us about climate change. Secondly, it identifies the main goals of laws and politics addressing climate change. Thirdly, it presents the elements constituting the international law on climate change. Lastly, it introduces the outlines of this textbook.

I. THE SCIENCE OF CLIMATE CHANGE

The international law on climate change relies on a scientific understanding of our influence on the climate. The following provides synthetic explanations of the "greenhouse" effect, the origin of anthropogenic greenhouse gas (GhG) emissions, their impact, projections of future changes and the possibilities to alter these projections.

A. The "Greenhouse" Effect

Life, as we know it, is possible only because the temperature in many places on Earth remains stable within a certain range. This temperature is determined by complex planetary equilibriums. The temperature on the Earth's surface depends on what is often called an "energy budget" – the difference between the quantity of energy that the Earth receives almost exclusively from the Sun and the quantity of energy it releases to space. Energy enters or leaves the Earth system through electromagnetic radiations such as infrared radiations, visible light and ultraviolet radiations. Like any other warm object, both the Sun and the Earth emit electronic radiations through which they release some energy. The Earth's surface gains energy by absorbing electromagnetic radiations from the Sun and thus warms. In turn, the Earth loses energy by emitting electromagnetic radiations toward the space and thus cools. When the Earth's surface absorbs as much energy as it releases, the Earth's energy budget is balanced and temperature remains stable.

Not all electromagnetic radiations are identical. Their wavelength depends on the temperature of the body from which these radiations are emitted. Extremely hot objects such as the Sun emit high-energy electromagnetic radiations characterized by a short wavelength, including ultraviolet radiations as well as visible light. Far cooler objects like the Earth's surface emit far less intense radiations with a longer wavelength, such as infrared radiations. The human eye does not perceive the infrared radiations emitted by the Earth day and night – all we can see is the visible light emitted by the Sun (or artificial sources) and its reflection by objects around us. Infrared cameras, however, perceive the radiations that objects around us emit days and nights, whose specific wavelength varies as a function of the object's temperature.

Our atmosphere is not totally transparent to all electromagnetic radiations. It acts like a tinted window – a selective filter which reflects or absorbs some electromagnetic radiations while letting others go through. In particular, our atmosphere lets some ultraviolet radiations and almost all visible light enter the Earth system, but it stops some longwave infrared radiations from leaving the Earth system (see Figure 1.1). Due to their physical properties, some gases tend to be opaque to infrared radiations: when these gases are present in the atmosphere, they reflect a proportion of the infrared radiations emitted by the Earth's surface back toward the Earth's surface. Thus, the presence of these gases in the atmosphere causes an additional warming of the Earth's surface, both day and night, by preventing some of the heat absorbed from the Sun during the day from being released to space. Without this partial opacity of its atmosphere to infrared radiations, the Earth's surface would – like the surface of Moon – instantly reach sub-freezing temperatures at night when it stops being exposed to sunlight. This would make many forms of life impossible.

This selective opacity to infrared radiations is called the *greenhouse effect*. The analogy with a greenhouse is somewhat inaccurate because the inside of a greenhouse warms through a different mechanism. A greenhouse operates by preventing airflow: it traps the air warmed by contact with the surface and reduces cooling by *convection* (the circulation of warm air away from the source of heat). By contrast, the atmosphere regulates the temperature of the Earth's system by filtering inputs and outputs of electromagnetic *radiations*. Although the analogy is slightly misleading, it is very common. The gases that contribute to the "greenhouse" effect are called greenhouse gases (GhGs). Most of our atmosphere is composed by nitrogen and oxygen that are completely transparent to infrared radiations, but it also contains relatively small amounts of such GhGs, in particular water vapor (H_2O), carbon dioxide (CO_2), methane (CH_4) and nitrous oxide (N_2O).

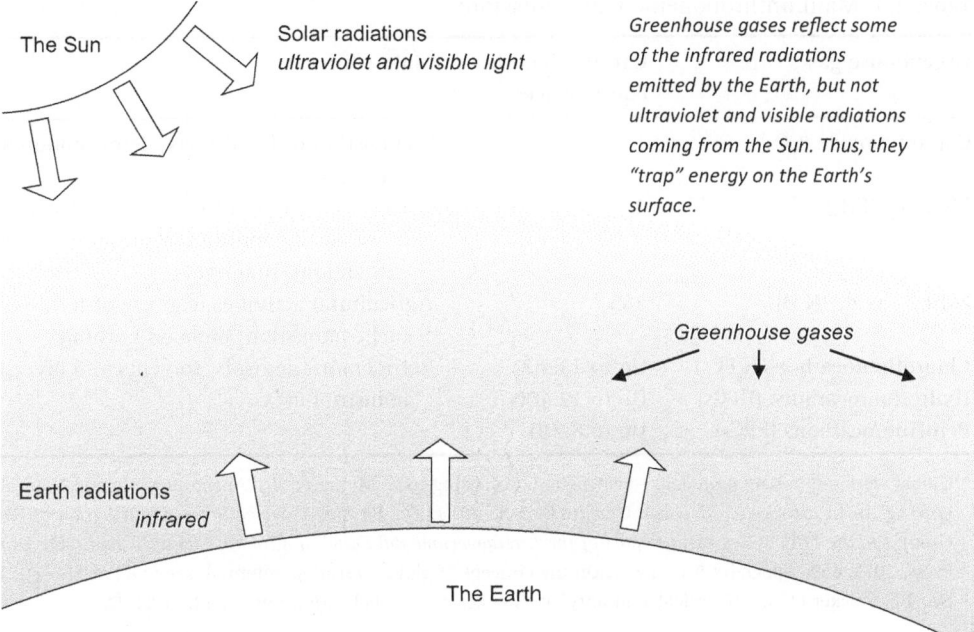

Figure 1.1 The greenhouse effect

B. Anthropogenic GhG Emissions

Since the Industrial Revolution in late eighteenth-century England, most societies have gradually turned to a model of industrial development which relies heavily on the combustion of fossil fuels – coal, then oil and gas – as a source of energy. Fossil fuels were formed gradually over millions of years through the decomposition of buried dead organisms. Today, they provide us with a convenient source of energy to produce electricity, propel vehicles and cook, among other things. Yet, burning fossil fuels releases GhGs into our atmosphere. Other human activities have released significant quantities of GhGs in our atmosphere, including deforestation, the decomposition of organic wastes, agricultural processes, the production of cement or the production of synthetic organic compounds used for fridges and air-conditioning systems.

Over time, the combustion of fossil fuels and other diverse human activities have caused the emission of tremendous amounts of GhGs into our atmosphere. Carbon dioxide is the main GhG produced by human activities. Since 1750, more than 2 teratons of carbon dioxide ($TtCO_2$) were emitted as a result of such industrial processes – this represents $2e^{+12}$ tons, or 2,000,000,000,000 tons. Other GhGs, such as methane, nitrous oxide and carbon compounds which have a much higher opacity to infrared radiations, have also been emitted in smaller quantities, but they often have a much stronger warming potential. A ton of CFC-13 emitted into the atmosphere, for instance, has as much global warming effect over a century as 13,900 tons of carbon dioxide (see Table 1.1).

Anthropogenic GhG emissions were historically concentrated in a few developed States, mainly in Europe and North America. Yet, many other countries are now following the same model of industrial development and GhG emissions have been rapidly increasing in emerging

Table 1.1 Main anthropogenic GhG emissions

Greenhouse gas	Carbon dioxide equivalence[a]	Main sources[b]
Carbon dioxide (CO_2)	1	Combustion of fossil fuels; deforestation and land-use change
Methane (CH_4)	28	Extraction and use of fossil fuels; decomposition of wastes in landfills; agriculture (e.g. rice cultivation); ruminants
Nitrous oxide (N_2O)	265	Agricultural activities (e.g. use of fertilizers); fossil fuel combustion; biomass burning
Chlorofluorocarbons (CFCs)	Up to 13,900	Refrigerants, aerosols, solvents and diverse industrial uses
Hydrofluorocarbons (HFCs)	Up to 12,400	
Perofluorocarbons (PFCs)	Up to 8,210	

[a] Global warming potential on a 100-year period, according to G. Myhre *et al.*, "Anthropogenic and Natural Radiative Forcing" in T.F. Stocker *et al.* (eds.), *Climate Change 2013: The Physical Science Basis. Contribution of Working Group I to the Fifth Assessment Report of the Intergovernmental Panel on Climate Change* (Cambridge University Press, 2013) 659, Appendix 8.1, at 731. On the concept of global warming potential, see *ibid.*, at 710–712.

[b] See T.F. Stocker *et al.*, "Technical summary" in Stocker *et al.* (eds.), *supra* note 1, 33, at 53–59.

economies and in the developing world generally. China became the largest GhG emitter in 2005 and it is currently responsible for as much GhG emissions as the United States and the European Union combined.[1] Yet, China's population is significantly larger than the United States and the European Union taken together, and per capita emissions in China remain lower than in many developed States.

The pursuit of industrial development by an increasing number of countries, in a context of rapid global demographic growth, has gradually increased the pace of anthropogenic GhG emissions. A first teraton of anthropogenic carbon dioxide was emitted over more than two centuries, between 1750 and 1970. The second teraton was emitted in 40 years, between 1970 and 2010. At the current rate of 40 gigatons of carbon dioxide ($GtCO_2$) emissions per year, it would take another 25 years to emit another $TtCO_2$. Yet, the pace of anthropogenic GhG emissions is increasing, and the third teraton may have been entirely emitted before the end of the 2020s.[2]

Anthropogenic emissions of GhGs into our atmosphere have been so massive that the chemical composition of the Earth's atmosphere has been significantly affected. This had the effect of altering the opacity of our atmosphere to shortwave electromagnetic radiations, thus amplifying the greenhouse effect by reducing the release of energy into space. This is causing a steady increase in the global average temperature, which can be observed and measured year after year at a speed unprecedented in geological history.

The impact of anthropogenic emissions on the concentration of GhGs in the atmosphere has been measured in many ways. The Mauna Loa observatory in Hawaii started to measure the

[1] WRI, CAIT Climate Data Explorer, "Total GHG emissions excluding land-use change and forestry" (2013).

[2] R.K. Pachauri *et al.*, *Climate Change 2014: Synthesis Report. Contribution of Working Groups I, II and III to the Fifth Assessment Report of the Intergovernmental Panel on Climate Change* (IPCC, 2015) 45.

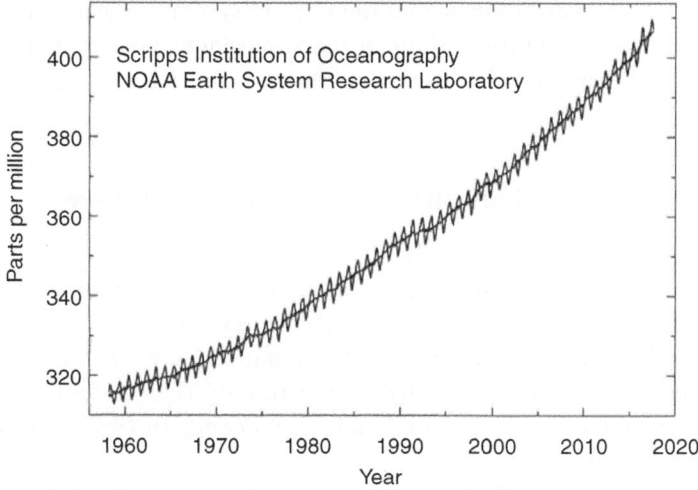

Figure 1.2 Concentration of carbon dioxide (CO2) in the atmosphere at Mauna Loa Observatory

Note: Reproduced with permission from the US National Oceanic and Atmospheric Administration.

concentration of carbon dioxide in 1958. Apart from seasonal variations caused by photosynthesis in the Northern hemisphere, its measures revealed a steady increase in the concentration of carbon dioxide, from 315 parts per million (ppm) in 1958 to more than 400 ppm today (see Figure 1.2). Many other direct measurements have been made to the concentration of diverse GhGs in the atmosphere since the 1960s, first from land-based observatories and increasingly now from satellite systems, evidencing similar increases in the concentration of methane, nitrous oxide and other GhGs in the atmosphere. Indirect measures were also made, for instance, through analyzing air bubbles trapped for centuries in deep ice sheets in Greenland and Antarctica, in order to retrace the historical evolution in the composition of our atmosphere.[3]

Based on such information, scientists with the relevant expertise are unanimous in concluding that human activities such as the massive combustion of fossil fuels result in changes in the composition of our atmosphere which increase the greenhouse effect and warm the Earth's surface beyond the pre-industrial range of temperatures. The Intergovernmental Panel on Climate Change (IPCC), an institution created to assess available scientific knowledge in a transparent manner in order to inform policy-makers, has concluded that "warming of the climate system is unequivocal"[4] and that there is clear evidence that this warming is caused by the increase in the atmospheric concentrations in GhGs, in particular carbon dioxide.[5] Although scientific evidence is different in nature from mathematical conclusions, the amount of data that have been collected over the last decades provides compelling evidence of an ongoing change in our climate caused by anthropogenic GhG emissions. As the first working

[3] See generally Dennis L. Hartmann *et al.*, "Observations: atmosphere and surface" in T.F. Stocker *et al.* (eds.), *Climate Change 2013: The Physical Science Basis. Contribution of Working Group I to the Fifth Assessment Report of the Intergovernmental Panel on Climate Change* (Cambridge University Press, 2013), 159, at 165–180.

[4] L.V. Alexander *et al.*, "Summary for policymakers" in Stocker *et al.* (eds.), *supra* note 3, 3, at 4.

[5] *Ibid.*, at 13.

group of the IPCC concluded, "human influence on the climate system is clear" and "evident from the increasing greenhouse gas concentrations in the atmosphere, positive radiative forcing [i.e. increase in the greenhouse effect], observed warming, and understanding of the climate system."[6]

C. The Impacts of Increased GhG Concentrations in the Atmosphere

The IPCC has further noted that "since the 1950s, many of the observed change are unprecedented over decades to millennia."[7] By altering the chemical concentration of the Earth's atmosphere, human societies are causing far-reaching, often irreversible changes in multiple complex planetary systems. The nature and ambit of these changes, and their consequences for us as a civilization and species, cannot be predicted with certainty. There is no certainty that life on Earth will be able to adjust to a substantial and rapid change in GhG concentrations and its far-reaching consequences.

The first, most simple consequence of an increase in the concentration of GhGs in our atmosphere is an increase in the global average temperature. As of mid-2017, the best estimate is that the global average surface temperature on Earth increased by around 1.1 degrees Celsius between 1880 and 2016.[8] Heat records were repeatedly broken in recent years. As of mid-2016, 2015 stood as the hottest year on record, immediately followed by 2014; the ten hottest years on record were all after 1998. The IPCC noted that "[e]ach of the last three decades has been successively warmer at the Earth's surface than any preceding decade since 1850,"[9] when the first reliable global temperature measurements were made.

Yet, climate change does not only affect temperatures. Referring to climate change as "global warming" belittles the long chain of consequences that a slight change in global average temperature is having on multiple complex planetary systems. Warming, for instance, increases evaporation, thus altering the water cycle, with consequences on precipitation patterns throughout the world. Floods *and* droughts occur more often, and so do some extreme weather events, such as tornadoes, in some parts of the world. Wildfires are also occurring more often in certain regions of the world as a consequence of changes in temperature, rain and wind systems. The level of the sea is rising as the consequence of the melting of glaciers and ice sheets, but also because, just like the mercury in a thermometer, the water of the ocean expands when it warms. At the same time, oceans are becoming more acidic because of the absorption of some carbon dioxide from the atmosphere as soluble carbon acid (H_2CO_3). Ocean acidification makes it more difficult for some plankton to develop; this alone could wreak havoc in marine biology throughout the food chain. On land too, ecosystems are deeply affected by changes in temperature and aridity, which exacerbate risks of extinction for many vegetal and animal species.[10]

---

[6] *Ibid.*, at 15.

[7] *Ibid.*, at 4.

[8] World Meteorological Organization, *Statement on the State of the Global Climate in 2016* (WMO, 2017) 2. See also Alexander *et al.*, *supra* note 4, at 5, noting an increase of 0.85 degrees Celsius between 1880 and 2012.

[9] Alexander *et al.*, *supra* note 4, at 5.

[10] See generally Christopher B. Field *et al.*, "Summary for policymakers" in Christopher B. Field *et al.* (eds.), *Climate Change 2014: Impacts, Adaptation, and Vulnerability. Part A: Global and Sectoral Aspects. Working Group II Contribution to the Fifth Assessment Report of the Intergovernmental Panel on Climate Change* (Cambridge University Press, 2014) 1.

These transformations in planetary systems are and will be impacting human societies very severely. Food production could be hindered by adverse climatic conditions in large production regions. Human health is affected by heat waves as well as by the diffusion of some pathogens, such as malaria and dengue fever, to regions where they did not use to be prevalent – and where populations may have lower immunity.[11]

The most severe adverse impacts of climate change occur in some of the least developed States, affecting those populations who live under the less forgiving climate of tropical regions, depend more directly on natural resources (in particular, subsistence agriculture) and have the least resources to cope with changes in their environment. The notion of "climate justice," further discussed throughout this book, sheds light on the disconnect between the industrial nations which benefit most from GhG emissions and the least developed nations which suffer the most from the impacts of climate change.[12] Yet, developed States are not immune from the consequences of climate change, some of which – such as food crises, epidemics, economic recession, migration or conflicts – will stop at no borders.[13]

D. Projections

While the impacts of climate change can already be observed throughout the world, they will inevitably become more severe in the coming years, decades and centuries. This would be the case even if all anthropogenic GhG emissions were to stop suddenly, as planetary systems pursue their slow adjustment to a different atmospheric chemistry. Sea-level, for instance, would continue to rise over centuries by probably one to several meters in the absence of any additional anthropogenic GhG emissions. Continuing GhG emissions, however, raise the risk of a sea-level rise in the scale of tens of meters over several centuries.[14] A "large fraction of species" will face extinction in the coming centuries.[15] The impacts on food production, livelihood, economic growth, health, migration and conflicts cannot be predicted with certainty, but they are likely to be critical.

Overall, the severity of climate change largely depends on GhG emissions in the coming years and decades. Large amounts of additional GhG emissions will result in more dramatic consequences. Global average surface temperature will likely increase by more than 4 degrees Celsius by the end of the century, and further in the following centuries, if the growth in global GhG emissions goes unimpeded. While most debates on the impacts of climate change focus on the twenty-first century, scientists estimate that current anthropogenic greenhouse gas emissions will have a profound impact on the climate system, ecosystems and human societies "for the next ten millennia and beyond."[16]

Predictions come, however, with a certain degree of uncertainty – and catastrophic scenarios cannot be ruled out. The planetary systems at stake, from the climate system to the evolution of ice sheet and ocean streams, are extremely complex. One of the greatest sources of scientific

---

[11] *Ibid.*

[12] See e.g. Teresa M. Thorp, *Climate Justice: A Voice for the Future* (Palgrave Macmillan, 2014).

[13] See A. Guzman, *Overheated: The Human Cost of Climate Change* (Oxford University Press, 2013).

[14] Pachauri *et al.*, *supra* note 2, at 16.

[15] *Ibid.*, at 13.

[16] Peter U. Clark *et al.*, "Consequences of twenty-first-century policy for multi-millennial climate and sea-level change" (2016) 6:4 *Nature Climate Change* 360.

uncertainty regarding the reaction of the climate system to continuing anthropogenic GhG emissions stems from the feedbacks of planetary systems to GhG emissions. A feedback is a natural reaction in a system that either amplifies (positive feedback) or mitigates (negative feedback) the consequences of an interference with this system. Some "negative" feedbacks could reduce the impact of anthropogenic GhG emissions. For instance, an increase in carbon dioxide concentrations in the atmosphere could accelerate photosynthesis and vegetation growth, which could remove some carbon dioxide from the atmosphere and mitigate climate change, although this phenomenon is likely to have a fairly insignificant impact at the global scale.[17] By contrast, "positive feedbacks" refer to phenomena that could amplify the impact of anthropogenic GhG emissions. Multiple positive feedbacks have been documented. For instance, the melting of glaciers, ice sheets or ice seas changes the color of large regions of the Earth. Seas and rocks are darker than ice: they reflect less sunlight and warm faster. Ice melting thus decreases in the Earth's "albedo" (or "whiteness"), which hastens the pace of warming, especially in the polar regions. Another positive feedback relates to the increase in wildfires, which release large amounts of GhGs, thus amplifying climate change.[18]

Some powerful positive feedbacks, or "tipping points," could have abrupt consequences leading to catastrophic "runaway" climate change – a scenario where our climate would continue to warm even without further anthropogenic GhG emissions due to self-sustaining positive feedback. Such a tipping point could occur as a consequence of a thawing of permafrost – soils that have been frozen for thousands of years, in particular in Siberia, the north of Canada and Alaska. The permafrost contains great quantities of frozen organic materials. The thawing of the permafrost is already setting free large amounts of methane and carbon dioxide, significantly amplifying climate change.[19] A runaway climate change scenario could cause extremely dangerous consequences for humankind. The likelihood of such a scenario is difficult to assess due to the lack of any historical precedent. We are, in other words, in uncharted territory.

E. Ways Forward

The impact of GhG emissions on the atmosphere has not always been known. Empirical evidence of climate change started to emerge in the early 1960s; by the mid-1980s, a strong scientific consensus had been achieved.[20] Thirty years later, even though the impacts of climate change are becoming more obvious and more dramatic every year, global GhG emissions continue to grow steadily. Depending on response measures taken in the coming years and decades, global GhG emissions could continue to increase and reach catastrophic levels, or they could soon peak and quickly decrease.

To inform decision-makers of the consequences of their decisions, the IPCC developed different scenarios of GhG emission pathways, or Representative Concentration Pathways (RCPs), for the twenty-first century. Each of these scenarios is characterized by a particular radiative forcing – an intensity of the greenhouse effect – by the end of the twenty-first century.[21]

[17] Stocker *et al.*, *supra* note 3, 33 at 57–58.
[18] *Ibid.*, at 58.
[19] Alexander *et al.*, *supra* note 4, at 27.
[20] See S.R. Weart, *The Discovery of Global Warming* (Harvard University Press, 2008).
[21] See Pachauri *et al.*, *supra* note 2, at 21. A complete explanation of the representative concentration pathways can be found in U. Cubasch *et al.*, "Introduction" in Stocker *et al.* (eds.), *supra* note 3, 119, at 147–150, Box 1.1. See also

The current annual rate of anthropogenic GhG emissions is about 50 $GtCO_2eq$,[22] which represents an aggregate impact equivalent to the emission of 50 billion tons of carbon dioxide. If no or little action is taken to mitigate climate change, as per scenario "RCP8.5," the rate of global GhG emissions will continue to increase. Scenario RCP8.5 assumes a rate of about 100 $GtCO_2eq$ per year by the end of the twenty-first century. It is estimated that the global average temperature would increase by more than 4 degrees Celsius before the end of the twenty-first century and, if the same course of action is maintained, by 8 degrees Celsius by 2300. The summer Arctic ice sea would have totally disappeared by 2060 and one could sail over the North Pole. Even on the basis of the most conservative estimates, the sea level would possibly exceed one meter by 2100, reaching several meters by 2500, possibly even tens of meters in the circumstance of a catastrophic collapse of massive ice sheets in Antarctica and Greenland. Sea-water acidity would rapidly increase, with great impacts on marine ecosystems. Heatwaves, floods, droughts, wildfires, famines, conflicts for access to natural resources, mass migration and possibly epidemics would make the world a much, much more difficult place to live for future generations. Tipping points could trigger a run-away climate change scenario. Things could get totally out of control. The possibility of a global civilizational collapse – the end of humankind within a few centuries – could certainly not be excluded.[23]

Alternative scenarios exist where serious efforts are made to mitigate climate change. Such efforts would not prevent further impacts of climate change as planetary systems slowly adjust to changes in atmospheric chemistry, but they would most likely avoid some of the worst consequences of a runaway climate change. The IPCC's most optimistic scenario, scenario "RCP2.6," assumes that drastic efforts are made to mitigate climate change. In particular, it assumes that a peak in global GhG emissions is reached in the 2020s and that it is followed by a very rapid decrease in global GhG emissions thereafter. Net global GhG emissions by the end of the twenty-first century would be null or even negative as efforts would be made to remove GhG from the atmosphere through reforestation and geoengineering. In these circumstances, the global average temperature would possibly be held under 2 degrees Celsius above pre-industrial levels by the end of the twenty-first century and could remain relatively stable thereafter. Sea-level rise would perhaps not increase by more than a further 50 cm before the end of the century and it could be limited to between one and a few meters by 2500. Some sea ice would be maintained all year round over the Arctic Ocean. Even then, however, many great risks would remain for many populations throughout the world. Biodiversity losses, extreme weather events, conflicts and other social impacts of climate change would not all be avoided. Some low-lying coastal regions and island States, for

O. Edenhofer et al., "Summary for policymakers" in R. Pichs-Madruga et al. (eds.), *Climate Change 2014: Mitigation of Climate Change. Contribution of Working Group III to the Fifth Assessment Report of the Intergovernmental Panel on Climate Change* (Cambridge University Press, 2014) 1, at 11–13.

[22] This includes carbon dioxide emissions as well as emissions of other greenhouse gases, weighted on the basis of their global warming potential over one hundred years. On carbon dioxide equivalence of other GhGs, see above, Table 1.1.

[23] On the notion of civilizational collapse, see Jared M. Diamond, *Collapse: How Societies Choose to Fail or Succeed* (Penguin, 2011). Although many civilizations have collapsed in the past, a global civilizational collapse only becomes possible in the age of globalization. See a discussion of the risk of such catastrophic crisis in M.L. Weitzman, "On modeling and interpreting the economics of catastrophic climate change" (2009) 91:1 *Review of Economics & Statistics* 1.

instance, would become uninhabitable as a consequence of sea-level rise. The risk of a global civilizational collapse would be significantly reduced, but a succession of crises is certainly unavoidable. Even this very "best" scenario involves great harm to human societies and ecosystems.

II. CLIMATE CHANGE LAWS AND POLICIES

Laws and policies can foster processes through which societies address climate change. They can promote efforts to decrease GhG emissions or even perhaps efforts to remove GhGs from the atmosphere. Moreover, laws and policies can also foster preparedness to the impacts of climate change. Thus, two categories of action are generally distinguished: efforts toward mitigating climate change and efforts toward adapting to its impacts. As climate change is already occurring and no amount of climate change mitigation will avoid all harms, climate change mitigation and adaptation are both necessary. In the long run, however, adaptation efforts will become more challenging if mitigation efforts are not successful.

A. Climate Change Mitigation

Laws and policies have been adopted to mitigate global anthropogenic climate change. Such efforts consist primarily in reducing sources of GhG emissions, but they can also seek to remove GhGs from the atmosphere through enhancing sinks and reservoirs.

Reducing sources of GhG emissions is the most obvious way to promote climate change mitigation. This can be done through sector-specific measures ranging from power generation to automobile traffic, cement production and waste disposal. Reductions in GhG emissions can often be achieved through improved efficiency. For instance, coal plants can be required to use the latest available technology in order to emit less GhG when producing as much power. More efficient power grids can reduce energy waste in transmission. Better air traffic control can reduce fuel consumption for each flight. Gains in efficiency result in savings which, sometimes, balance the cost of the particular measure. For instance, purchasing a more efficient car would lead to savings in fuel expenses which could balance the additional investment. Yet, while gains in efficiency may come at no or little cost, their outcome is limited to incremental reduction in GhG emissions. A more efficient car still burns oil.

Sources of GhG can also be reduced through more structural changes, often involving behavioral change – alterations to the way we live and consume. Walking or even taking public transportation instead of driving a car is an example of behavioral change leading to a reduction in GhG emissions. Such changes can also be induced at a collective level, for instance, through laws and policies. Shifting power generation away from a reliance on fossil fuels by investing in renewable or nuclear energy is a typical way to reduce GhG emissions through changes in production patterns. Building a national railway system to divert passengers from air or car transportation could also lead to substantial reductions in GhG emissions. Such measures are typically more costly than gains in efficiency and require more time for implementation, but they also lead to much greater climate benefits. While gains in efficiency may help reduce GhG emissions in the short term, more structural changes are necessary to ensure a turn to carbon-neutral ways of life.

Besides reducing sources of GhG emissions, measures can also be taken to enhance sinks. Sinks of GhGs are processes through which GhG are removed from the atmosphere and durably stored in reservoirs. Forests, for instance, are reservoirs of GhG because vegetation develops through photosynthesis – a process through which carbon dioxide is absorbed to produce nutrients. Reforestation can thus be promoted in order to remove some carbon dioxide from the atmosphere. Yet, reforestation may come into conflict with other possible uses of land, for instance, food production. Some alternative techniques have been developed to sequestrate more carbon in soils, for instance through the use of biochar, a charcoal produced through the pyrolysis of biomass, as soil fertilizer. Oceans are also a substantial carbon reservoir as seawater naturally absorbs some of the carbon dioxide contained in the air. Yet, enhancing the use of oceans as a sink of GhG would also exacerbate ocean acidification (as carbon dioxide is absorbed in seawater as an acid), with serious consequences for marine ecosystems.

In addition to enhancing natural sinks of GhGs, debates have developed on the possibility of developing artificial reservoirs through geoengineering. Carbon capture and storage (CCS) involves diverse techniques to isolate carbon dioxide from large industrial facilities (e.g. power plants) and to inject it in deep geological storage locations – which could include, somewhat ironically, some of the same locations from which oil and gas are being extracted. Yet, such techniques face significant obstacles regarding the cost of capturing and storing massive quantities of carbon dioxide as well as the ability to ensure durable sequestration.

B. Adaptation to Climate Change

Climate change is already under way and its impacts are affecting populations throughout the world. It is necessary for laws and policies to mitigate climate change through reducing sources and enhancing sinks of GhG – but this is insufficient. Response measures also need to be taken to protect anything we value – populations, culture, economic production, ecosystems, etc. – from harm caused by the impacts of climate change.

Climate change adaptation is usually defined as the "process of adjustment to actual or expected climate and its effect" in order "to moderate or avoid harm or exploit beneficial opportunities."[24] Adaptation measures are extremely varied and context-specific. Building dikes can reduce risks of floods. Developing and introducing drought-resistant food crops can reduce the risk that a drought translates into significant loss of livelihood or even into a famine. Preparing shelters can protect populations exposed to an increased risk of extreme weathers.

Adaptation to climate change is often mainstreamed in diverse laws, policies, programs and projects, most obviously through development and disaster risk reduction policies or in the design of large development projects. It is often difficult to single out adaptation efforts within such broader laws, policies, programs and projects. In this sense, climate change adaptation can be much more abstract than climate change mitigation.

The impacts of climate change seem to reinforce economic inequalities at all scales. The countries and populations most affected by climate change are often those who benefit the least from industrial development and those who are the least able to afford adaptation efforts.

---

[24] Pachauri *et al.*, *supra* note 2, at 118.

Arguments have thus been made in favor of support to adaptation, in particular toward developing countries, as assistance but also as a matter of causal responsibility. While the idea has partly been endorsed by international negotiations on climate change, little support has yet been provided to the States and populations most affected by and most vulnerable to the impacts of climate change. As a consequence, to date, most of the cost of adapting to a changing climate continues to be borne by developing States themselves, notwithstanding the responsibilities of industrial nations for emitting colossal quantities of GhGs.

III. THE INTERNATIONAL LAW ON CLIMATE CHANGE

Laws and policies relating to climate change have appeared in multiple forums. Concrete measures are generally defined by States and subnational authorities, or sometimes through regional cooperation – most obviously in the European Union (EU). International law plays a leading role, in particular in promoting climate change mitigation and providing support for climate change adaptation.

The international law on climate change consists of specific treaty rules as well as general norms of customary international law (see Table 1.2). Treaties establish rules that their parties expressly recognize as binding upon themselves. Specific treaties have been adopted to address climate change, including the United Nations Framework Convention on Climate Change (UNFCCC) and treaties negotiated under the UNFCCC.[25] Other treaty-based regimes have also developed rules relating to climate change, whether with regard to GhG emissions from specific sectors such as international shipping or international aviation, or with regard to particular GhGs.[26] Customary norms, by contrast, do not require the express agreement of any particular State. They form, instead, when a general practice becomes widely accepted as law.[27] Some customary norms of international law apply to climate change in spite of the existence of more specific treaty rules.[28]

A. The UNFCCC Regime

International negotiations over the last quarter of a century established a regime of international cooperation specifically dedicated to addressing climate change. This regime was created by the adoption of the UNFCCC. It was further developed through the negotiation of other treaties and decisions under these treaties.

Firstly, the UNFCCC was adopted on May 9, 1992 and entered into force on March 21, 1994. As of early 2018, the parties to the UNFCCC include 196 States – virtually all of them – as well as the EU. The UNFCCC established institutions, including a Conference of the Parties (COP) which meets at least once a year, as well as a permanent secretariat which facilitates

---

[25] United Nations Framework Convention on Climate Change, May 9, 1992, 1771 *UNTS* 107 (hereinafter UNFCCC). See also Kyoto Protocol to the United Nations Framework Convention on Climate Change, December 11, 1997, 2303 *UNTS* 162 (hereinafter Kyoto Protocol); and Paris Agreement, December 12, 2015, in the annex of decision 1/CP.21, "Adoption of the Paris Agreement" (December 12, 2015). See generally Chapter 3.

[26] See generally Chapter 4.

[27] Statute of the International Court of Justice, June 26, 1945, 3 Bevans 1179 (hereinafter ICJ Statute), art. 38.1(b).

[28] See generally Chapter 5.

Table 1.2 The components of the international law on climate change

Source[a]	International custom, as evidence of a general practice accepted as law	International conventions establishing rules expressly recognized by States	
Contents	Norms of general international law	Collective objectives, national commitments, as well as procedural and institutional provisions	
		UNFCCC regime	Other treaty regimes
Illustrations	e.g. no-harm principle, law of State responsibility	e.g. UNFCCC, Kyoto Protocol and Paris Agreement, as well as decisions adopted by the COP/CMP/CMA decisions	e.g. Montreal Protocol on Substances that Deplete the Ozone Layer, International Convention for the Prevention of Pollution from Ships, Convention on International Civil Aviation, World Heritage Convention, Convention on Biological Diversity

[a] In the terminology of the ICJ Statute, *supra* note 27, art. 38.1.

negotiations. It also contains some obligations for States to promote climate change mitigation and adaptation, albeit in rather vague language.

Secondly, the Kyoto Protocol to the UNFCCC was adopted by a COP decision on December 11, 1997. After ratification by a critical number of States, it entered into force on December 16, 2005. As of early 2018, its parties included 191 States as well as the European Union. The United States never ratified the Kyoto Protocol, whereas Canada withdrew from it with effect on December 15, 2012. The Kyoto Protocol defined specific national commitments on climate change mitigation which were applicable to developed States for a commitment period running from 2008 to 2012. The Doha Amendment to the Kyoto Protocol, which was adopted on December 8, 2012 and had not yet entered into force as of early 2018, seeks to define a second commitment period running from 2013 to 2020.[29]

Thirdly, the Paris Agreement was adopted on December 12, 2015 and entered into force on November 4, 2016.[30] As of early 2018, it had been signed by 194 States and ratified by 171 States as well as the European Union, with many other States still engaged in the process of ratification.[31] The Paris Agreement calls on every party to communicate and make efforts towards the realization of its Nationally Determined Contribution (NDC) to the global response to climate

[29] Doha Amendment to the Kyoto Protocol, December 8, 2012, in the annex of decision 1/CMP.8, "Amendment to the Kyoto Protocol pursuant to its Article 3, paragraph 9 (Doha Amendment)" (December 8, 2012) (hereinafter Doha Amendment).

[30] Paris Agreement, *supra* note 25.

[31] The United States is, as of early 2018, a party to the Paris Agreement, having ratified it on September 3, 2016. On June 1, 2017, newly elected US President Trump announced that the United States would withdraw from the Paris Agreement. According to the conditions in arts. 28.1 and 28.2, the US withdrawal from the Paris Agreement could not take effect less than four years after its entry into force, i.e. November 4, 2020.

change. Besides climate change mitigation, it promotes climate change adaptation as well as international support to developing countries.

In addition to climate change treaties, hundreds of decisions have been adopted by the COP (under the UNFCCC) by the Conference of the Parties serving as the Meeting of the Parties to the Kyoto Protocol (CMP) and most recently by the Conference of the Parties Serving as the Meeting of the Parties to the Paris Agreement (CMA) at their joint annual sessions. In particular, the 2010 Cancún Agreements call all developed States and voluntary developing States to pledge mitigation efforts applicable from 2013 to 2020.

B. Relevant Provisions in Other Regimes

Efforts to address climate change are not confined to the UNFCCC regime. They have taken place in a variety of forums,[32] in particular under treaties for the protection of the ozone layer and forums facilitating international transport.

Even though this was not its main objective, the regime on the protection of the ozone layer has played a significant role in mitigating climate change. The ozone regime aims to control and reduce the emissions of chemical substances which damage a stratospheric layer of ozone which filters some dangerous ultraviolet radiations emitted by the Sun. Some of these gases, such as chlorofluorocarbons (CFCs), some halons and hydrochlorofluorocarbons (HCFCs), are also very potent GhGs. The 1987 Montreal Protocol on Substances that Deplete the Ozone Layer successfully coordinated efforts to phase out the production of most of such ozone-depleting substances, thus immensely contributing to climate change mitigation.[33] To avoid duplication of efforts, the UNFCCC and the Kyoto Protocol did not apply to substances already controlled by the Montreal Protocol.

Moreover, in October 2016 the Parties to the Montreal Protocol agreed to extend its application to substitute substances which, although not damaging to the ozone layer, are also potent GhGs.[34] This utilization of the regime on the protection of the ozone layer to foster cooperation on climate change mitigation reflects the growing frustration of many stakeholders with the slow pace of negotiations conducted under the UNFCCC.

The Kyoto Protocol excluded emissions from aviation and marine bunker fuels from its scope, preferring to refer these questions to specialized forums. Negotiations were conducted under the aegis of the International Maritime Organization (IMO) for an amendment to the 1973 International Convention for the Prevention of Pollution from Ships (MARPOL). They led in 1997 to the adoption of an Annex on the Prevention of Air Pollution from Ships,[35] which was amended in 2011 to impose energy efficiency standards on all new ships and energy efficiency management plans on all ships.[36] Likewise, negotiations on civil aviation were held

---

[32] See Robert O. Keohane and David G. Victor, "The regime complex for climate change" (2011) 9:1 *Perspectives on Politics* 7.

[33] Montreal Protocol on Substances that Deplete the Ozone Layer, September 16, 1987, 1522 *UNTS* 3.

[34] Amendment to the Montreal Protocol on Substances that Deplete the Ozone Layer (October 15, 2016), (2017) 56 ILM 196 (hereinafter Kigali Amendment).

[35] 1997 Protocol to amend the 1973 International Convention for the Prevention of Pollution from Ships, as modified by the 1978 Protocol, adopted in London on September 26, 1997.

[36] MARPOL resolution MEPC.203(62), "Amendments to the Annex of the Protocol of 1997 to Amend the International Convention for the Prevention of Pollution from Ships, 1973, as modified by the Protocol of 1978 relating thereto" (July 15, 2011), in IMO document MEPC 62/24/Add.1. See Chapter 4, section II.A.

under the aegis of the International Civil Aviation Organization (ICAO), leading in 2016 to an agreement on the creation of the Carbon Offsetting and Reduction Scheme for International Aviation (CORSIA).[37]

Other international legal regimes contain provisions relevant either to climate change mitigation or to climate change adaptation. This includes general provisions which can be approached in a new light in the context of climate change. Thus, the notion of "pollution of the marine environment" in the UN Convention on the Law of the Sea (UNCLOS) is defined sufficiently broadly to include some impacts of climate change. The parties to UNCLOS must accordingly "take, individually or jointly as appropriate, all measures consistent with this Convention that are necessary to prevent, reduce and control pollution of the marine environment from any source."[38] Human rights law may also be considered to imply a broad mandate for States to address climate change, whose impacts affect the enjoyment of human rights in many ways.[39]

Increasingly, cooperation to tackle climate change has involved actors other than States and forums other than intergovernmental organizations. Many important actions were carried out by subnational governments, businesses and non-profit organizations. While the action of "non-party stakeholders" was recognized and encouraged by decisions adopted by the COP to the UNFCCC,[40] it gradually emerged spontaneously from civil society's initiatives taking place outside of the UNFCCC regime, implementing commitments, procedures and standards defined and developed by non-State actors themselves. The international responses to climate change – mitigation as well as adaptation – have thus been scattered, not only between different legal regimes, but also between classical State-centered public international law and more innovative transnational initiatives led by private enterprises.

C. Relevant Norms of General International Law

The UNFCCC and other multilateral regimes establish rules applicable to specific issue-areas. In addition to these rules, international law comprises some norms of a customary nature and of a general application. These norms of general international law apply to any issue-area in international law, unless their exclusion is explicitly provided or necessarily implied by special rules. According to Christian Tomuschat, general international law includes "axiomatic premises of the international legal order" such as the principle of equal sovereignty, systematic features that derive almost automatically from these premises such as the law of State responsibility, and possibly some widely shared values such as the right to life.[41]

As a corollary of the principle of equal sovereignty, States have an obligation not to cause serious harm to the territory of other States or to areas beyond their jurisdiction, and not

---

[37] ICAO Assembly Resolution A39-3, "Consolidated statement of continuing ICAO policies and practices related to environmental protection – Global Market-Based Measure (MBM) scheme" (September 27–October 7, 2016), para. 5. See Chapter 4, section II.B.

[38] United Nations Convention on the Law of the Sea, December 10, 1982, 1833 *UNTS* 3, art. 194.1. For a definition of pollution of the marine environment, see art. 1.1(4).

[39] See, for instance, CESCR, "Concluding observations on the fifth periodic report of Australia" (June 23, 2017), doc. E/C.12/AUS/CO/5, para. 12.

[40] Decision 1/CP.21, *supra* note 25, at paras. 134–135.

[41] See C. Tomuschat, "What is general international law?" http://legal.un.org/avl/ls/Tomuschat.html# (accessed December 30, 2017).

to allow activities under their jurisdiction to cause such harm. The "no-harm" principle, which forms part of customary international law, is widely recognized as the "cornerstone" of international environmental law.[42] This norm of general international law is relevant to the circumstances of climate change, whereby a few States bear most of the responsibility for the massive GhG emissions which are causing significant harm to others – threatening the territorial integrity of low-lying small-island developing States, the prosperity of many others and possibly the very survival of humankind.

The general international law on the responsibility of States for internationally wrongful acts could also be relevant in the context of climate change. An authoritative interpretation by the International Law Commission stated that "[e]very internationally wrongful act of a State entails the international responsibility of that State."[43] Inasmuch as excessive GhG emissions constitute a breach of an international obligation attributable to industrial States – a breach of the no-harm principle, for instance – these States bear secondary obligations under the general law of State responsibility, including, in particular, obligations of ceasing the continuing internationally wrongful act and to make adequate reparation.[44]

Other norms of general international law may be relevant to climate change or responses to it. There is arguably a general international law obligation for States to cooperate in good faith in addressing an issue of a global nature such as climate change. Furthermore, general international law involves an obligation for each State to protect everyone within its jurisdiction and to provide some humanitarian assistance and development aid to other States which drastically need them, as may be the case when a State is severely affected by impacts of climate change of an increasing severity. Lastly, the principle of State sovereignty suggests particular deference to the national determination of the most appropriate mitigation or adaptation measures.

Yet, the relevant norms of general international law are often ill-defined. Because of their customary nature, they lack a unique authoritative written statement which would provide clarity on their content. Many questions arise when trying to define the modalities of application, for instance, of the no-harm principle in relation to climate change.

The application of general international law can be precluded by the adoption of special rules under particular circumstances. Yet, the mere existence of treaty provisions seeking to address climate change does not preclude the application of a general norm of general international law to the same subject matter.[45] As interpreted by the International Law Commission, the principle according to which special laws derogate from general laws (*lex specialis derogate lege generali*) only applies when there is "some actual inconsistency" between the special rule and the general norm or else "a discernible intention that one provision is to exclude

---

[42] Philippe Sands and Jacqueline Peel, *Principles of International Environmental Law*, 3rd edn (Cambridge University Press, 2012), at 191. See also ICJ, *The Legality of the Threat or Use of Nuclear Weapons*, Advisory Opinion of July 8, 1996, para. 29.

[43] ILC, *Draft Articles on Responsibility of States for Internationally Wrongful Acts with Commentaries* in (2001) *Yearbook of the International Law Commission*, vol. II, part two (hereinafter *Articles on State Responsibility*), art. 1.

[44] *Ibid.*, arts. 30–31.

[45] See, for instance, ICJ, *Military and Paramilitary Activities in and against Nicaragua (Nicaragua v. United States)*, judgment of June 27, 1986, para. 174.

the other."[46] The UNFCCC regime does not involve any such inconsistencies, and there is no consensual intention to exclude the application of the no-harm principle or that of the general law on State responsibility. Therefore, as will be further discussed later in the book,[47] it seems more accurate to approach the UNFCCC regime as a collective attempt to promote compliance with the no-harm principle and, to a lesser extent, with the general international law on State responsibility.

IV. OUTLINE OF THE BOOK

This book aims to provide a comprehensive overview of the international law on climate change, including various treaty rules and norms of general international law as well as developments taking place in diverse forums.

Chapter 2 examines the rationales which justify and orient international cooperation on climate change. Beyond scientific information and economic arguments, it highlights the diverse cultural and moral narratives which justify and orient international cooperation on climate change. Chapters 3, 4 and 5 then delve into the sources of the international law on climate change. Chapter 3 starts with an overview of the UNFCCC regime, while Chapter 4 reviews the developments which took place within other treaty regimes. Chapter 5 turns to the relevance of general international law. Finally, Chapter 6 is dedicated to a discussion of differentiation, a central concept in the international law on climate change. It engages with the grounds for differentiation before documenting their impact on the international law on climate change.

Chapters 7 to 12 explore goal-specific aspects of the international law on climate change through discussions which place technical developments within the UNFCCC regime in perspective with developments taking place in other regimes and norms of general international law. Thus, Chapter 7 discusses efforts to promote climate change mitigation generally, documenting in particular national commitments under different treaty regimes. Chapter 8 looks more specifically at the flexibility mechanisms through which a State can fulfill its mitigation commitment through action conducted in another State's territory. Chapter 9 reviews normative debates on geoengineering – attempts made to create artificial carbon sinks or otherwise to cool the Earth system as a whole. Chapter 10 turns to efforts to promote climate change adaptation and identifies the emerging principles of international cooperation in this field. Chapter 11 examines ongoing negotiations on "approaches to address loss and damage," an as-yet ill-defined concept which, more often than not, orbits around the idea of compensation for the impacts of climate change. Chapter 12 discusses the role of international support, including financial support, transfer of technology, and capacity-building.

The last group of chapters engages with classical themes of international law which unfold somewhat differently in the context of climate change. Chapter 13 explores the question of

---

[46] *Articles on State Responsibility, supra* note 43, Commentary under Article 55, para. 4. See also ILC, *Fragmentation of International Law: Difficulties Arising from the Diversification and Expansion of International Law* (April 13, 2006), doc. A/CN.4/L.682, para. 88.

[47] See, more specifically, Chapter 5, section IV and Chapter 13, section I.

ambition and compliance, arguing that the UNFCCC regime can be understood as an effort to encourage States to comply with their obligations under general international law, in particular under the no-harm principle. Chapter 14 assesses the role of adjudication at the international level as well as before domestic courts. Chapter 15 highlights the increasing role of non-State actors in international cooperation on climate change. Lastly, Chapter 16 concludes with a general discussion of the transformations of international law in times of climate change, highlighting that international law is unlikely to remain unchanged while addressing this major global crisis.

2

The Rationale for International Action on Climate Change

Before further analyzing the actual laws applicable to climate change in the rest of the book, this chapter proposes a preliminary reflection regarding the rationale for international cooperation relating to climate change. A rationale is a narrative that is able to justify and to guide a particular course of action. The need to do something about climate change may be quite obvious at this point, but *what* precisely should be done, *by whom*, *how* and *at what cost* largely depend on one's understanding of the rationale for such action. The discussions introduced in this chapter have far-reaching implications related to more technical questions explored in the following chapters, for instance, concerning the level of ambition of mitigation action, differentiation between developed and developing States, and the need for international support to adaptation or to approaches to address loss and damage.

Although climate change is certainly a legitimate concern for governments as well as individuals, there is no single objectively "best" technical fix to it. Rather, there are multiple ways to justify, and many alternative considerations that could guide multiple choices, for instance, between different levels of ambition or different methods for reducing GhG emissions, enhancing sinks, possibly through geoengineering, and adapting to the impacts of climate change. As Mike Hulme wrote:

> Climate change is not "a problem" waiting for "a solution". It is an environmental, cultural and political phenomenon which is reshaping the way we thing about ourselves, our societies and humanity's place on Earth.[1]

To provide a more detailed overview of these questions, the following sections explore the insights of climate science, economics and moral theories on possible justifications for international action on climate change, as well as psychological, social and political reception processes.

---

[1] M. Hulme, *Why We Disagree about Climate Change* (Cambridge University Press, 2009) v.

I. THE ROLE OF SCIENCE

The previous chapter detailed the extensive impacts that GhG emissions have had and will be having on the global average temperature as a whole. Based on this knowledge, an oft-heard political argument calls for action against climate change as simply "necessary." From this perspective, the scientific knowledge of climate change determines action on climate change. This view is frequently conveyed not only in the media and in advocacy documents, but also in international policy documents. For instance, the Copenhagen Accord – a document adopted in December 2009 and accepted by most States – recognized "the *scientific view* that the increase in global temperature should be below 2 degrees Celsius."[2]

This quotation displays a logical fallacy – a pattern of reasoning rendered invalid by a flaw in its structure.[3] Science alone cannot compel any specific action. Science is a method of study to understand the universe. It seeks to determine how the universe *is* and thus possibly how it would react to any particular course of action. Science alone cannot determine what *should* be; it cannot assess whether the result of any course of action is desirable or not or how much so. Climate science cannot tell us what increase in global temperature is acceptable; such a determination is only possible based on particular values and interests.[4]

British philosopher David Hulme has demonstrated that a normative statement (a statement about what *should* be done) cannot be deduced only from a factual premise (an understanding of what *is*); it also needs to build upon normative premises or values (an understanding of what *should* be).[5] For instance, simply asserting that smoking harms does not suffice to justify a ban on smoking: to justify such a ban, a moral argument would need to rely on values, for instance, the understanding that public health ought to be protected rather than, say, individual self-determination. A decision to ban smoking requires a balancing of conflicting values.

Just like science has long demonstrated the adverse health effects of smoking, it has also revealed the consequences that current GhG emissions have on the global average temperature. This, however, does not tell us whether these consequences are desirable or not. Values are needed to interpret these scientific facts and to assess them as desirable or undesirable. Some values may seem quite obvious and consensual, but they may be in conflict with other values. For instance, every society recognizes the value of public health, but different responses have been given to questions such as smoking, as the value of public health conflicts with other values.

Likewise, it seems obvious that something must be done to tackle climate change in order to avoid serious harms to human societies and ecosystems, and to reduce the risk of a cataclysmic runaway climate change scenario. Virtually any possible value suggests that such a scenario

---

[2] Copenhagen Accord, in the annex of decision 2/CP.15 (December 18–19, 2009), para. 2. A similar instance can be found in decision 1/CP.13, "Bali Action Plan" (December 14–15, 2007), recital 5, which refers to an IPCC report as a justification of the "urgency" to address climate change.

[3] The inclusion of this fallacy in the Copenhagen Accord obviously stems from the nature of this agreement, hastily drafted in the last hours of protracted negotiations and unedited.

[4] Saleemal Huq is thus reported saying that: "While two degrees is safe for many countries, and many ecosystems, it is not safe for all countries, and all ecosystems." See Helen Burley, "The climate negotiator" (International Institute for Environment and Development, February 15, 2016), www.iied.org/climate-negotiator (accessed December 30, 2017).

[5] See D. Hume, *A Treatise of Human Nature* (Clarendon Press, 1896), book III, part I, section I.

should be avoided, whether for the sake of our own life or for that of future generations, for our own country or for all, and for humankind or other forms of lives on Earth. Yet, efforts to tackle climate change come at a price. The resources that are invested in tackling climate change are diverted from other objectives, some of which may also address legitimate concerns such as global poverty or education. Such diverging concerns require a balancing of values, a task for which climate science is unsuited. Climate science reveals that the likelihood of more dramatic consequences increases with the severity and the rapidity of climate change, but this alone does not justify any particular threshold that should not be surpassed. Contrary to the assessment of the Copenhagen Accord, science alone cannot possibly tell us that "the increase in global temperature *should* be below 2 degrees Celsius,"[6] no more than it tells us how much exposure to cigarette smoke is tolerable.

A valid normative argument to justify climate action should take the form of a syllogism. As David Hume pointed out, this syllogism needs to involve normative as well as factual premises if it is to conclude in a normative statement – it needs to involve values as well as science. Reverting to the example of smoking in public places, a syllogism could be expressed as follows: "We should protect human health, and smoking in public harms human health: we should therefore not smoke in public places." This formal expression of a syllogism does not automatically reconcile people of different views. Someone could retort with another syllogism based on a different value, for instance: "We should protect individual freedom, and smoking is an exercise of individual freedom: we should therefore protect the right to smoke in public places." Yet, formal reasoning helps to clarify the terms of the debate by identifying the source of disagreement – a divergence in the values on which normative arguments are grounded. An agreement regarding smoking in public places will not be found in better science, but in a better understanding of what we really care about: either public health or individual freedom. Compromises will often be necessary.

Likewise, what action we desire against climate change largely depends on what we take to be morally right. Based on the same understanding of climate science, alternative values can justify very different approaches of action on climate change, ranging on a continuum from almost complete inaction to stringent immediate action to cease most GhG emissions at all costs.

To illustrate this point, let us consider two somewhat extreme perspectives, both based on the same understanding of climate science, but on totally different moral intuitions (see Table 2.1):

- One extreme perspective builds on a doctrine of ethical egoism, a radical ethical doctrine according to which "it is necessary and sufficient for an action to be morally right that it maximize one's self-interest."[7] According to this value system, we should only take harm affecting others into account if, and inasmuch as, it will affect us too as a consequence. Individuals could decide to cooperate for the promotion of their own self-interest and thus take *some* steps to mitigate climate change. Yet, the impacts of climate change beyond those able to

---

[6] On the construction of the two-degree target, see Piero Morseletto, Frank Biermann and Philipp Pattberg, "Governing by targets: reduction ad unum and evolution of the two-degree climate target" (2017) 17:5 *International Environmental Agreements: Politics, Law & Economics* 655; Reto Knutti *et al.*, "A scientific critique of the two-degree climate change target" (2016) 9 *Nature Geoscience* 13.

[7] Robert Shaver, "Egoism," *Stanford Encyclopedia of Philosophy* (Spring 2015), http://plato.stanford.edu/archives/spr2015/entries/egoism (accessed December 20, 2017).

Table 2.1 Alternative normative arguments on climate change

	Ethical egoism	Deep ecology
Normative premise	One should maximize one's self interest	One should respect the equal right of all creatures to live and blossom
Factual premise	Anthropogenic GhGs affect the global atmospheric system	
Conclusion	Very little, if any, mitigation action	Radical efforts to cease anthropogenic GhG emissions as soon as possible

cooperate, including future generations and non-intelligent life, will be entirely disregarded. Although young individuals could be willing to cooperate in order to promote their individual interest to prevent a cataclysmic climate change scenario from unfolding within their lifetime, they would have no interest for the long-term implications of current GhG emissions on even the possibility of life on Earth.

- Another somewhat radical perspective builds on the much more demanding moral intuition of the "deep ecology" movement. Deep ecology rejects the anthropocentrism of most ethical doctrines. Some theories of deep ecology endorse the concept of "biospherical egalitarianism," according to which all living organisms have the same intrinsic value. This perspective precludes human activities causing any significant damage which would affect the equal right of all creatures to live and blossom.[8] Accordingly, no significant anthropogenic interference with the climate system could be accommodated within this system of values, and radical efforts would be prescribed to immediately cease virtually any industrial GhG emissions.

These moral intuitions are the extremes of a broad continuum. More reasonable normative premises lead to more moderate conclusions whereby we should decrease GhG emissions rapidly, but without disproportionately affecting human development. Nevertheless, these examples illustrate the moral neutrality of science: based on the same scientific understanding of climate change, different values are able to shape radically different visions of the climate regime. Science is necessary but insufficient to defining the rationale for climate action.

II. THE ROLE OF ECONOMICS

Rather than a mere scientific question, economists often approach climate change as a collective action problem. A collective action problem is one where individual interests push us to adopt a course of conduct which is collectively irrational, and where it would accordingly be better for everyone if we were all to act differently.[9]

A particular collective action problem was theorized by Garrett Hardin in an article published in 1968 on the "tragedy of the commons."[10] Hardin asks us to imagine a common parcel of grassland where several herders can send their cattle to graze. Each individual herder is

---

[8] See in particular Arne Naess, "The shallow and the deep, long-range ecology movement: a summary" (1973) 16:1–4 *Inquiry (Oslo)* 95.

[9] See Mancur Olson, *The Logic of Collective Action* (Harvard University Press, 1965).

[10] Garrett Hardin, "The tragedy of the commons" (1968) 162:3859 *Science* 1243.

interested in increasing his cattle in order to maximize his gain. As all do so, however, there is soon more cattle grazing the common field than the field can sustain, resulting in a deterioration of the land and, possibly, loss of livestock. The "tragic" part of the fable is that the herders' individual rationality leads them to act in a way which threatens their individual interest. Each herder would be better off if everyone were to limit the size of their cattle or their use of the common land.

The theory of the tragedy of the commons applies readily to many situations of environmental degradation which occur when many individual actors pursue their individual interests without sufficient consideration of collective interests, resulting, for instance, in overhunting, overfishing, exhaustion of freshwater resources or accumulation of wastes of different sorts in the environment. As Hardin noted in 1968, "we are locked into a system of 'fouling our own nest,' so long as we behave only as independent, rational, free-enterprisers."[11]

Climate change can be approached as such a collective action problem. As individuals, we emit large amounts of GhGs when pursuing our own interests, buying SUVs or flying to other parts of the world, without really considering the implication that our conduct has on planetary systems. Likewise, each State tends to act in a similar way, pursuing short-sighted national interests without giving sufficient consideration to the cumulative consequences.

Thus, two American law and economics scholars, Eric Posner and David Weisbach, described climate change as "perhaps the most important case of a widespread problem known as the tragedy of the commons."[12] They explain the problem in the following terms:

> Climate scientists have taught us that the atmosphere is a limited resource similar to roads or fisheries ... Therefore, whenever people engage in activities that emit carbon [dioxide] ... they deplete the resource but do not pay a price for the harm they impose on others. The result is overuse. In economic terms, carbon emissions are an "externality" – the effects of emissions on other people are not included in the price. The price people pay – zero – is below the true cost, so people use too much.[13]

There are different ways of addressing a collective action problem, all of which seek to impose collective rationality on individual conduct. Hardin himself called for "a fundamental extension in morality"[14] – an effort of everyone toward more self-restraint and altruism. Regulation can also be an efficient way of imposing collective rationality, for instance – in Hardin's example – by limiting the number of cattle that each herder can rear. Such regulation can come from central authorities or from local ones, or even from communities themselves through social regulation.[15] In addition to spontaneous individual or national efforts to reduce our GhG emissions, international agreements followed by domestic measures of implementation could seek to control individual or national GhG emissions in order to reduce the impact of human societies on the climate system.

[11] *Ibid.*, at 1245.
[12] Eric Posner and David Weisbach, *Climate Change Justice* (Princeton University Press, 2010) 42.
[13] *Ibid.*
[14] Hardin, *supra* note 10.
[15] See Elinor Ostrom *et al.*, "Revisiting the commons: local lessons, global challenges" (1999) 284:5412 *Science* 178.

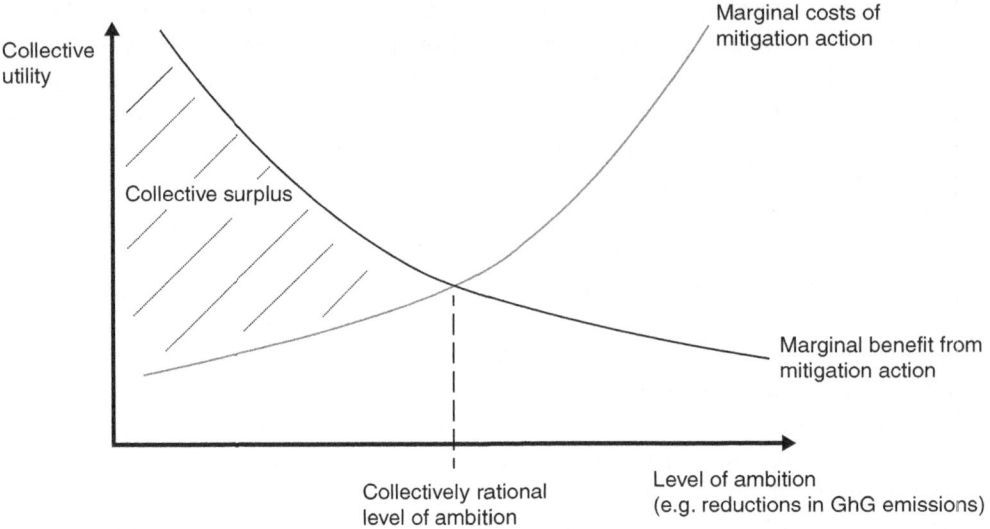

Figure 2.1 Determination of a rational level of ambition in climate change mitigation

In this view, international cooperation on climate change should seek to limit global GhG emissions to a level which is collectively rational. It is rational, from a collective perspective, to seek to maximize the sum of everyone's satisfaction – what economists call collective utility. The first measures to curb GhG emissions are relatively inexpensive when compared with the benefits they provide, but additional measures become more expensive or bring fewer benefits. Maximum collective utility is reached when every effort worth making to advance collective interests has been made: any additional measure would cost more to the community than it would benefit to the community. In other words, maximal collective utility is characterized by a level of ambition whereby the marginal costs of action to mitigate climate change equal the marginal benefits from such action (see Figure 2.1). The costs of mitigation action comprise any loss suffered as a consequence of such action, for instance, by industrial companies that need to invest in new equipment, consumers compelled to pay higher prices or anyone losing a chance to have a higher level of economic development. The benefits from mitigation action, on the other hand, reflect whatever costs of climate change are avoided through such mitigation action: lives which are saved, livelihoods which are not affected, disasters which do not occur, food production which can continue unhindered, future generations which can thrive, biodiversity which can be maintained, among others. Mitigation action, economists say, should be pursued as long as its costs are inferior to the costs of the harms it avoids.[16]

Classical economic analyses suggests that the rationale for international action of climate change could be defined by this point of equal marginal costs and benefits, where all efforts which increase global utility have been made.[17] But while this reasoning is irrefutable in principle – and even rather tautological – what it means concretely is far from clear. Trying to determine whether a specific mitigation action increases global utility requires a valuation of both its cost and its benefit in a common unit. While the cost of a mitigation action may be relatively easy to determine, much

---

[16] See for instance Nicholas Stern, *The Economics of Climate Change* (Cambridge University Press, 2007) 26.
[17] See for instance *ibid.*, Part I, Chapter 2.

greater difficulties are to be met in trying to assess the benefit of such action. This requires not just determining the nature of the impacts of climate change that a particular measure contributes to avoid, but also ascribing an objective economic value to these impacts – a nearly impossible task.

Nicholas Stern, the former chief economist of the World Bank, published an authoritative study of the economics of climate change which came to be known as the "Stern review."[18] This report estimates that "if we don't act, the overall costs and risks of climate change will be equivalent to losing at least 5% of global GDP each year, now and forever."[19] This number suggests that an economic value of climate change can objectively be assessed, making it possible to assess the benefit drawn from a measure which mitigates climate change. Yet, Stern immediately adds that "[i]f a wider range of risks and impacts is taken into account, the estimates of damage could rise to 20% of GDP or more."[20] By comparison, the IPCC's last assessment report suggests that "global annual economic losses for temperature increases of ~2.5°C above pre-industrial levels are between 0.2 and 2.0% of income," while recognizing that such estimates are "partial and affected by important conceptual and empirical limitations."[21] These figures are based on such acrobatic methodological assumptions that one may doubt that they indicate anything useful at all beyond the general understanding that climate change will significantly disrupt human societies.[22]

Economists often assume that everything, including avoiding climate change, has or can objectively be ascribed a particular economic cost. Yet, determining the cost of climate change faces at least four formidable obstacles. Firstly, determining the cost of climate change would require us to ascribe an economic value to things which are not and arguably should not be on the market. Most of the harm caused by climate change is likely to consist not in economic losses, but in "non-economic" human or environmental losses. Losses in human lives or in living conditions (health, culture, identity, security, etc.) and environmental degradation cannot readily be ascribed an economic value because they are not tradable. The sum that one person is ready to pay to save one's life or that a government is ready to pay to reduce mortality (e.g. by investing in road safety) depends on individual and national wealth; this does not necessarily reveal the true value of life. Most of us would justly find repulsive the idea that the life of the poor is worth less simply because they or their government are able to pay less to protect them. This goes against a common deontological understanding that human life and dignity have an intrinsic value and cannot be traded.[23] Likewise, non-human lives, biological diversity and culture do not harm an objective economic value: they are not to be bought or sold.

Secondly, determining the cost of climate change would require us to aggregate individual losses. Even economic losses cannot be aggregated by simple addition when they are suffered by different individuals because they simply do not mean the same thing – the sum of US$100 has a different meaning for an American billionaire or a Bangladeshi farmer. A simple aggregation of economic losses would fail to recognize the implications of climate change for global

[18] *Ibid.*

[19] *Ibid.*, at xv.

[20] *Ibid.*, at xv.

[21] R.K. Pachauri *et al.*, *Climate Change 2014: Synthesis Report. Contribution of Working Groups I, II and III to the Fifth Assessment Report of the Intergovernmental Panel on Climate Change* (IPCC, 2015) 79, box 3.1.

[22] See Richard Mendelsohn, "Economic estimates of the damages caused by climate change" in John S. Dryzek, Richard B. Norgaard and David Schlosberg (eds.), *The Oxford Handbook of Climate Change and Society* (Cambridge University Press, 2011) 177.

[23] See e.g. Henry Shue, "Environmental change and the varieties of justice" in Fen Osler Hampson and Judith Reppy (eds.), *Earthly Goods: Environmental Change and Social Justice* (Cornell University Press, 1996) 9.

inequalities. The impacts of climate change are arguably much more morally significant because they often affect poor populations throughout the world most severely.

Thirdly, determining the cost of climate change would require taking future losses into account – losses that will continue to unfold over decades, centuries and even tens of millennia.[24] As we prefer a loss to occur later rather than earlier, economists usually apply a discount rate on future losses. Defining this discount rate requires us to assess the relative value of present and future losses. There is no objective way to assess this rate, or even to determine whether this should be a fixed rate or one evolving over time. Yet, the question is of the highest consequence to any estimate of the price of climate change. Different assumptions regarding the discount rate to be applied on future loss largely explain differences in order of magnitude among the economic studies trying to ascribe an economic value to climate change.[25]

Lastly, determining the cost of climate change requires us to account for fundamental uncertainties and existential risks. There are risks whose likelihood can be assessed: heads and tails, for instance, are equally likely outcomes of the flipping of a coin. Yet, there are also circumstances where the likelihood of a particular event cannot be determined in any objective and reliable way. Despite decades of intensive research, scientists have not yet been able to predict with reasonable certainty the extent of some positive feedbacks or the likelihood of particular tipping points.[26] These are not assessible risks where the likelihood of occurrence is known, but fundamental uncertainties where the probability cannot reliably be assessed. But some such unknown risks could have cataclysmic consequences. It cannot be excluded that anthropogenic GhG emissions would trigger a scenario of runaway climate change or impacts of such a severity as to threaten our existence as a civilization or our survival as a species. There is no objective way to ascribe a cost to our aversion to poorly assessed existential risk.

The studies that attempt to ascribe an economic value to climate change often satisfy themselves with taking into account as much as can be taken into account. As the IPCC noted, many such estimates "do not account for the possibility of large-scale singular events and irreversibility, tipping points and other important factors, especially those that are difficult to monetize, such as loss of biodiversity."[27] As a consequence, these studies give only a very truncated vision of the benefits of tackling climate change. A more complete valuation of climate change – one that would address the four sets of issues detailed above – would need to consider more than economic data. It would need to look at the way we value things such as life and living conditions, some forms of equality, the future of civilization and life on Earth, and fundamental uncertainties about all of the above. Such questions cannot be solved by

---

[24] Peter U. Clark *et al.*, "Consequences of twenty-first-century policy for multi-millennial climate and sea-level change" (2016) 6:4 *Nature Climate Change* 360.

[25] See William D. Nordhaus, "A review of the Stern Review on the Economics of Climate Change" (2007) 45(3) *Journal of Economic Literature* 686, at 689. See generally Dieter Helm and Cameron Hepburn (eds.), *The Economics and Politics of Climate Change* (Oxford University Press, 2011); Navraj Singh Ghaleigh, "Economics and international climate change law" in Cinnamon P. Carlarne, Kevin R. Gray and Richard Tarasofsky (eds.), *The Oxford Handbook of International Climate Change Law* (Oxford University Press, 2016) 72; Richard S.J. Tol, *Climate Change: Economics Analysis of Climate, Climate Change and Climate Policy* (Edward Elgar, 2014).

[26] T.F. Stocker *et al.*, "Technical summary" in T.F. Stocker *et al.* (eds.), *Climate Change 2013: The Physical Science Basis. Contribution of Working Group I to the Fifth Assessment Report of the Intergovernmental Panel on Climate Change* (Cambridge University Press, 2013), 33, at 50.

[27] Pachauri *et al.*, *supra* note 21, at 79, box 3.1.

economic studies alone; they should rely on genuine political deliberations involving various cultural and moral perspectives.

III. MORAL PERSPECTIVES

Most systems of values recognize the importance of preserving the environment, at least for our sake and that of future generations, while also protecting the most vulnerable. Yet, these legitimate considerations can be articulated in various ways, with more or less emphasis on each element, resulting in varied approaches to the need for international action on climate change. There is, for instance, no objective way of balancing considerations for future generations with concerns for the protection of the environment's intrinsic values. While wealthy populations have the opportunity to think serenely about the fate of future generations, this would be a luxury for those populations trapped in extreme poverty or who are victims of bloody wars or recurrent natural disasters – for someone with an empty stomach, more immediate priorities legitimately prevail over climate change. Therefore, while moral philosophers and multiple civil society organizations have attempted to forge and define a concept of "climate justice," there appears to be various alternative, sound approaches to justice in the context of climate change.[28] Rather than a definitive prescription, this debate has highlighted four sets of moral issues which should be considered, relating respectively to corrective justice, distributive justice, intergenerational justice and environmental justice.

A. Corrective Justice

Climate change raises questions of corrective justice. Corrective justice refers to the widely held principle that a person has the duty to repair the harm caused to others as a consequence of his or her wrongful conduct. This principle is not only recognized in moral systems throughout the world, but also in laws, in particular as one of the foundations of the laws of criminal responsibility and tort. The French Civil Code asserts in the broadest possible terms that "[a]ny act whatever of man, which causes damage to another, obliges the one by whose fault it occurred, to compensate it."[29] A similar principle is recognized in the customary international law of State responsibility.[30]

Climate change results from colossal anthropogenic GhG emissions, most of which are caused by just a handful of industrial nations. Its impacts disproportionately affect the most vulnerable populations which often benefit the least from these GhG emissions. Therefore, a moral argument can be made calling those responsible for colossal GhG emissions to make reparation to those affected by the impacts of climate change. This argument can alternatively

---

[28] See e.g. Stephen Gardiner and David A. Weisbach, *Debating Climate Ethics* (Oxford University Press, 2016); Denis Arnold (ed.), *The Ethics of Global Climate Change* (Cambridge University Press, 2011); Lukas H. Meyer, "Climate justice and historical emissions" (2010) 13:1 *Critical Review of International Social & Political Philosophy* 229; Chukwumerije Okereke, "Climate justice and the international regime" (2010) 1(3) *Wires Climate Change* 462; Alyssa R. Bernstein, "Climate justice" in Deen K. Chatterjee (ed.), *Encyclopedia of Global Justice* (Springer, 2011) 144; Lukas H. Meyer and Pranay Sanklecha (eds.), *Climate Justice and Historical Emissions* (Cambridge University Press, 2017).

[29] French Civil Code (Georges Rouhette and Anne Rouhette-Berton trans.), art. 1382.

[30] ILC, *Draft Articles on Responsibility of States for Internationally Wrongful Acts with Commentaries* in (2001) *Yearbook of the International Law Commission*, vol. II, part two (hereinafter *Articles on State Responsibility*).

be made at the individual level by ascribing individual responsibilities for individual harm, at the level of corporations by ascribing responsibilities to some key industrial corporations, or at the national level by ascribing responsibilities to States for the harm suffered by or within other States. These arguments could be made on moral grounds, but also on legal grounds, either at the municipal level based on the law of tort or its functional equivalents, or at the international level based on the law of State responsibility.

These arguments generally assume that the conduct causing the harm is wrongful.[31] It is self-evident that emitting colossal amounts of GhGs that cause great harm to others throughout the world while affecting future generations and ecosystems is morally wrongful; it may also be inconsistent with States' obligations under general international law.[32] Nevertheless, there are difficulties in defining the extent of the wrong and the respective responsibilities of concurrent GhG emitters. One could reasonably argue that *some* GhG emissions either are not wrongful or else are excusable in certain circumstances. Thus, one cannot reasonably be blamed for breathing, for instance, even though we constantly produce carbon dioxide through this process. Besides, GhG emissions justified by the pursuit of a basic level of development could also appear to be justified. Henry Shue suggested a distinction between excusable "subsistence emissions" and wrongful "luxury emissions."[33] It would, however, be particularly difficult to draw a line between what is justified for subsistence and what is not. Rather than a dichotomy, it may be more reasonable to consider that any additional emission unit is just slightly more wrongful than the previous unit.

One could also question the wrongfulness of historical GhG emissions which occurred prior to the discovery of climate change or at a time where there was no clear scientific consensus on the cause and the severity of climate change. The validity of this argument is uncertain. Knowledge is not always considered as a necessary element for wrong-based responsibility, whether in morals or in laws. The law of tort, by analogy, often recognizes negligence or risk-taking as sufficient grounds to ascribe responsibility in certain circumstances. If one accepts that no responsibility could occur absent knowledge of the consequence, determining the starting point of responsibilities for historical GhG emissions would be difficult because a scientific consensus does not appear all at once. Here again, there is no objective ground for tracing a line at a particular point in time.

B. Distributive Justice

In addition to corrective justice, climate change raises or, often, exacerbates pre-existing issues of distributive justice. Distributive justice relates to the distribution of goods among individuals within a community. In the era of globalization, considerations for distributive justice have increasingly been transposed beyond the State to questions of global justice, which relate to the distribution of goods among peoples, and international justice, regarding the

---

[31] Wrongdoing, however, is not the only ground for remedial obligations. Many legal regimes recognize that responsibility (sometimes called "liability") can stem from taking risk which could affect others even without a more specific "wrong." See e.g. ILC, *Draft Principles on the Allocation of Loss in the Case of Transboundary Harm Arising out of Hazardous Activities*, in (2006) *Yearbook of the International Law Commission*, vol. II, part two.

[32] See Chapter 5, section I.

[33] Henry Shue, "Subsistence emissions and luxury emissions" (1993) 15:1 *Law & Policy* 39.

distribution of goods among States. In a context of increasing global inequalities, the world's wealthy have onerous duties toward those who have insufficient access to the basic resources necessary for survival and flourishing.[34] Redistribution is essential to ensure that all enjoy their most fundamental rights and that everyone's human dignity is duly respected.

Distributive justice questions could be raised in relation to the impacts of climate change which, often, disproportionately affect developing States with few resources to protect their populations. Distributive justice would suggest some duty of assistance to the victims of disasters or more broadly to populations in need.[35] These arguments are not specific to climate change – the duty of assistance in question applies equally to anyone in need because, unlike corrective justice arguments, it does not relate to any causal attribution to a wrongdoing. It is, however, an understatement that the practice of international solidarity remains limited to date. Based on historical experience, distributive justice arguments alone appear unlikely to trigger massive international support.[36]

Distributive justice is also relevant to determining the respective obligations of each actor – States, companies or individuals – in relation to climate change mitigation. It may for instance be argued that every individual should be entitled to emit an equal amount of GhGs. Some distinctions could be based on relevant considerations, for instance, when considering States' duties, to geographic or climatic circumstances requiring particular energy consumptions (e.g. long-distance traveling, heavy use of heating/air conditioning) or to the availability of sources of renewable energy. The financial capacity of a State and its ability to take rapid mitigation measures without disproportionately affecting its development should arguably also be taken into account.[37] Here again, however, these arguments can be developed in myriad ways, as is illustrated by the protracted international negotiations on differentiation in international cooperation in response to climate change.[38]

C. Intergenerational Justice

In addition, climate change raises questions regarding the duties of present generations toward future generations. In fact, there is hardly any clearer example of an intergenerational injustice than climate change, which Stephen Gardiner called the "Tyranny of the Contemporary."[39] Most GhGs emitted today will remain in the atmosphere for centuries or millennia. As planetary systems are only beginning a slow process of adjustment to a sudden change in the atmospheric chemistry, most of the impacts of today's GhG emissions will only unfold in the decades or centuries to come. Consequently, future generations will bear more

---

[34] See e.g. Thomas Pogge, *World Poverty and Human Rights: Cosmopolitan Responsibilities and Reforms* (Blackwell, 2002).

[35] Thus, distributive justice arguments have often played an important role in international negotiations, despite the nearly constant refusal of developed States to recognize the validity of any corrective justice argument. See e.g. Chapters 10 (adaptation) and 12 (support).

[36] Thus, most of the support provided in relation to climate change seeks to enhance international action on climate change mitigation rather than assisting adaptation efforts in developing countries. See Chapter 12.

[37] See e.g. Henry Shue, "Global environment and international inequality" (1999) 75:3 *International Affairs* 531.

[38] See Chapter 6.

[39] Stephen Gardiner, *A Perfect Moral Storm: The Ethical Tragedy of Climate Change* (Oxford University Press, 2011) 143.

of the costs of today's GhG emissions. Their very existence is threatened by the prospect of a cataclysmic runaway climate change scenario.

What are our duties to future generations? Different societies appear to have different approaches to this question. Traditional societies are generally depicted as putting greater emphasis on future generations than modern societies. The discourse on a "sustainable development" originates from an attempt to bring future generations back to the debate in modern societies. Thus, sustainable development was defined as "development that meets the needs and aspirations of the present without compromising the ability of future generations to meet their own needs."[40] Yet, this again appears rather vague and abstract, subject to different interpretations.

Multiple biases act in favor of present generations; future generations do not participate in current debates. The application of a discount rate on the assessment of future losses caused by climate change reflects the difficulty of accounting for the losses that are inflicted on future generations.[41] Likewise, faith in technological progress is sometimes invoked in favor of postponing measures to tackle climate change, on the elusive ground that some alternative sources of energy, for instance, could soon become economically viable. While GhG is accumulating in the atmosphere, current generations are creating an increasingly difficult task for future generations not just to mitigate climate change, but also to adapt to its impacts.

D. Environmental Justice

Lastly, climate change raises questions relating to environmental justice. These include not only questions of corrective and distributive justice applied to environmental goods, but also questions relating to our relation to the environment, in particular to other forms of life on Earth. Some scholars in environmental ethics suggest that we have certain duties to protect or, at least, not to harm our environment more than is strictly necessary.[42] Accordingly, animals and plants, as well as ecosystems as a whole, have not only an instrumental value (one related to the role they play, or could play, for human societies) but also an intrinsic value; they should be protected, or at least respected, for their own sake.

This aspect of the ethical debate is often overseen. Thus, the research on the impact of climate change is typically centered on human societies. More often than not, the environment is approached through anthropocentric concepts such as "ecosystem goods" or "ecosystem services,"[43] being only recognized for its instrumental value. This often conceals the fact that climate change is taking a terrible toll on non-human life on Earth, and will increasingly do so, as many species are unable to adapt to changes in their environment. Warren and colleagues suggest that, absent climate change mitigation, 57 percent of plant species and 34 percent of animal species would lose more than half of their climatic range by the 2080s, "amounting to

---

[40] WCED, *Our Common Future* (Oxford University Press, 1987). On the principle of sustainable development, see Chapter 5, section II.A.

[41] See *supra* note 25 and accompanying text.

[42] See e.g. Naess, *supra* note 8.

[43] See Pachauri *et al.*, *supra* note 21, at 122, defining ecosystem services as "[e]cological processes or functions having monetary or non-monetary value to individuals or society at large." This approach artificially distinguishes useful parts of nature from other parts of nature, ignoring the fact that nature is an interdependent whole. Ecosystem services are likely to be disturbed in direct and unpredictable ways when other aspects of the natural environmental are being harmed.

a substantial global reduction in biodiversity and ecosystem services by the end of this century."[44] No study could be found on what would happen to non-human life on Earth after the end of the twenty-first century in the absence of appropriate mitigation efforts.[45]

IV. PSYCHOLOGICAL, SOCIAL AND POLITICAL OBSTACLES

Scientific, economic and moral arguments on the need to address climate change have an influence on the conduct of international affairs, but they do not directly frame the international law on climate change. Rather, these arguments are conveyed and filtered through a series of psychological, social and political processes.[46] Like arguments for peace, arguments for addressing climate change are often disregarded through these complex processes.

In some societies like the United States, climate change denial remains influential, reinforced by powerful economic interests.[47] Like elsewhere, arguments on climate change are also filtered through psychological processes of denial – it is simply easier to "forget" about an issue when there is no simple fix and no immediate individual threat.[48] As a slow change in average weather conditions on a global scale, climate change does not get in our way on a daily basis. Jared Diamond documented the tendency of crises occurring through "creeping normalcy" to remain unaddressed, leading to the collapse of many past civilizations.[49] As a slow process through which GhGs accumulate in the atmosphere, the climate change crisis "lacks a sense of urgency" which would trigger support to action.[50] While States devote considerable resources to address the imminent risks of terrorism or ensure national security, the significant risks raised by climate change are far too often overlooked in public debates due to the lack of a salient incident unequivocally attributable to climate change.[51] A creeping crisis does not make the news on any particular day and does not "jump" into the political agenda.

Thus, political arguments for action on climate change need to overcome many barriers if they are to succeed. As Stephen Gardiner noted, "even if we initially accept that we face a serious moral challenge, our resolve remains vulnerable to corrupt mechanisms of persuasion."[52] The populations most affected by climate change, often marginalized, tend to be among those that are the least able to influence global GhG emissions and processes through

---

[44] R. Warren et al., "Quantifying the benefit of early climate change mitigation in avoiding biodiversity loss" (2013) 3:7 Nature Climate Change 678.

[45] See, for instance, Stocker et al., supra note 26, at 1029, which contains no discussion of the impact of "long-term" climate change (i.e. until 2300) on biodiversity.

[46] See Neta Crawford, "Homo politicus and argument (nearly) all the way down: persuasion in politics" (2009) 7:1 Perspectives on Politics 103; Martha Finnemore and Kathryn Sikkink, "International norm dynamics and political change" (1998) 52:4 International Organization 887; Thomas Risse, "'Let's argue!': communicative action in world politics" (2000) 54:1 International Organization 1.

[47] Riley E. Dunlap and Aaron M. McCright, "Organized climate change denial" in John S. Dryzek, Richard B. Norgaard and David Schlosberg (eds.), The Oxford Handbook of Climate Change and Society (Oxford University Press, 2011) 144.

[48] Karie Marie Norgaard, "Climate denial: emotion, psychology, culture, and political economy" in Dryzek et al. (eds.), supra note 47, 399.

[49] Jared M. Diamond, Collapse: How Societies Choose to Fail or Succeed (Penguin, 2011) 425.

[50] Anthony Leiserowitz, "Climate change risk perception and policy preferences: the role of affect, imagery, and values" (2006) 77:1–2 Climatic Change 45.

[51] See Cass R. Sunstein, "On the divergent American reactions to terrorism and climate change" (2007) 107:2 Columbia Law Review 503.

[52] Gardiner, supra note 39, at 301.

which they are regulated. Electoral processes in democractic countries structurally favor candidates who defend national interests rather than those who seek a more responsible approach to international cooperation. Future generations and non-human life are never directly represented. This unequal argumentative field makes it possible for a complacent, self-serving, ahistorical narrative on climate change to develop, putting great emphasis on the willingness of Western States to do *something* when they could do *nothing* at all – overseeing their moral duties to do *much more*. It is only through a gradual process of awareness-raising and education that arguments for more ambitious climate action could one day succeed.

V. CONCLUSION

The rationale for international action on climate change is not defined by scientists, economists or moral philosophers in isolation. Nor is there a single best technical fix to climate change which could be determined by experts of any sort. Determining the right course of action in response to climate change requires a balancing of values – environmental protection and development, the interest of present and future generations, advancing human development and respecting the intrinsic value of the non-human environment, or accepting small risks of cataclysmic futures or preferring to remain on the safe side. Such choices must be based on appropriate scientific and economic considerations, but science or economics alone cannot determine what our societies ought to do about climate change. Instead, political deliberations based on cultural and moral perspectives determine the willingness of each community to act. These deliberations take place in non-ideal circumstances and they are hindered by various psychological and political obstacles

Each of the ethical arguments noted above – on corrective justice, distributive justice, future generations and environmental justice – is contested by some and interpreted by others in many different ways. For some, tackling climate change is about addressing "an act of aggression by the rich world against the poor one."[53] For others, it is about trying to avoid a "dangerous anthropogenic interference with the climate system"[54] which "may adversely affect natural ecosystems and humankind"[55] and "undermine the ability of all countries to achieve sustainable development."[56] For yet others, it is an infringement of the inherent rights of "Mother Earth," which needs to be ceased and remedied.[57] There is not one single but many concurrent moral narratives capable of justifying international action against climate change. This variety of views should be acknowledged and, as far as possible, accommodated in the international law on climate change.

---

[53] Y. Museveni, President of Uganda, cited in "Drying up and flooding out" *The Economist* (May 10, 2007).

[54] United Nations Framework Convention on Climate Change, May 9, 1992, 1771 *UNTS* 107 (hereinafter UNFCCC), art. 2.

[55] *Ibid.*, recital 3.

[56] UN General Assembly Resolution 70/1, "Transforming our world: the 2030 Agenda for Sustainable Development" (September 25, 2015), para. 14.

[57] World People's Conference on Climate Change and the Rights of Mother Earth, Universal Declaration on the Rights of Mother Earth (April 22, 2010). The Universal Declaration on the Rights of Mother Earth is a declaration adopted by a global gathering of civil society and governments, organized by the government of Bolivia, on behalf of "the peoples and nations of Earth."

3

The UNFCCC Regime, from Rio to Paris

The international law on climate change comprises, first of all, a series of international climate agreements. This chapter provides a general overview of the regime constituted under the UNFCCC, including the Kyoto Protocol and the Paris Agreement as well as the Copenhagen Accord and the Cancún Agreements, among others. The next chapter turns to relevant developments under other, non-specific treaty regimes, while Chapter 5 delves into norms of general international law.

The UNFCCC regime is the most evident component of the international law on climate change. It is constituted by treaties adopted with the aim of responding to climate change, and decisions taken by the parties to these treaties to promote their implementation. The UNFCCC regime has developed through difficult and protracted international negotiations, trying to achieve consensus despite the incompatible expectations of multiple key stakeholders – European States keen to push for strong mitigation efforts, developing States pointing at developed States' responsibilities, the United States rejecting any such responsibility, oil-producing States often hostile to any substantial cut in fossil fuel consumption and the small island developing States calling for more ambition, among others. A quarter of a century of protracted international negotiations achieved only partial success, and frustration grew among partakers and observers. While the Paris Agreement has fueled new hopes, its effectiveness remains uncertain and many doubts remain regarding its modalities of application. In June 2017, US President Trump's decision to pull the United States out of the Paris Agreement shows, once again, the difficulty of orchestrating international cooperation on climate change.

Adopting a chronological approach, the present chapter explores the main achievements of successive international climate agreements: the UNFCCC, the Kyoto Protocol and its Doha Amendment, the Copenhagen Accord and the Cancún Agreements, and finally the Paris Agreement. Only the most prominent features of successive agreements are presented here; other chapters discuss specific areas of cooperation, including mitigation, adaptation, loss and damage, and support.[1] Table 3.1 provides a general overview of the chronology of the different developments discussed in this chapter.

---

[1] See respectively Chapters 7, 10, 11 and 12.

Table 3.1 Milestones in the development of the UNFCCC regime

Date	Event
October 26, 1990	Publication of *Climate Change: The IPCC Scientific Assessment*,[a] a report prepared by the first working group of the Intergovernmental Panel on Climate Change as part of the First Assessment Report
December 21, 1990	The UN General Assembly establishes the Intergovernmental Negotiating Committee for a Framework Convention on Climate Change (INC/FCCC)
May 9, 1992	The INC/FCCC adopts the UN Framework Convention on Climate Change (UNFCCC), which entered into force on March 21, 1994
December 11, 1997	The third session of the Conference of the Parties to the UNFCCC (COP3) adopts the Kyoto Protocol, which enters into force on February 16, 2005
November 10, 2001	COP7 adopts the Marrakesh Accords, a series of decisions which paved the road to the entry into force of the Kyoto Protocol
December 14–15, 2007	COP13 adopts the Bali Action Plan
December 18, 2009	COP15 takes note of the Copenhagen Accord, a text negotiated by a group of critical States and soon accepted by 141 parties
December 10–11, 2010	COP16 adopts the Cancún Agreements, essentially taking stock of the Copenhagen Accord and defining modalities of application
December 8, 2012	The eighth session of the Conference of the Parties serving as the meeting of the Parties to the Kyoto Protocol (CMP8) adopts the Doha Amendment to the Kyoto Protocol, which had not yet entered into force as of early 2018
December 12, 2015	COP21 adopts the Paris Agreement, which enters into force on November 4, 2016. The Paris Rulebook is expected to be adopted in December 2018

[a] See J.T. Houghton, G.J. Jenkins and J.J. Ephaums (eds.), *Climate Change: The IPCC Scientific Assessment. Report Prepared for Intergovernmental Panel on Climate Change by Working Group I* (Cambridge University Press, 1990). See also W.J. McG. Tegart, G.W. Sheldon and D.C. Griffiths (eds.), *Climate Change: The IPCC Impacts Assessment. Report Prepared for Intergovernmental Panel on Climate Change by Working Group II* (Australian Government, 1990); F. Bernthal *et al.*, *Climate Change: The IPCC Response Strategies. Report Prepared for Intergovernmental Panel on Climate Change by Working Group III* (Island Press, 1990).

I. THE UNFCCC

Evidence that human activities affect the climate system were progressively gathered during the second half of the twentieth century.[2] In the late 1950s, measurements taken from the Mauna Loa observatory in Hawaii revealed a steady increase in carbon dioxide concentration in the atmosphere. In the 1980s, enough empirical evidence had been gathered to leave no reasonable doubt: colossal GhG emissions from the combustion of fossil fuels and other human activities were impacting the global climate system with largely unknown consequences on human and non-human life. Successful international negotiations had been conducted to reduce the production of ozone-depleting substances during the second half of the 1980s.[3]

[2] For a historical recount, see Spencer Weart, *The Discovery of Global Warming*, 2nd edn (Harvard University Press, 2008).

[3] The experience of the regime on the protection of the ozone layer and its contribution to reducing global GhG emissions is recounted in Chapter 4, section I.

A strong global momentum was building up in favor of similar efforts toward the limitation of global GhG emissions.

In 1988, the World Meteorological Organization (WMO) and the United Nations Environmental Programme set up the IPCC in order to take stock of a growing number of scientific studies to inform policy-makers based on the best available science. The IPCC's first report, published in 1990, confirmed the scientific bases of arguments for an international action against climate change. Accordingly, in December 1990, the UN General Assembly established the Intergovernmental Negotiating Committee for a Framework Convention on Climate Change (INC/FCCC).[4] From February 1991 to May 1992, intense negotiations took place under the aegis of the INC/FCCC. To achieve consensus among actors with very different expectations, the negotiations often resorted to constructive ambiguities and vague provisions with little prospects of enforcement.[5]

The UNFCCC was formally adopted on May 9, 1992 and opened for signatures at the Earth Summit, which took place in Rio de Janeiro, Brazil, the following month. It entered into force on March 21, 1994. To date, it has been ratified by 196 States and the European Union, making it one of the most widely ratified treaties.[6]

The UNFCCC establishes a general framework. It defines an "ultimate" objective and some principles for cooperation, outlines some vague national commitments to promote mitigation and adaptation, and establishes institutions which could facilitate further negotiations. It was clearly understood in 1992 that more specific obligations would need to be defined in a subsequent instrument. In the minds of the negotiators was the experience of the 1985 Vienna Convention on the Protection of the Ozone Layer, followed two years later by the adoption of its Montreal Protocol.[7] Like the Vienna Convention, the UNFCCC was seen as a first step toward laying the ground for a new regime.

A. The Ultimate Objective

Article 2 of the UNFCCC defines the "ultimate objective" of the UNFCCC and any related legal instruments:

> to achieve, in accordance with the relevant provisions of the Convention, stabilization of greenhouse gas concentrations in the atmosphere at a level that would prevent dangerous anthropogenic interference with the climate system.[8]

---

[4] UN General Assembly Resolution 45/212, "Protection of global climate for present and future generations of mankind" (December 21, 1990).

[5] For a historical recount, see Irving M. Mintzer and J. Amber Leonard (eds.), *Negotiating Climate Change: The Inside Story of the Rio Convention* (Cambridge University Press, 1994).

[6] As a comparison, there are currently 193 Member States of the United Nations.

[7] The main instruments of the ozone regime are the Vienna Convention for the Protection of the Ozone Layer, March 22, 1985, 1513 *UNTS* 293, which provides a general framework, and the Montreal Protocol on Substances that Deplete the Ozone Layer, September 16, 1987, 1522 *UNTS* 3, which defines more specific national commitments. See Chapter 4, section I.

[8] United Nations Framework Convention on Climate Change, May 9, 1992, 1771 *UNTS* 107 (hereinafter UNFCCC), art. 2. See generally David Freestone, "The United Nations Framework Convention on Climate Change: the basis for the climate change regime" in Cinnamon P. Carlarne, Kevin R. Gray and Richard Tarasofsky (eds.), *The Oxford Handbook of International Climate Change Law* (Oxford University Press, 2016) 97.

This provision defines the UNFCCC regime as a goal-oriented, transitory regime. This suggests that the UNFCCC and any further international climate agreement would become obsolete once atmospheric GhG concentrations are stabilized at a safe level. Nevertheless, the UNFCCC does not contain any provision on its termination following the achievement of this goal, which in any case remains a distant prospect.

The ultimate objective of the UNFCCC places emphasis on climate change mitigation. Nevertheless, adaptation efforts are mentioned several times in the Convention and they could be considered as a complementary way of reducing the danger of anthropogenic interference with the climate system.

What would constitute a "dangerous anthropogenic interference with the climate system" is far from clear. The UNFCCC provides no definition of what level of risk should be considered "dangerous." One could argue that any interference with the climate system is dangerous per se, given our limited understanding of how the climate system and other planetary systems will react. The lack of a more specific definition of the UNFCCC's ultimate objective reflects the difficulty of reaching a consensus among States, which, at this point, had radically different expectations from international cooperation on climate change.

B. The Principles

Article 3 of the UNFCCC states a list of principles. The inclusion of the term "principle," however, was opposed by the United States. While "principles" was agreed to be the title of Article 3, a footnote was added under the title of Article 1 to specify that "Titles of articles are included solely to assist the reader,"[9] thus suggesting that they should not be used to interpret the provisions. It is difficult, however, to distinguish between a word included "to assist the reader" and one that may have an interpretative value.

The first paragraph of Article 3 defines the principle of Common but Differentiated Responsibilities and Respective Capabilities, providing the ground for differentiation among States. It reads as follows:

> The Parties should protect the climate system for the benefit of present and future generations of humankind, on the basis of equity and in accordance with their common but differentiated responsibilities and respective capabilities. Accordingly, the developed country Parties should take the lead in combating climate change and the adverse effects thereof.[10]

The Rio Declaration on Environment and Development is another document adopted during the 1992 Earth Summit. Principle 7 of the Rio Declaration affirms the principle of Common but Differentiated Responsibilities in slightly different terms, without mentioning "respective capabilities":

> In view of the different contributions to global environmental degradation, States have common but differentiated responsibilities.[11]

---

[9] UNFCCC, *supra* note 8, footnote under art. 1. See Daniel Bodansky, "The United Nations Framework Convention on Climate Change: a commentary" (1993) 18:2 *Yale Journal of International Law* 451, at 502.

[10] UNFCCC, *supra* note 8, art. 3.1.

[11] UNCED, Rio Declaration on Environment and Development (June 3–14, 1992), available in (1992) 31 *ILM* 874 (hereinafter *Rio Declaration*), principle 7. See generally Jorge E. Viñuales (ed.), *The Rio Declaration on Environment and Development: A Commentary* (Oxford University Press, 2015).

The principle of Common but Differentiated Responsibilities (and Respective Capabilities) has remained ill-defined due to a lack of consensus among negotiating States. There is no specification, in the UNFCCC or in documents adopted later, of the ground, extent or ambit of differentiation. Developing States promoted the insertion of elements of language that would recognize the historical responsibility of developing nations and the consequential obligations of these nations. Developed States, by contrast, accepted that they had greater moral obligations on the ground of their financial capacity, but rejected any mention of historical responsibilities. For example, the United States transmitted the following written statement to the UN Secretary General upon the adoption of the Rio Declaration:

The United States understands and accepts that principle 7 highlights the special leadership role of the developed countries, based on our industrial development, our experience with environmental protection policies and actions, and our wealth, technical expertise and capabilities.

The United States does not accept any interpretation of principle 7 that would imply a recognition or acceptance by the United States of any international obligations or liabilities, or any diminution in the responsibilities of developing countries.[12]

In addition to the principle of Common but Differentiated Responsibilities (and Respective Capabilities), Article 3 also acknowledges "the specific needs and special circumstances of developing country Parties."[13] It notes the relevance of "precautionary measures" in application of the principle of precaution.[14] Finally, it highlights the "right" of the parties "to ... promote sustainable development"[15] and their obligation to "promote a supportive and open international economic system,"[16] thus suggesting that mitigation measures should not impose disproportionate constraints on national development policies.

C. National Commitments

Article 4 defines the particular efforts that States agreed to carry on. In particular, Article 4.1 defines national commitments that all parties should implement, although taking their common but differentiated responsibilities into account. Article 4.2 contains more demanding commitments applicable only to developed countries as listed in Annex I, which includes the States which were then Members of the Organisation for Economic Co-operation and Development (OECD). The obligations contained in Article 4, paragraphs 1 and 2, include in particular obligations to make a national inventory of GhG emissions and obligations to take some measures to mitigate climate change and to facilitate adaptation.

---

[12] Written statement of the United States on Principle 7 of the Rio Declaration, in *Report of the United Nations Conference on Environment and Development*, UN document A/CONF.151/26 (Vol. IV) (September 28, 1992), para. 16.

[13] UNFCCC, *supra* note 8, art. 3.2.

[14] *Ibid.*, art. 3.3.

[15] *Ibid.*, art. 3.4.

[16] *Ibid.*, art. 3.5.

These provisions are formulated in a vague, almost incantatory language. They involve some obligations of conduct, for instance, the obligation for all States to "formulate, implement, publish and regularly update ... programmes containing measures to mitigate climate change"[17] and for developed States to "adopt national policies and take corresponding measures on the mitigation of climate change."[18] Yet, these commitments do not define any obligation to achieve a particular result, such as any quantified targets of emission reduction.

Article 4, paragraphs 3–5 suggest that developed States shall provide financial and techno-logical assistance to the implementation of mitigation and adaptation policies by developing States. Article 4.8 recognizes the particular vulnerability of nine categories of States, including small island states and countries with low-lying coastal areas as well as countries with areas of high urban atmospheric pollution and "countries whose economies are highly depending on income generated from the production, processing and export, and/or on consumption of fossil fuels and associated energy-intensive products."

D. Institutional Developments

Other articles in the UNFCCC establish institutions which will become central to the UNFCCC regime, being later endorsed by the Kyoto Protocol and the Paris Agreement. Article 7 establishes the COP, an assembly of the representatives of all parties meeting at annual sessions to take decisions on the implementation of and if need be on amendments or protocols to the UNFCCC. Articles 9 and 10 establish two specialized organs to hold inter-sessional multilateral negotiations, the Subsidiary Body for Scientific and Technological Advice (SBTA) and the Subsidiary Body for Implementation (SBI), both of which report to the COP. The COP was directed to adopt its own rules of procedures at its first session,[19] but no agreement could be reached on voting rules for substantive matters. While the reminder of the draft rules of procedure has been applied provi-sionally since 1995, the COP has followed the customary usage in international negotiations of taking substantial decisions by consensus only.[20] This means that, in principle at least, any party can always block a decision by objecting to it.[21]

Article 8 establishes a permanent Secretariat. The first session of the Conference of the Parties (COP1) agreed that the UNFCCC Secretariat would be "institutionally linked to the United Nations, while not being fully integrated in the work programme and management structure of any particular department or programme."[22] The head of the Secretariat, with the title of Executive Secretary, is appointed by the Secretary General of the United Nations (UN).[23] It was also decided that the Secretary would be located in Bonn, the former capital of West Germany.[24] The UNFCCC Secretariat provides support to the negotiations and follow-up on

---

[17] *Ibid.*, art. 4.1(b).

[18] *Ibid.*, art. 4.2(a).

[19] *Ibid.*, art. 7.3.

[20] UNFCCC, *Draft Rules of Procedure of the Conference of the Parties and its Subsidiary Bodies*, UNFCCC document FCCC/CP/1996/2 (May 22, 1996). See also *Report of the Conference of the Parties on its First Session*, UNFCCC document FCCC/CP/1995/7 (May 24, 1995), para. 14.

[21] See generally Antto Vihma, "Climate of consensus: managing decision making in the UN climate change negoti-ations" (2015) 24:1 *Review of European Community & International Environmental Law* 58.

[22] Decision 14/CP.1, "Institutional linkage of the Convention secretariat to the United Nations" (April 7, 1995), para. 2.

[23] *Ibid.*, para. 7.

[24] Decision 16/CP.1, "Physical location of the Convention secretariat" (April 7, 1995), para. 1.

the implementation of the Convention and subsequent instruments. It is fairly small-scale in structure, currently employing around 500 people.

The final articles of the Convention contain general provisions. Article 11 establishes a Financial Mechanism. Article 14 relates to the settlement of disputes. Articles 15 and 16 provides rules on the adoption of amendments to the Convention and revisions to its Annexes. Article 17 defines the procedure for the adoption of Protocols to the Convention. Articles 19–23 relates to the adoption and entry into force of the UNFCCC. Article 24 excludes reservations to any provision of the Convention. Article 25 regulates withdrawal of a party from the Convention. Article 26 clarifies that the versions in the six UN official languages are equally authentic.

II. THE KYOTO PROTOCOL

A few months after the entry into force of the UNFCCC, the first decision adopted by the COP at its first session (decision 1/CP.1) established the "Berlin Mandate" for the negotiation of "a protocol or another legal instrument."[25] The protocol would aim to specify the obligation of developed states under Article 4.2 of the UNFCCC, in particular by "set[ting] quantified limitation and reduction objectives with specified time-frames."[26] It was agreed that the Berlin Mandate would "[n]ot introduce any new commitments" for developing States.[27]

Negotiations under the Berlin Mandate took place in the two years that followed, leading to the adoption of the Kyoto Protocol at COP3 on December 11, 1997. These were among the most difficult international negotiations ever undertaken. Defining national Quantified Emission Limitation and Reduction Commitments (QELRCs) required negotiators to take position on two extremely controversial questions. Firstly, an overall objective had to be agreed upon, implicitly at least, in a more specific way than in the UNFCCC. Secondly, respective national commitments had to be determined based on an interpretation of grounds of differentiation.

A. Quantified Emission Limitation and Reduction Commitments (QELRCs)

Article 3.1 of the Kyoto Protocol reads as follows:

The Parties included in Annex I shall, individually or jointly, ensure that their aggregate anthropogenic carbon dioxide equivalent emissions of the greenhouse gases listed in Annex A do not exceed their assigned amounts, calculated pursuant to their quantified emission limitation and reduction commitments inscribed in Annex B and in accordance with the provisions of this Article, with a view to reducing their overall emissions of such gases by at least 5 per cent below 1990 levels in the commitment period 2008 to 2012.[28]

---

[25] Decision 1/CP.1, "The Berlin Mandate" (April 7, 1995), recital 4. The Berlin Mandate was named after the city where the first COP was held.

[26] *Ibid.*, para. 2(a).

[27] *Ibid.*, para. 2(b).

[28] Kyoto Protocol to the United Nations Framework Convention on Climate Change, December 11, 1997, 2303 *UNTS* 162 (hereinafter Kyoto Protocol), art. 3.1.

This provision refers to two annexes. Annex A contains a list of six GhGs – carbon dioxide, methane, nitrous oxide, hydrofluorocarbons (HFCs), perfluorocarbons (PFCs) and sulfur hexafluoride – and a list of relevant sectors and source categories. Annex B contains a list of 38 developed parties essentially identical to Annex I of the UNFCCC. For each of these parties, a second column defines a QELRC expressed as a percentage of the party's 1990 emissions.[29] "Annex B parties" is commonly used as shorthand to refer to the parties to the Kyoto Protocol which, being included in the Annex B, have agreed to emission limitation and reduction commitments under the Kyoto Protocol.

The QELRCs under the Kyoto Protocol are defined as a percentage of a State's emissions in 1990 that States should not exceed during a "commitment period" extending from 2008 to 2012.[30] Unlike the obligations of conduct defined in Articles 4.1 and 4.2 of the UNFCCC, QELRCs are defined as obligations of result – an obligation to "ensure" the realization of a specified outcome. These specific mitigation commitments vary from a reduction of GhG emissions by 8 percent for the European Community and 7 percent reduction for the United States, to a limitation to an increase by 10 percent for Iceland and by 8 percent for Australia. These different commitments were agreed as a recognition of differences in national circumstances, in particular in energy needs and capacity to develop alternative sources of energy.[31] The Protocol was drafted in such a way as to allow further commitment periods to be added through subsequent amendments. As will be shown below, the Doha Amendment would – subject to its entry into force – add a second commitment period.[32]

In combination, the individual commitments included in Annex B aimed to achieve a reduction by 5 percent below 1990 levels in the GhG emissions of developed States.[33] Yet, no similar commitment applied to developing States, which were not included in Annex B. This slight reduction in the GhG emissions of developed States was insufficient to prevent a total increase in global GhG emissions due to a rapid increase in emerging economies.

B. Flexibility Mechanisms

To facilitate compliance with QELRCs, the Kyoto Protocol established three so-called "flexibility mechanisms." These mechanisms allow an Annex B party to report mitigation outcome occurring beyond its own territory as if it was its own mitigation outcome. In particular:

- Joint Implementation permits an Annex B party to receive "Emission Reduction Units" from a mitigation project taking place within the territory of another Annex B party and to use these Emission Reduction Units for the purpose of fulfilling its own QELRC.[34] This allows an Annex B party to invest in cheaper mitigation projects abroad if it faces difficulties limiting domestic GhG emissions, while exploiting opportunities for further mitigation action in another Annex B party.
- The Clean Development Mechanism (CDM) allows an Annex B party to receive "Certified Emission Reductions" from a mitigation project taking place within the territory of a non-Annex B party (a developing State) and to use these Certified Emission Reductions for the

---

[29] "Reduction" means a decrease in the overall GhG emissions from a State, while "limitation" confines increase in the overall GhG emissions from a State by a certain percentage.

[30] Kyoto Protocol, *supra* note 28, art. 3.1.

[31] UNFCCC, COP3, *Adoption of a Protocol or Another Legal Instrument: Fulfilment of the Berlin Mandate: Revised Text under Negotiation* (November 12, 1997), doc. FCCC/CP/1997/2, at 31.

[32] See below, section III.B.

[33] Kyoto Protocol, *supra* note 28, art. 3.1.

[34] *Ibid.*, art. 6.

purpose of achieving compliance with its QELRC.[35] Like Joint Implementation, this allows an Annex B party to invest in cheaper mitigation projects abroad if it faces difficulties limiting domestic GhG emissions, while also spurring initial consideration for climate change mitigation in developing countries.

- An Emission Trading system allows Annex B parties to sell and buy "Assigned Amount Units" to and from one another in order to fulfill their respective QELRCs.[36] This allows Annex B parties that have achieved mitigation outcomes beyond their commitment to sell some of these outcomes to other Annex B parties that have achieved less than their commitment.

A more complete discussion of the Flexibility Mechanisms of the Kyoto Protocol can be found in Chapter 8.

C. The Reception of the Kyoto Protocol

Adopted on December 11, 1997, the Kyoto Protocol was opened for signature on March 16, 1998. There were already significant doubts, however, about the possibility of the United States ratifying the Kyoto Protocol. Under the law of the United States, ratification of the Kyoto Protocol would require a vote by two-thirds of the Senate – a condition that has excluded the United States from several major historical treaties.[37] On July 25, 1997, the US Senate adopted by 95 votes and 5 abstentions the Byrd-Hagel resolution, where it announced its sense that the United States should not be party to a treaty which would:

(A) mandate new commitments to limit or reduce greenhouse gas emissions for the Annex I Parties, unless the protocol or other agreement also mandates new specific scheduled commitments to limit or reduce greenhouse gas emissions for Developing Country Parties within the same compliance period, or

(B) would result in serious harm to the economy of the United States.[38]

Even though President Clinton signed the Kyoto Protocol on November 12, 1998, the US Senate refused to ratify it and, as a consequence, the United States did not become bound by any emission limitation and reduction commitment. The United States was then the largest GhG emitter (now surpassed by China) and its GhG emissions represented a large share of the emissions of developed States that the Kyoto Protocol sought to regulate.

The United States' non-participation raised questions about the opportunity for other Annex I Parties to implement the Kyoto Protocol. A political agreement was finally achieved at COP7, held in Marrakesh in 2001. The "Marrakesh Accords,"[39] constituted by 23 concomitant COP decisions, marked an agreement on numerous modalities of application of

---

[35] Kyoto Protocol, *supra* note 28, art. 12.

[36] Kyoto Protocol, *supra* note 28, art. 17.

[37] These include the 1919 Treaty of Versailles, the 1982 United Nations Convention on the Law of the Sea and the 1998 "Rome" Statute of the International Criminal Court, among others.

[38] US Senate Resolution 98, 105th Cong., 143 Cong. Rec. S8138–39 (July 25, 1997) (hereinafter the Byrd-Hagel Resolution).

[39] Decisions 2/CP.7 to 24/CP.7 (November 10, 2001). These decisions were formally adopted by the COP serving as the Meeting of the Parties to the Kyoto Protocol at its first session in 2005.

the Kyoto Protocol, in particular regarding its flexibility mechanisms, and indicated an agreement of all other developed States to continue the ratification process. Ratified by virtually every State other than the United States, the Kyoto Protocol finally entered into force on February 16, 2005. Yet, after years of inaction and growing national GhG emissions, Canada withdrew from the Kyoto Protocol with effect on December 15, 2012, two weeks before the expiration of the commitment period, to avoid a finding of non-compliance. Without Canada and the United States, only thirty-six Annex B States implemented their emission limitation and reduction commitments.

The implementation of QELRCs was facilitated by rapid emission reductions in economies in transition (i.e. countries formerly from the Eastern Bloc) during the 1990s and by the progressive offshoring of greenhouse gas-intensive industries to emerging economies. The 2009 global economic crisis also had the effect of reducing GhG emissions in developed States, although temporarily, making it easier for several States to comply with their commitment during the 2008–2012 period. As a result, most of the 36 participating Annex B States overachieved their commitment.[40] This, however, did not prevent global GhG emissions from continuing to increase due to fast-paced industrial development in emerging economies, the non-participation of the United States and the withdrawal of Canada. Consequently, the developed parties to the Kyoto Protocol showed little enthusiasm for a second commitment period, preferring to engage in negotiations for an agreement which would include some mitigation commitment also applicable to emerging economies.

III. THE 2020 HORIZON

On December 16, 2005, as the Kyoto Protocol entered into force, many were already wondering about what would happen after the expiry of the first commitment period, extending from 2008 to 2012. Given that seven years had been necessary for the entry into force of the Kyoto Protocol and that the first commitment period would end in 2012, the time was already ripe for negotiations toward further climate action.

One option was to extend the application of the Kyoto Protocol to a second commitment period. Accordingly, formal negotiations on a second commitment period under the Kyoto Protocol were initiated in December 2005, at the first meeting of the CMP. An Ad Hoc Working Group on Further Commitments for Annex I Parties under the Kyoto Protocol (AWG-KP) was set up in order to organize consultations and seek to build a consensus.[41] Yet, the prospects of these negotiations under the Kyoto Protocol were limited. The United States had made it clear that they would not participate in the Kyoto Protocol; accordingly, it did not formally join the negotiations for a second commitment period. As GhG emissions from emerging economies were rapidly increasing, a second commitment period limited to developed States and without the participation of the United States would do little to curb the growth in global GhG emissions. It appeared increasingly desirable to open a second track of negotiations not under the Kyoto Protocol, but directly under the UNFCCC, with the participation of the United States,

---

[40] See Romain Morel and Igor Shishlov, "Ex-post evaluation of the Kyoto Protocol: four key lessons for the 2015 Paris Agreement" (Climate Report: Research on the Economics of Climate Change, No. 44, CDC Climate, May 2014).

[41] Decision 1/CMP.1, "Consideration of commitments or subsequent periods for Parties included in Annex I to the Convention under Article 3, paragraph 9, of the Kyoto Protocol" (December 9–10, 2005), para. 2.

and seek to convince emerging economies to make specific commitments on climate change mitigation.

Thus, in December 2007, COP13 adopted the "Bali Action Plan" initiating a second track of negotiations under the Convention. An Ad Hoc Working Group on Long-Term Cooperative Action under the Convention (AWG-LCA) was set up to conduct "a comprehensive process to enable the full, effective and sustained implementation of the Convention through long-term cooperative action, now, up to and beyond 2012."[42] In a political compromise between developed and developing countries at the end of tortuous political negotiations, the Bali Action Plan defined "a shared vision for long-term cooperative action" which included but, crucially, was not limited to mitigation action. On mitigation, the Plan called for "quantified emission limitation and reduction objectives" by all developed States as well as "nationally appropriate mitigation actions" by developing States.[43] In exchange, developing States obtained the inclusion of a call for "enhanced action on adaptation,"[44] including "means to address loss and damage,"[45] as well as enhanced action "on technology development and transfer to support action on mitigation and adaptation,"[46] and for enhanced action "on the provision of financial resources and investment to support action on mitigation and adaptation."[47]

Thus, negotiations on the post-2012 horizon took place in two parallel tracks. The AWG-LCA track initiated by the Bali Action Plan, placed directly under the UNFCCC, led to the adoption of the Copenhagen Accord and the Cancún Agreements. The AWG-KP track, under the Kyoto Protocol, led to the adoption of the Doha Amendment. As a result, two parallel systems of national commitments are applicable between 2013 and 2020: pledges made by States under the Cancún Agreements and, pending the entry into force of the Doha Amendment, QELRCs of Annex B parties in a second commitment period.

A. The AWG-LCA Track: The Copenhagen Accord and the Cancún Agreements

The negotiations conducted within the AWG-LCA were supposed to lead to the adoption of an instrument at the 15th session of the Conference of the Parties to the UNFCCC (COP15), held in Copenhagen in December 2009.[48] Yet, the negotiations were plagued by the vexing question of the mitigation obligations applicable to developing States. The United States and China had incompatible views about what would constitute "appropriate" mitigation actions by developing States. The Copenhagen Summit was a very public failure for international negotiations on climate change. No formal decision could be taken on long-term cooperative action, and the mandate of the AWG-LCA was extended by an additional year.[49]

At the last minute, however, informal consultations between some of the main contenders reached an agreement on a text called the "Copenhagen Accord." Because many States protested that they had not been consulted during the drafting of this document, no consensus

---

[42] Decision 1/CP.13, "Bali Action Plan" (December 14–15, 2007), para. 1.
[43] *Ibid.*, para. 1(b)(i) and (ii).
[44] *Ibid.*, para. 1(c).
[45] *Ibid.*, para. 1(c)(iii).
[46] *Ibid.*, para. 1(d).
[47] *Ibid.*, para. 1(e).
[48] Copenhagen Accord, in the annex of decision 2/CP.15 (December 18–19, 2009).
[49] Decision 1/CP.15, "Outcome of the work of the Ad Hoc Working Group on Long-Term Cooperative Action under the Convention" (December 18–19, 2009), para. 1.

could be reached for its adoption as a COP decision. Instead, the Copenhagen Accord was annexed to a COP decision, which merely took note of its existence.[50] During the Copenhagen conference and in the following weeks, 141 of the 197 Parties to the UNFCCC notified the UNFCCC Secretariat that they agreed with the content of the Copenhagen Accord.[51]

Despite its controversial development through informal consultations, the Copenhagen Accord defined important orientations that would be pursued in the following years. In particular, it adopted the objective of limiting an increase in global average temperature to 2 degrees Celsius beyond pre-industrial temperatures. It adopted the outlines of a regulatory regime where each State would formally be free to define its own mitigation commitment, in contrast to the Kyoto Protocol, where respective commitments had been at the core of international negotiations. Furthermore, confirming the Bali Action Plan, the Copenhagen Accord invited Annex I parties to submit "quantified economy-wide emissions targets for 2020," but it also invited developing States to submit "nationally appropriate mitigation actions"[52] – all States were to make some specific commitments on climate change mitigation, although developing States' mitigation commitments would not need to be quantified. Lastly, developed States committed to "new and additional resources" in support of adaptation and mitigation, which should reach a value "approaching USD 30 billion for the period 2010–2012" and with "a goal of mobilizing jointly USD 100 billion dollars a year by 2020 to address the needs of developing countries."[53]

The following year, all of these orientations were formally adopted and clarified in the Cancún Agreements adopted at COP16. In particular, the Cancún Agreements took note of the Quantified Economy-Wide Emissions Targets for 2020 and Nationally Appropriate Mitigation Actions by Developing Country Parties called upon by the Copenhagen Accord,[54] which became known as the "Cancún pledges." The Cancún Agreements have also placed great emphasis on adaptation, affirming in particular that it "must be addressed with the same priority as mitigation" and calling for "appropriate institutional arrangements to enhanced adaptation action and support."[55] Lastly, the Cancún Agreements endorsed the financial provisions of the Copenhagen Accord and established the Green Climate Fund (GCF) to channel some of this funding.

Every developed State has pledged quantified economy-wide emission reduction targets for 2020 under the Cancún Agreements,[56] while more than 50 developing States, accounting for three-quarters of the total GhG emissions of developing States, have submitted nationally appropriate mitigation actions.[57] Yet, many pledges are not particularly ambitious; some

---

[50] Copenhagen Accord, *supra* note 48.

[51] "Copenhagen Accord," http://unfccc.int/meetings/copenhagen_dec_2009/items/5262.php (accessed December 31, 2017).

[52] Copenhagen Accord, *supra* note 48, paras. 4 and 5.

[53] *Ibid.*, para. 8.

[54] Decision 1/CP.16, "The Cancún Agreements: outcome of the work of the Ad Hoc Working Group on Long-Term Cooperative Action under the Convention" (December 10–11, 2010), paras. 36 and 49.

[55] *Ibid.*, para. 2(b).

[56] For a compilation, see UNFCCC Technical Paper, *Quantified Economy-Wide Emission Reduction Targets by Developed Country Parties to the Convention: Assumptions, Conditions, Commonalities and Differences in Approaches and Comparison of the Level of Emission Reduction Efforts*, UNFCCC document FCCC/TP/2014/8 (November 18, 2014).

[57] UNFCCC Revised Note by the Secretariat, *Compilation of Information on Nationally Appropriate Mitigation Actions to Be Implemented by Developing Country Parties*, UNFCCC document FCCC/SBI/2013/INF.12/Rev.3 (January 19, 2015).

do not significantly differ from business as usual (i.e. predictable emission scenarios without mitigation action).[58] Moreover, the legal nature of these "pledges" is unclear. China's communication on its nationally appropriate mitigation action, for instance, insists that this document is "voluntary in nature."[59] Yet, a sovereign State is able to voluntarily undertake binding commitments. Depending on the particular circumstances of each communication, some Cancún "pledges" could constitute unilateral declarations from which arise international obligations. The added value of the Cancún pledges is the fact that they apply to States which would have no emission limitation and reduction commitments under a second commitment period of the Kyoto Protocol. They involve developing States such as China and India as well as the United States and Canada.

B. The AWG–KP Track: The Doha Amendment to the Kyoto Protocol

The adoption of the Cancún Agreements did not interrupt the negotiations toward a second commitment period under the Kyoto Protocol.[60] Unlike the Cancún pledges, a second commitment period would define mitigation commitments of a clearly binding legal nature, taking place within the much more rigorous institutional context, including flexibility mechanisms as well as verification and compliance processes, established by the Kyoto Protocol.

A second commitment period to the Kyoto Protocol was formally adopted at the eighth session of the Conference of the Parties Serving as the Meeting of the Parties to the Kyoto Protocol (CMP8), held in Doha, Qatar, in December 2012, and is known as the Doha Amendment.[61] The Doha Amendment amends Annex B of the Kyoto Protocol to include additional QELRCs applicable to a second commitment period. Yet, the 38 parties with a QELRC under the second commitment period, constituted mostly of European States, only represented an eighth (12.5 percent) of global GhG emissions in 2013.[62] Together, these States seek to reduce their aggregate GhG emissions to 18 percent below 1990 levels.

To enter into force, the Doha Amendment needs to be accepted by three-quarters of the parties to the Kyoto Protocol (i.e. 144 parties).[63] As of early 2018, only 108 States Parties had communicated their instrument of acceptance. Pending the entry into force of the Doha Amendment, the parties to the Kyoto Protocol are called upon to "implement their commitments and other responsibilities in relation to the second commitment period, in a manner consistent with their national legislation or domestic processes."[64] While the Doha Amendment can thus be implemented with the assistance of the institutions of the Kyoto Protocol, including the flexibility mechanisms, it has not yet received the legal authority of a treaty in force.

---

[58] See UNEP, *The Emissions Gap Report 2016* (November 2016), Chapter 2.

[59] UNFCCC Revised Note by the Secretariat, *supra* note 57, at 15.

[60] The word "agreement" is used in plural because a second agreement (though less important) was reached under the Kyoto Protocol. See decision 1/CMP.6, "The Cancún Agreements: outcome of the work of the Ad Hoc Working Group on Further Commitments for Annex I Parties under the Kyoto Protocol at its fifteenth session" (March 15, 2011).

[61] Doha Amendment to the Kyoto Protocol, December 8, 2012, in the annex of decision 1/CMP.8, "Amendment to the Kyoto Protocol pursuant to its Article 3, paragraph 9 (Doha Amendment)" (December 8, 2012) (hereinafter the Doha Amendment).

[62] Data from WRI, CAIT Climate Data Explorer, "Total GHG emissions excluding land-use change and forestry" (2013).

[63] Kyoto Protocol, *supra* note 28, art. 20.4.

[64] Doha Amendment, *supra* note 61, para. II.6.

IV. THE PARIS AGREEMENT

With the lack of political support for a second commitment period under the Kyoto Protocol, the hypothesis of a third commitment period did not seem attractive. Instead, negotiations on international cooperation in the longer term, extending beyond 2020, took place under the UNFCCC rather than under the Kyoto Protocol. In 2011, COP17 established the Ad Hoc Working Group on the Durban Platform for Enhanced Action (AWG-DP) as "a process to develop a protocol, another legal instrument or an agreed outcome with legal force under the Convention applicable to all Parties,"[65] which would "come into effect and be implemented from 2020."[66]

This instrument was adopted four years later on December 12, 2015 at COP21 in Paris.[67] The Paris Agreement was opened for signature on April 22, 2016 and was immediately signed by 175 States. It entered into force on November 4, 2016, in record time for a multilateral convention, having fulfilled the condition of ratification by 55 parties accounting for more than 55 percent of the total global GhG emissions by that date.[68] As of early 2018, it had been signed by 194 States and ratified by 171 States as well as the European Union, with many other States still engaged in the process of ratification.

On June 1, 2017, US President Trump announced his intention of pulling the United States out of the Paris Agreement – a decision which could not be implemented less than four years after the entry into force of the treaty, that is, November 4, 2020.[69] The United States would likely then be the only large State not to participate in an international climate agreement. Unlike the Kyoto Protocol, however, participation in the Paris Agreement is largely supported by the American political class,[70] making the re-entry of the United States to the Paris Agreement in the post-Trump era likely.

A. Legal Nature

The legal nature of the instrument was a contentious question during most of the work of the AWG-DP. European States and most non-governmental organizations insisted on a legally binding treaty. Yet, there was a concern that ratification by the US Senate could be an obstacle to a formal treaty. A consensus was found on an instrument which is unquestionably a treaty under international law,[71] yet may be approved by Executive Order under US constitutional

---

[65] Decision 1/CP.17, "Establishment of an Ad Hoc Working Group on the Durban Platform for Enhanced Action" (December 11, 2011), para. 2.

[66] *Ibid.*, para. 4.

[67] Paris Agreement, December 12, 2015, in the annex of decision 1/CP.21, "Adoption of the Paris Agreement" (December 12, 2015). See Daniel Klein *et al.* (eds.), *The Paris Climate Agreement: Analysis and Commentary* (Cambridge University Press, 2017); Christine Bakker, "The Paris Agreement on Climate Change: balancing 'legal force' and 'geographical scope'" (2016) 25:1 *Italian Yearbook of International Law* 299; Laurence Boisson de Chazournes, "One swallow does not a summer make, but might the Paris Agreement on Climate Change a better future create?" (2016) 27:2 *European Journal of International Law* 253; and the special issue in (2016) 6:1–2 *Climate Law* 1.

[68] Paris Agreement, *supra* note 67, art. 21.1.

[69] *Ibid.*, arts. 28.1 and 28.2.

[70] See e.g. Scott Clement and Brady Dennis, "Post-ABC poll: nearly 6 in 10 oppose Trump scrapping Paris Agreement" *Washington Post* (June 5, 2017).

[71] Under art. 2.1(a) of the Vienna Convention on the Law of Treaties, May 23, 1969, 1155 *UNTS* 331, a "treaty" is defined as "an international agreement concluded between States in written form and governed by international law, whether embodied in a single instrument or in two or more related instruments and whatever its particular designation."

law for lack of virtually any direct and specific provision. This, ironically, ended up making the US participation more vulnerable to a change of administration.

As a result of this constraint, the Paris Agreement relies largely on modalities of implementation which are to be determined by decisions adopted by the COP. Decision 1/CP.21 on the Adoption of the Paris Agreement contains 139 paragraphs in which some of the modalities of the Paris Agreement are defined.[72] The Ad Hoc Working Group on the Paris Agreement (AWG-PA) was established to prepare for the entry into force of the Agreement through the adoption of other modalities of application at the first session of the Conference of the Parties Serving as the Meeting of the Parties to the Paris Agreement (CMA1).[73] Following the entry into force of the Paris Agreement on November 4, 2016, it was decided that CMA1 would be held in three parts, coinciding with COP22, COP23 and COP24, from 2016 to 2018, and that the complete "Paris rulebook" would be adopted at the third part of CMA1.3 at the end of 2018.[74]

B. The Objective

Article 2.1 of the Paris Agreement specifies the "ultimate objective" of the UNFCCC, reflecting the consensus which had been reached in the intervening two decades. In particular, it defines the following three sub-objectives on mitigation, adaptation and support:

(a) Holding the increase in the global average temperature to well below 2°C above pre-industrial levels and to pursue efforts to limit the temperature increase to 1.5°C above pre-industrial levels, recognizing that this would significantly reduce the risks and impacts of climate change;

(b) Increasing the ability to adapt to the adverse impacts of climate change and foster climate resilience and low greenhouse gas emissions development, in a manner that does not threaten food production;

(c) Making finance flows consistent with a pathway towards low greenhouse gas emissions and climate resilient development.

Further, Article 4 of the Paris Agreement mentions the:

aim to reach global peaking of greenhouse gas emissions as soon as possible ... and to undertake rapid reductions thereafter ... so as to achieve a balance between anthropogenic emissions by sources and removals by sinks of greenhouse gases in the second half of this century.[75]

C. Nationally Determined Contributions (NDCs)

Article 3 provides that: "As nationally determined contributions to the global response to climate change, all Parties are to undertake and communicate ambitious efforts ... with the view to achieving the purpose of this Agreement." This approach differs from the Kyoto

---

[72] Decision 1/CP.21, *supra* note 67.
[73] *Ibid.*, paras. 7–11.
[74] Decision 1/CMA.1, "Matters relating to the implementation of the Paris Agreement" (November 18, 2016), para. 5.
[75] Paris Agreement, *supra* note 67, art. 4.1.

Protocol in three substantial ways. Firstly, there is no general dichotomy between developed and developing States, even though developed States are clearly expected to make more significant contributions in applying the principle of common but differentiated responsibilities.[76]

Secondly, respective national commitments are not negotiated and included in the instrument, as in the Kyoto Protocol, but are to be determined unilaterally by each State and communicated to the UNFCCC Secretariat. While this approach makes it significantly easier for States to participate in the Paris Agreement, it also makes it more difficult to ensure that national commitments are sufficiently ambitious. Rather than relying on the authority of legal rules, the Paris Agreement bets on bottom-up social and political processes, such as naming and shaming, to decide States to make ambitious commitments.

Lastly, while the Kyoto Protocol was largely centered on climate change mitigation, the Paris Agreement seeks to define specific national commitments in other areas of international climate change cooperation. In particular, it requires States to communicate their contribution to adaptation,[77] means to address loss and damage,[78] financial support,[79] technological assistance,[80] capacity building[81] and support to reporting activities.[82]

The Paris Agreement establishes a procedural obligation for each party to communicate its contribution and to update it every five years. It also requires that such contributions, in particular in relation to climate change mitigation and financial support, "represent a progression over time."[83] Regarding mitigation, Article 4.2 requires each party to "pursue domestic mitigation measures, with the aim of achieving the objectives of such contributions."[84] Thus, a State's nationally determined contribution (NDC) creates an obligation of conduct with regard to the target conveyed in NDCs, though not necessarily in relation to the modalities of implementation which are detailed in some NDCs.[85] Some other provisions of the Paris Agreement are rather vague or aspirational, stating, for instance, that "Parties should strengthen their cooperation on enhancing action on adaptation."[86]

The Paris Agreement recognizes the principle of common but differentiated responsibilities and respective capabilities as well as differences in "national circumstances" more generally.[87] Thus, it reaffirms the principle of the UNFCCC according to which "[d]eveloped country Parties should continue taking the lead" and it admits that the nationally determined contributions of developing States may initially consist in mitigation efforts which are not quantified.[88] Strong emphasis is placed on the provision of financial and technical support by developed parties (or, voluntarily, by other parties)[89] to mitigation and adaptation in developing States.[90]

---

[76] Ibid., arts. 2.2 and 3 (second sentence). See Daniel Bodansky, "The Paris Climate Change Agreement: a new hope?" (2016) 110:2 American Journal of International Law 288.

[77] Paris Agreement, supra note 67, art. 7.

[78] Ibid., art. 8.

[79] Ibid., art. 9.

[80] Ibid., art. 10.

[81] Ibid., art. 11.

[82] Ibid., art. 13.

[83] Ibid., art. 3. See also arts. 4.3 and 9.3.

[84] Ibid., art. 4.2.

[85] See Benoit Mayer, "Obligations of conduct in the international law on climate change: a defence" Review of European, Comparative and International Environmental Law (forthcoming).

[86] Paris Agreement, supra note 67, art. 7.7.

[87] Ibid., art. 2.2.

[88] Ibid., art. 4.4.

[89] Ibid., art. 9.2.

[90] Ibid., arts. 3, 4.5, 7.6, 7.13, 9.1, 10.6 and 11.3.

In the run-up to the Paris Agreement, most parties submitted the NDCs they intended to communicate and implement until 2025 or 2030 under the future instrument. A technical synthesis by the Secretariat of the UNFCCC revealed that although some of these intended nationally determined contributions (INDCs) were somewhat ambitious, their aggregate effect would fall very short of what the IPCC identified as the likely lowest-cost way to limit an increase in global average temperature to 2 degrees Celsius. Rather, INDCs appear more likely to be consistent with scenarios that would lead to at least 3 degrees Celsius of warming by the end of the twenty-first century.[91] The challenge for the implementation of the Paris Agreement will be to initiate a process that is able to spur the willingness of every State to commit to more significant climate action.[92]

D. Boosting Parties' Commitment

The success of the Paris Agreement will depend on the willingness of the parties to review and increase their commitments while also making the necessary efforts to implement them. Three mechanisms are established in the Paris Agreement to encourage parties in this sense.

Firstly, the Paris Agreement establishes "an enhanced transparency framework for action and support."[93] This mechanism aims to facilitate public oversight and pressure on governments, taking advantage of the strong involvement of civil society organizations in the UNFCCC regime. It will seek to provide up-to-date and reliable information on each party's GhG emissions and actual mitigation efforts, as well as on the support provided by developed States, based on regular reports from the parties and a technical expert review. This mechanism will "be implemented in a facilitative, non-intrusive, non-punitive manner, respectful of national sovereignty, and avoid placing undue burden on Parties."[94]

Secondly, the Paris Agreement establishes a mechanism through which the CMA will "take stock of the implementation of this Agreement to assess the collective progress towards achieving the purpose of this Agreement and its long-term goals."[95] Following the 2018 "Talanoa dialogue,"[96] this exercise will be conducted in 2023 and every five years thereafter. This mechanism will review national efforts on mitigation, adaptation, financial and technological support, and capacity building. The Paris Agreement provides that the outcome of this exercise "shall inform Parties in updating and enhancing, in a nationally determined manner, their actions and support."[97] This process is another attempt at promoting ambitious national commitments through international socialization.

Thirdly, the Paris Agreement establishes a mechanism to facilitate implementation and compliance.[98] Like the transparency mechanism, this mechanism will be expert-based and facilitative, and it will function "in a manner that is transparent, non-adversarial and non-punitive."[99]

---

[91] UNFCCC Secretariat, *Aggregate Effect of the Intended Nationally Determined Contributions: An Update*, doc. FCCC/CP/2016/2 (May 2, 2016), para. 41. See also Joeri Rogelj *et al.*, "Paris Agreement climate proposals need a boost to keep warming well below 2°C" (2016) 534:7609 *Nature* 631.

[92] See Chapter 13, section III.

[93] Paris Agreement, *supra* note 67, art. 13.1.

[94] *Ibid.*, art. 13.3.

[95] *Ibid.*, art. 14.1.

[96] See Decision 1/CP.23, "Fiji Momentum for Implementation" (November 2017), para. 11.

[97] Paris Agreement, *supra* note 67, art. 14.3.

[98] *Ibid.*, art. 15.1.

[99] *Ibid.*, art. 15.2.

As such, it is not clear how this mechanism will differ from the transparency mechanism,[100] but it could provide one more chance for social and political processes that could push national governments to boost national commitment.

V. CONCLUSION

From Rio to Paris, the development of the UNFCCC regime involves some of the longest and most complex international negotiations. The adoption of the Paris Agreement is certainly a significant achievement, but it is also a new starting point – an institutional basis for continuing international negotiations over the next few decades. The UNFCCC regime has achieved significant results. Virtually every nation now agrees to take some action to tackle climate change.[101] However, much more international cooperation is needed if the objective adopted in the Paris Agreement of limiting the increase in global average temperature by the end of the twenty-first century to 2 – let alone 1.5 – degrees Celsius above pre-industrial levels is to be met.

---

[100] Alexander Zahar, "A bottom-up compliance mechanism for the Paris Agreement" (2017) 1:1 *Chinese Journal of Environmental Law* 69.

[101] The United States is no exception. Despite the decision of US President Trump to withdraw the United States from the Paris Agreement, participation is supported by a strong majority of the American people and a strong majority of their representatives in the Congress. See Clement and Dennis, *supra* note 70.

4

Relevant Developments in Other Regimes

The previous chapter retraced the international negotiations conducted under the UNFCCC, whose very aim is to tackle climate change. Yet, important initiatives have also taken place outside the UNFCCC regime. These developments, scattered in other diverse treaty regimes, complement the UNFCCC regime in different ways. Some initiatives seek to tackle GhG emissions from particular sectors or to tackle emissions of specific types of GhGs; others promote adaptation or climate change through innovative approaches. Along with the UNFCCC regime, these initiatives under other treaty regimes form part of what Robert O. Keohane and David G. Victor call "the regime complex for climate change" – "a loosely-coupled set of specific regimes"[1] dealing in different ways with particular aspects of climate change.

This chapter provides a general overview of the most relevant legal developments taking place outside of the UNFCCC regime. Within the regime on the protection of the ozone layer, rules which were devised primarily to phase out ozone-depleting substances have also contributed to mitigating climate change – so successfully that the ozone regime has been extended to other GhGs which have no effect on the ozone layer. Rules have also been developed under the aegis of the IMO and the ICAO to reduce GhG emissions from international transportation. Some clubs of States have played a pioneering role in climate change mitigation. Lastly, consideration for climate change mitigation have also been mainstreamed in diverse international regimes, for instance under the World Heritage Convention and under the Convention on Biological Diversity.

I. THE REGIME ON THE PROTECTION OF THE OZONE LAYER

The regime on the protection of the ozone layer is of particular relevance to the international law on climate change for three reasons. Firstly, the architecture of this regime – the development of a framework convention soon followed by a protocol containing more substantive rules – had a certain influence on the historical development of the UNFCCC regime. Secondly, the measures that have been taken to reduce the consumption and production of ozone-depleting substances have largely contributed to climate change mitigation because many of these substances are also GhGs. Thirdly, a recent amendment has extended the application of relevant instruments in order to reduce the use of GhGs which are being used to replace ozone-depleting substances.

---

[1] Robert O. Keohane and David G. Victor, "The regime complex for climate change" (2011) 9:1 *Perspectives on Politics* 7.

The ozone layer refers to a concentration of ozone (O_3) in the lower stratosphere, from 15 to 25 kilometers of altitude.[2] The ozone layer filters particular ultraviolet solar radiations which are harmful to human life as well as animals and plants. Scientific research in the 1970s and 1980s has revealed a reduction in the concentration of stratospheric ozone.[3] This "hole" in the ozone layer (in fact rather a thinning) is particularly significant in the polar regions, where the ozone layer is naturally thinner. This results in an increased exposure of the populations of Australia and Northern Europe, in particular, to harmful ultraviolent radiations, with public health consequences.

The depletion of the ozone layer was caused by anthropogenic emissions of certain gases, in particular CFCs and halons, mostly from industrial States. Because of their non-flammable, non-toxic, non-corrosive and thermodynamic properties, CFCs and halons were commonly used as refrigerants, air conditioner coolants and aerosols. When released, reaching the stratosphere and being exposed to higher-frequency solar radiations, these molecules break down into free chlorine or bromine, which, in turn, react with ozone molecules and thus deplete the ozone layer.[4]

A complex international regime has been developed since the mid-1980s to regulate the consumption and production of ozone-depleting substances. The Vienna Convention for the Protection of the Ozone Layer, which was adopted on March 22, 1985 and entered into force on September 22, 1988, established the framework of a regime "to protect human health and the environment against adverse effects resulting or likely to result from human activities which modify or are likely to modify the ozone layer."[5] The Vienna Convention defines some principles and sets up some institutions, but it includes little substantive provision.[6]

Negotiations continued, spurred by the publication of clearer scientific evidence. The Montreal Protocol on Substances that Deplete the Ozone Layer was adopted on September 16, 1987, before the Vienna Convention had entered into force.[7] The Montreal Protocol establishes control measures on the consumption and production of certain CFCs and halons.[8] It imposes a less stringent regime on developing States with a limited consumption of controlled substances.[9]

Four subsequent amendments to the Montreal Protocol were adopted in the 1990s to add new controlled substances and reinforce control measures.[10] Table 4.1 provides an overview of the substances addressed by the original Montreal Protocol and subsequent amendments.

[2] The stratospheric ozone layer should not be confused with non-natural ground-level ozone, created by industrial pollution, which is a serious public health issue in many cities.

[3] See World Meteorological Organization, *Scientific Assessment of Ozone Depletion: 2014* (WMO *et al.*, 2014).

[4] See generally Edward A. Parson, *Protecting the Ozone Layer: Science and Strategy* (Oxford University Press, 2003); Susan Solomon, "Stratospheric ozone depletion: a review of concepts and history" (1999) 37:3 *Review of Geophysics* 275.

[5] Vienna Convention for the Protection of the Ozone Layer, March 22, 1985, 1513 *UNTS* 293, art. 2.1.

[6] See generally Peter H. Sand, "Protecting the ozone layer: the Vienna Convention is adopted" (1985) 27:5 *Environment: Science & Policy for Sustainable Development* 18; Jørgen Wettestad, "The Vienna Convention and Montreal Protocol on Ozone-Layer Depletion" in Edward L. Miles *et al.* (eds.), *Environmental Regime Effectiveness: Confronting Theory with Evidence* (MIT Press, 2002); Stephen O. Andersen and K. Madhava Sarma, *Protecting the Ozone Layer: The United Nations History* (UNEP and Earthscan, 2002).

[7] Montreal Protocol on Substances that Deplete the Ozone Layer, September 16, 1987, 1522 *UNTS* 3 (hereinafter Montreal Protocol).

[8] *Ibid.*, art. 2 and Annex A.

[9] *Ibid.*, art. 5. See generally Peter M. Morrisette, "The evolution of policy responses to stratospheric ozone depletion" (1989) 29:3 *Natural Resources Journal* 793.

[10] Amendments to the Montreal Protocol on Substances that Deplete the Ozone Layer adopted in London on June 29, 1990, 1598 *UNTS* 469; in Copenhagen on November 25, 1992, 1785 *UNTS* 517; in Montreal on September 17, 1997, 2054 *UNTS* 522; and in Beijing on December 3, 1999, 2173 *UNTS* 183.

Table 4.1 Substances controlled under the Montreal Protocol and its amendments

Instrument	Controlled substances
Montreal Protocol	Some chlorofluorocarbons (CFCs), some halons
1990 London Amendment	Other CFCs, carbon tetrachloride, methylchloroform
1992 Copenhagen Amendment	Halomethane (HCFCs), hydrobromofluorocarbons (HBFCs), methyl bromide
1997 Montreal Amendment	No new controlled substances
1999 Beijing Amendment	Bromochloromethane
2016 Kigali Amendment	Hydrofluorocarbons (HFCs)

The regime on the protection of the ozone layer has been, by and large, a success. It has significantly reduced the depletion of the ozone layer, which has now started to "heal."[11] This experience inspired early negotiations on climate change. An attempt was made to follow a similar pattern to address climate change with the adoption of a framework convention defining general objectives and principles for international cooperation followed by a protocol containing more specific national commitments, which could then be revised as needed. Yet, the UNFCCC regime has faced much greater difficulties because of the strong reliance of modern economies on GhG emissions and the reluctance of States to make the sort of structural changes that mitigating climate change would require – revamping energy systems, transportation, industrial production, construction, food production and even our way of life, among others. Whereas phasing out ozone-depleting substances only involves some minor technological shifts in very specific economic sectors, climate change mitigation requires a complete overhaul of current patterns of production and consumption.

Beyond this influence of the ozone regime on the structure of negotiations on climate change, the ozone regime has also contributed to a reduction of GhG emissions. Most ozone-depleting substances regulated under the Montreal Protocol happen to be, at the same time, extremely powerful GhGs. CFC-13, for instance, has an estimated 100-year global warming potential 13,900 times greater than the same mass of carbon dioxide.[12] Consumption and production of CFC-13 was phased out under the Montreal Protocol, whose preamble took note, almost as an aside, of "the potential climatic effects of emissions" of ozone-depleting substances.[13] To avoid an overlap between the two regimes, the UNFCCC and the Kyoto Protocol do not apply to the GhGs controlled by the Montreal Protocol.[14]

By phasing out the consumption and production of ozone-depleting substances which had a very high global warming potential, the Montreal Protocol has significantly contributed

---

[11] See e.g. Susan Solomon et al., "Emergence of healing in the Antarctic ozone layer" (2015) 353:6296 Science 269.

[12] G. Myhre et al., "Anthropogenic and natural radiative forcing" in T.F. Stocker et al. (eds.), Climate Change 2013: The Physical Science Basis. Contribution of Working Group I to the Fifth Assessment Report of the Intergovernmental Panel on Climate Change (Cambridge University Press, 2013), Appendix 8.1, at 731. See Table 1.1.

[13] Montreal Protocol, supra note 7, recital 5.

[14] See, for instance, United Nations Framework Convention on Climate Change, May 9, 1992, 1771 UNTS 107 (hereinafter UNFCCC), arts. 4.1(b) and 4.2(a); Kyoto Protocol to the United Nations Framework Convention on Climate Change, December 11, 1997, 2303 UNTS 162 (hereinafter Kyoto Protocol), art. 2(a)(ii). By contrast, the Paris Agreement, December 12, 2015, in the annex of decision 1/CP.21, "Adoption of the Paris Agreement" (December 12, 2015), does not specify to which gases it applies.

to climate change mitigation. Absent the Montreal Protocol, it is likely that the use of such substances would have increased sharply, as refrigeration and air-conditioning systems expanded quickly in the developing world. Guus J.M. Velders and colleagues suggested that the Montreal Protocol had made a larger contribution to climate change mitigation than the Kyoto Protocol.[15] It is ironic that, despite the most intense negotiations on climate change in the last quarter of a century, the greatest contribution to climate change mitigation came from an instrument which pursued a different objective. This is partly due to the fact that preventing significant increases in the production of ozone-depleting substances was a low-hanging fruit, facing fewer obstacles than substantial reductions in more diffuse sources of carbon dioxide emissions, methane or nitrous oxide.[16] Nevertheless, the great success of the Montreal Protocol in phasing out the consumption and production of ozone-depleting substances led to some consideration regarding the opportunity of gas-specific action on climate change mitigation.[17]

Some of the substances controlled by the Montreal Protocol and its amendments have progressively been replaced by hydrofluorocarbons (HFCs). HFCs do not affect the ozone layer; therefore, they were not originally controlled by the Montreal Protocol. Yet, like the substances they replace, HFCs are extremely powerful GhGs, with a 100-year global warming potential up to 12,400 times greater than the same mass of carbon dioxide.[18] Already in the late 1990s, the COP expressed great concern about the rapid growth in the emissions of HFCs as a substitute for ozone-depleting substances.[19] Emissions of HFCs were regulated under the Kyoto Protocol and its Doha Amendment,[20] and they are expressly included in some – but not all – of the INDCs that parties have communicated in the run-up to the Paris Summit.[21] Yet, given the relatively specific use of these gases in only a few economic sectors, their extremely high global warming potential and the existence of alternatives, discussions were held on the possibility of a new Amendment to the Montreal Protocol to phase out HFCs.

In 2010, 108 parties to the Montreal Protocol adopted a declaration at the twenty-second Meeting of the Parties to the Montreal Protocol in 2010 to recognize the Montreal Protocol as "well-suited to making progress in replacing hydrochlorofluorocarbons (HCFCs) and

---

[15] Guus J.M. Velders *et al.*, "The importance of the Montreal Protocol in protecting climate" (2007) 104:12 *Proceedings of the National Academy of Sciences of the United States of America* 4814.

[16] O. Edenhofer *et al.*, "International cooperation: agreements & instruments" in R. Pichs-Madruga *et al.* (eds.), *Climate Change 2014: Mitigation of Climate Change. Contribution of Working Group III to the Fifth Assessment Report of the Intergovernmental Panel on Climate Change* (Cambridge University Press, 2014) 13, at 1050.

[17] See, for instance, Guus J.M. Velders *et al.*, "Preserving Montreal Protocol climate benefits by limiting HFCs" (2012) 335:6071 *Science* 922.

[18] Stocker *et al.* (eds.), *supra* note 12, Appendix 8.1, at 732.

[19] See decisions 13/CP.4, "Relationship between efforts to protect the stratospheric ozone layer and efforts to safeguard the global climate system: issues related to hydrofluorocarbons and perfluorocarbons" (November 11, 1998); 17/CP.5, "Relationship between efforts to protect the stratospheric ozone layer and efforts to safeguard the global climate system" (November 4, 1999); and 12/CP.8, "Relationship between efforts to protect the stratospheric ozone layer and efforts to safeguard the global climate system: issues relating to hydrofluorocarbons and perfluorocarbons" (November 1, 2002).

[20] See Kyoto Protocol, *supra* note 14, Annex A; Doha Amendment to the Kyoto Protocol, December 8, 2012, in the annex of decision 1/CMP.8, "Amendment to the Kyoto Protocol Pursuant to its Article 3, paragraph 9 (Doha Amendment)" (December 8, 2012) (hereinafter Doha Amendment). See also discussions in Chapter 7, section III.A.

[21] UNFCCC Secretariat, *Aggregate Effect of the Intended Nationally Determined Contributions: An Update*, doc. FCCC/CP/2016/2 (May 2, 2016), para. 16.

chlorofluorocarbons (CFCs) with low-global warming potential alternatives."[22] The following year, a working group was set up to discuss the possible role of the Montreal Protocol in managing HFCs,[23] and formal negotiations of an amendment to the Montreal Protocol were initiated at the twenty-seventh Meeting of the Parties in November 2015.[24] On October 15, 2016, the twenty-eighth Meeting of the Parties to the Montreal Protocol adopted the Kigali Amendment to regulate the consumption and production of HFCs.[25] The Kigali Amendment will enter into force on January 1, 2019.[26] It provides a timeframe for developed and developing parties to phase out most of the consumption and production of HFCs over several phases, which will be implemented in several phases until 2036 for developed parties and 2045 or 2047 for developing ones.[27]

Since HFCs have no significant effect on the ozone layer, this Amendment to the Montreal Protocol stretches it beyond a strict reading of the objective of the Vienna Convention: "protect[ing] human health and the environment against adverse effects resulting or likely to result from human activities which modify or are likely to modify the ozone layer."[28] This formal inconsistency was accepted, with no objection being raised, because the Montreal Protocol appeared to offer a more convenient forum for more specific international cooperation aimed at phasing out a particular GhG.[29]

II. INTERNATIONAL TRANSPORTATION

As discussed in the previous chapter, the Kyoto Protocol defined emission limitation and reduction commitments applicable to Annex I parties. This required an attribution of GhG emissions to States. Such an attribution was understood to be based on the place where the emissions were taking place. This territorial basis for attribution is relatively straightforward when it applies to stationary facilities such as a power plant, but it creates difficulty when accounting for GhG emissions from international transportation, in particular by air or by sea. Should the GhG emissions from a plane or a ship transiting through multiple States and high seas be attributed to the country of origin or destination of a passenger or shipment, or to the country where the fuel is bought, or to that where the plane or ship is registered or where

---

[22] Declaration on the Global Transition away from Hydrochlorofluorocarbons (HCFCs) and Chlorofluorocarbons (CFCs), adopted by 108 parties at and around the Twenty-Second Meeting of the Parties (November 8–12, 2010), second recital.

[23] Twenty-Fifth Meeting of the Parties to the Montreal Protocol, *Report of the Co-chairs of the Discussion Group on Issues on the Management of Hydrofluorocarbons using the Montreal Protocol and its Mechanisms* (October 21–25, 2013).

[24] Twenty-Seventh Meeting of the Parties to the Montreal Protocol decision XXVII/1, "Dubai Pathway on Hydrofluorocarbons" (November 1–5, 2015).

[25] Amendment to the Montreal Protocol on Substances that Deplete the Ozone Layer, adopted in Kigali on October 15, 2016, reproduced in (2017) 56 *ILM* 196.

[26] *Ibid.*, art. IV.1.

[27] For a synthetic presentation, see e.g. *Kigali Amendment to the Montreal Protocol: A Crucial Step in the Fight against Catastrophic Climate Change* (Environmental Investigation Agency, 2016).

[28] Vienna Convention on the Law of Treaties, May 23, 1969, 1155 *UNTS* 331, art. 2.1.

[29] See D. Shindell *et al.*, "A climate policy pathway for near- and long-term benefits" (2017) 356:6337 *Science* 493; Margaret M. Hurwitz *et al.*, "Early action on HFCs mitigates future atmospheric change" (2016) 11:11 *Environment Research Letters* 114019.

the owner or operator is based?[30] Or should the nationality of the passengers, or that of the producers or consumers of the cargo, be relevant? The IPCC recommends that emissions from international transportation should not be included in national totals, but reported separately.[31] As such, emissions from international transportation are not included under any Quantified Emission Limitation and Reduction Commitments (QELRCs) under the Kyoto Protocol.

Rather than being attributed to any specific country, it appeared that emissions from international transportation ought to be addressed separately. Yet, reaching a political consensus appeared particularly difficult given the vested interests of particular States, and questions of institutional legitimacy could be raised. To avoid these issues, the Kyoto Protocol excluded GhG emissions from aviation and marine bunker fuels from its scope, and called upon its parties to cooperate through specialized agencies of the UN dealing specifically with regulation on international aviation and on maritime transportation. Therefore, Article 2.2 of the Kyoto Protocol calls on developed States to pursue the limitation and reduction of these emissions "working through the International Civil Aviation Organization and the International Maritime Organization."[32] Negotiations on climate change mitigation have been held in these two agencies with some success in recent years.

A. Developments in International Maritime Law

Negotiations on air pollution by ships took place under the aegis of the International Maritime Organization and led to an amendment to the 1973 International Convention for the Prevention of Pollution from Ships as modified by its 1978 Protocol (MARPOL).[33] Annex VI on the Prevention of Air Pollution from Ships was adopted in 1997 and entered into force in 2005 to regulate emissions of air pollutants from ships.[34] Yet, Annex VI only contains nitrogen oxides (NOx), some of which are GhGs, along with other common pollutants such as sulfur oxide (SOx).

To address carbon dioxide emissions, an amendment to Annex VI was adopted on July 17, 2011 and entered into force on January 1, 2013.[35] This amendment includes regulations on the intensity of GhG emissions from ships. It contains two mandatory measures relating to an Energy Efficiency Design Index (EEDI) and to a Ship Energy Efficiency Management Plan (SEEMP). The EEDI is a minimal efficiency factor, expressed in carbon dioxide emissions per ton transported and per mile, applicable to any new ship. This standard is set to be tightened every five years. By contrast, the SEEMP applies immediately to the operations of any existing ships. The IMO adopts guidelines on technical measures to be included in the SEEMP, including, for instance, measures

---

[30] See Sebastian Oberthür and Hermann E. Ott, *The Kyoto Protocol: International Climate Policy for the 21st Century* (Springer, 1999) 111 ff.

[31] Simon Eggleston *et al.*, *2006 IPCC Guidelines for National Greenhouse Gas Inventories* (IGES, 2006), para. 8.2.1. See also J.T. Houghton *et al.*, *Revised 1996 IPCC Guidelines for National Greenhouse Gas Inventories* (UK Meteorological Office, 1997), 1.2.

[32] Kyoto Protocol, *supra* note 14, art. 2.2.

[33] 1973 International Convention for the Prevention of Pollution from Ships, November 2, 1973, 1340 *UNTS* 184; and 1978 Protocol Relating to the 1973 International Convention for the Prevention of Pollution from Ships, February 17, 1978, 1340 *UNTS* 61 (hereinafter MARPOL).

[34] 1997 Protocol to Amend the 1973 International Convention for the Prevention of Pollution from Ships, as Modified by the 1978 Protocol, adopted in London on September 26, 1997.

[35] MARPOL resolution MEPC.203(62), "Amendments to the Annex of the Protocol of 1997 to Amend the International Convention for the Prevention of Pollution from Ships, 1973, as Modified by the Protocol of 1978 relating thereto" (July 15, 2011), in IMO document MEPC 62/24/Add.1.

relating to improved voyage planning, weather routing and speed optimization.[36] These measures imply extensive technical regulations taken at the global scale to reduce GhG emissions.

Negotiations did not stop there. Following the adoption of the 2011 amendment to Annex VI, further proposals have been made for the creation of a complementary market-based mechanism.[37] In October 2016, a Working Group on Reduction of GhG Emissions from Ships was established to carry consultations toward a possible next step.[38]

B. Developments in International Aviation Law

Negotiations were also held under the aegis of the ICAO[39] to address the impact of international civil aviation, a fast-growing economic sector, on global GhG emissions.[40] Although significant limitations in GhG intensity can be achieved through technological advances, civil aviation is likely to remain by nature a non-negligible contributor to global GhG emissions.

In September and October 1992, shortly after the Earth Summit, the 29th session of the ICAO Assembly recognized "growing concerns about environmental problems in the upper atmosphere such as global warming and depletion of the ozone layer."[41] Yet, the Assembly took a rather defensive posture, noting that "the extent to which international civil aviation contributes to these problems has not yet been ascertained."[42] Emphasizing the role of ICAO as "the primary organization responsible for ... developing policy guidance on possible means of minimizing any undesirable effects of international civil aviation on the environment,"[43] the ICAO Assembly called the Council of the ICAO to "co-operate closely" with the relevant organizations.[44]

Following the adoption of the Kyoto Protocol, the 32nd Session of the ICAO Assembly held in September and October 1998 requested the Council "to study policy options to limit or reduce the GhG emissions from civil aviation."[45] At the same session, the ICAO Assembly also noted that "the subject of environmental charges or taxes on air transport ha[d] been

---

[36] See MARPOL resolution MEPC.213(63), "2012 Guidelines for the Development of a Ship Energy Efficiency Management Plan (SEEMP)" (March 2, 2012), in IMO document MEPC 63/23.

[37] See IMO, Report of the Marine Environment Protection Committee on its Sixty-Third Session, doc. MEPC 63/23 (March 14, 2012), section 6. See generally Md. Saiful Karim, *Prevention of Pollution of the Marine Environment from Vessels: The Potential and Limits of the International Maritime Organisation* (Springer, 2015) 117–118; Harilaos N. Psaraftis, "Market-based measures for greenhouse gas emissions from ships: a review" (2012) 11:2 *WMU Journal of Maritime Affairs* 211; Yoshifumi Tanaka, "Regulation of greenhouse gas emissions from international shipping and jurisdiction of states" (2016) 25:3 *Review of European Community & International Environmental Law* 333.

[38] IMO, Report of the Marine Environment Protection Committee on its Seventieth Session, doc. MEPC 70/18 (November 11, 2016), para. 7.14.

[39] The ICAO is a specialized agency of the United Nations. It was established by the Convention on International Civil Aviation, December 7, 1944, 15 *UNTS* 295 (hereinafter Chicago Convention).

[40] See generally Md. Tanveer Ahmad, *Climate Change Governance in International Civil Aviation* (Eleven, 2016); Alejandro Piera Valdés, *GhG Emissions from International Aviation: Legal and Policy Challenges* (Eleven, 2015); Beatriz Martinez Romera, *Regime Interaction and Climate Change: The Case of International Aviation and Maritime Transport* (Routledge, 2018).

[41] ICAO Assembly Resolution A29-12, "Environmental impact of civil aviation on the upper atmosphere" (September 22–October 8, 1992), in ICAO, *Assembly Resolutions in Force (as of 8 October 1992)*, doc. 9602, at I-35, recital 1.

[42] *Ibid.*, recital 3.

[43] *Ibid.*, recital 5.

[44] *Ibid.*, para. 2.

[45] ICAO Assembly Resolution A32-8, "Consolidated statement of continuing ICAO policies and practices related to environmental protection" (September 22–October 2, 1998), Appendix F, para. 4.

raised, for example, in the context of controlling greenhouse gas emission."[46] It "urge[d] States to ... refrain from unilateral action to introduce emission-related levies"[47] until a further decision had been adopted. In 2001, the 33rd session of the ICAO Assembly concluded that "an open emissions-trading system was a cost effective measure to limit or reduce carbon dioxide emitted by civil aviation in the long-term,"[48] but it postponed the adoption of guidelines on such a system.[49] Furthermore, the ICAO Assembly reaffirmed its opposition to "unilateral action to introduce emission-related levies,"[50] allowing only "voluntary actions" under which "industry and governments agree to a target and/or to a set of actions to reduce emissions."[51]

Although negotiations were stalling during the following sessions, the Assembly at its 36th session in 2007 requested the Council to "ensure that ICAO exercises continuous leadership on environmental issues relating to international civil aviation, including GhG emissions,"[52] thus in fact precluding any development in other forums. In addition, the ICAO Assembly urged States, once again, "to refrain from unilateral implementation of greenhouse gas emissions charges" and "not to implement an emissions trading system on other Contracting States' aircraft operators except on the basis of mutual agreement between those States."[53]

EU Member States placed a reservation on this and other similar provisions. Instead of waiting for a possible agreement under the ICAO, the EU decided to impose mitigation measures on international aviation. Since 2005, the EU has implemented an Emissions Trading Scheme (ETS) on a range of stationary activities.[54] Like other forms of market-based mechanisms, an ETS requires economic actors to acquire an allowance for each unit of GhG emission. Allowances can be traded, creating an economic incentive to decrease GhG emissions.[55] Thus, in late 2008, the EU decided that it would extend its ETS to international flights from or to any airport situated within the European Union, including for sections of these flights taking place outside of the territory of EU Member States.[56] Despite a favorable judgment by the European Court of Justice,[57] this measure led to heated political and academic debate,[58] but it spurred new developments at ICAO, as third States preferred a coordinated policy to unilateral action.

---

[46] *Ibid.*, Appendix H, recital 1.

[47] *Ibid.*, Appendix H, para. 2(b).

[48] ICAO Assembly Resolution A33-7, "Consolidated statement of continuing ICAO policies and practices related to environmental protection" (September 25–October 5, 2001), Appendix I, recital 12.

[49] *Ibid.*, Appendix I, para.2(c)(2).

[50] *Ibid.*, Appendix I, para. 2(b)(3).

[51] *Ibid.*, Appendix I, recital 13 and footnote.

[52] ICAO Assembly Resolution A36-22, "Consolidated statement of continuing ICAO policies and practices related to environmental protection" (September 18–28, 2007) Appendix J, para. 1(a).

[53] *Ibid.*, Appendix L, paras. 1(a)(3) and 1(b)(1).

[54] EU Directive 2003/87/EC establishing a scheme for GhG emission allowance trading within the Community (October 13, 2003), doc. 32003L0087. The EU ETS is further discussed in Chapter 7, section IV.B.

[55] See Chapter 7, section IV.B.

[56] EU Directive 2008/101/EC of November 19, 2008 amending Directive 2003/87/EC so as to include aviation activities in the scheme for GhG emission allowance trading within the Community (January 13, 2009), doc. 32008L0101.

[57] See e.g. ECJ judgment (Grand Chamber), Case C-366/10, *ATA* v. *Secretary of State for Energy*, judgment of December 21, 2011, doc. 62010CJ0366.

[58] See e.g. Glan Plant's analysis in (2013) 107(1) *American Journal of International Law* 183; Snaja Bogojević, "Legalising environmental leadership: a comment on the CJEU's ruling in C-366/10 on the inclusion of aviation in the EU Emissions Trading Scheme" (2012) 24:2 *Journal of Environmental Law* 345; annotations in (2012) 49:3 *Common Market Law Review* 1113 (annotations by Benoit Mayer); Andrea Gattini, "Between splendid isolation and tentative imperialism: the EU's extension of its Emission Trading Scheme to international aviation and the ECJ's judgment in the ATA case" (2012) 61:4 *International & Comparative Law Quarterly* 977; Kati Kulovesi, " 'Make your

Thus, in 2010, the 37th session of the Assembly defined an aspirational objective of "a global annual average fuel efficiency improvement of 2 per cent until 2020 and an aspirational global fuel efficiency improvement rate of 2 per cent per annum from 2021 to 2050."[59] It also adopted "guiding principles for the design and implementation" of market-based instruments.[60] Lastly, it invited States to communicate national action plans, thus providing them with an opportunity to showcase national commitments to limit GhG emissions from the aviation sector. Three years later, the 38th session adopted the "global aspirational goal of keeping the global net carbon emissions from international aviation from 2020 at the same level,"[61] asking States to submit consistent action plans. This aspirational goal appeared rather ambitious, given the rapid expansion of aviation in the developing world.[62] Its actual implementation called for further developments, and expectations rose for a multilateral agreement on a market-based mechanism.[63]

Such an agreement on a market-based mechanism was finally reached at the 39th session of the ICAO Assembly held in September 2016. The ICAO Assembly confirmed its commitment to "a comprehensive approach, consisting of a basket of measures including technology and standards, sustainable alternative fuels, operational improvements and market-based measures to reduce emissions."[64] Overall, building on the preparatory work of the ICAO Council and its Environment Advisory Group, the ICAO Assembly established the Carbon Offsetting and Reduction Scheme for International Aviation (CORSIA).[65] This market-based mechanism will allow aircraft operators to make up for their excess emissions by subsidizing mitigation projects in other economic sectors. It will progressively enter into application from 2021 onward, first through a pilot phase (2021–2023), followed by an initial voluntary phase (2024–2026) and a second phase applicable to all States (2027–2035).[66] Countries with a very small share of international civil aviation activities are exempted along with all least developed States, small island developing States and landlocked developing countries, unless they volunteer otherwise.[67]

The significance of CORSIA as the first global market-based mechanism cannot be overstated. As of mid-2017, 70 States representing close to 90 percent of international aviation activities

own special song, even if nobody else sings along': international aviation emissions and the EU Emissions Trading Scheme" (2011) 2:4 *Climate Law* 535.

[59] ICAO Assembly Resolution A37-19, "Consolidated statement of continuing ICAO policies and practices related to environmental protection – climate change" (September 28–October 8, 2010), para. 4. This language has since been integrated into ICAO Assembly Resolution A39-2, "Consolidated statement of continuing ICAO policies and practices related to environmental protection – climate change" (September 27–October 7, 2016), para. 4.

[60] See Annex of ICAO Assembly Resolution A37-19, *supra* note 59. See also Annex of ICAO Assembly Resolution A39-2, *supra* note 59.

[61] ICAO Assembly Resolution A38-18, "Consolidated statement of continuing ICAO policies and practices related to environmental protection – climate change" (September 24–October 4, 2013), para. 7. See also ICAO Assembly Resolution A39-2, *supra* note 59, para. 6.

[62] O. Edenhofer *et al.*, "Transport" in Pichs-Madruga *et al.* (eds.), *supra* note 16, 599, at 646.

[63] See in particular the special issue on aviation and the impacts of climate change and on the design of a global market-based measure in (2016) 10:2 *Carbon & Climate Law Review* 91–163.

[64] ICAO Assembly Resolution A39-2, *supra* note 59, recital 7.

[65] ICAO Assembly Resolution A39-3, "Consolidated statement of continuing ICAO policies and practices related to environmental protection – Global Market-Based Measure (MBM) Scheme" (September 27–October 7, 2016), para. 5.

[66] *Ibid.*, para. 9.

[67] *Ibid.*, para. 9(e).

had expressed their intention to participate in the pilot phase.[68] However, the modalities of application of this scheme remain to be defined by the Council of the ICAO before the end of 2018.[69]

III. OTHER PIONEERING INITIATIVES ON CLIMATE CHANGE MITIGATION

Some initiatives, often limited to a handful of like-minded States or sometimes involving non-State actors, have sought to go further in promoting climate change mitigation, including by promoting innovation and simply through exercising leadership toward greater ambition. While some of these initiatives used pre-existing forums of international cooperation, new forums have also emerged. It is impossible to list all such pioneering initiatives here, but some of the most prominent examples deserve at least a brief mention.

The EU could be considered as one such pioneering initiative exercising a global leadership, as illustrated by its role in the development of a market-based mechanism under the ICAO. In 2003, even before the Kyoto Protocol entered into force, the EU established an ETS applicable to about half of its economic production.[70] The diplomatic influence of the European Union and its Member States has been used to promote mitigation and adaptation throughout the world, for instance, by integrating relevant provisions in virtually any bilateral or multilateral agreement.[71] The EU mitigation policies are discussed in more depth in Chapter 7, along with national measures of implementation.[72]

Clubs of States which do not necessarily span a particular region have also taken initiatives in relation to climate change. The Group of Seven or Eight (G7/G8) has repeatedly considered climate change and emphasized the need to tackle it.[73] The Group of Twenty (G20) agreed on innovative provisions, calling on its Member States to "rationalize and phase out over the medium term inefficient fossil fuel subsidies that encourage wasteful consumption."[74] Bilateral negotiations have helped States reach a consensus for international cooperation, as illustrated by a series of US–China joint announcements and joint presidential statements in the run-up to and immediate follow-up of the Paris Agreement under the Obama administration.[75]

In addition to the engagement of existing institutions, new clubs of States have been formed to promote particular initiatives. The Major Economies Forum on Energy and Climate (MEF) was launched in 2009 to facilitate international negotiations on climate change, for instance, within the UNFCCC regime and the ICAO. The MEF includes 17 States which account for about

---

[68] At the time of concluding this book, it was not clear how the election of President Trump and his decision to pull out of the Paris Agreement would affect the participation of the United States in CORSIA.

[69] ICAO Assembly Resolution A39-3, *supra* note 65, para. 20.

[70] See Directive 2003/87/EC, *supra* note 54; and also Rüdiger K.W. Wurzel and James Connelly (eds.), *The European Union as a Leader in Climate Change Politics* (Routledge, 2011); Sebastian Oberthür and Claire Roche Kelly, "EU leadership in international climate policy: achievements and challenges" (2008) 43:3 *International Spectator* 35.

[71] See e.g. Stephan Keukeleire and Tom Delreux, *The Foreign Policy of the European Union*, 2nd edn (Palgrave Macmillan, 2014) 227; Diarmuid Torney, *European Climate Leadership in Question* (MIT Press, 2015).

[72] See Chapter 7, section IV.

[73] See e.g. G7 Ise-Shima Leaders' Declaration, May 27, 2016.

[74] G20 Leaders Statement: The Pittsburgh Summit (September 24–25, 2009), para. 29. See also G20 Hamburg Climate and Energy Action Plan for Growth (July 7–8, 2017), with unspecific reservations of the United States.

[75] See U.S.-China Joint Announcement on Climate Change, November 12, 2014; U.S.-China Joint Presidential Statement on Climate Change, September 25, 2015; U.S.-China Joint Presidential Statement on Climate Change, March 31, 2016.

three-quarters of global GhG emissions; its membership is largely similar, though not identical, to that of the G20. By contrast, the Cartagena Dialogue for Progressive Action includes about 30 developed or developing States, encompassing some small island developing States and some of the least developed countries, with the notable absence of the United States. It seeks to promote ambitious, comprehensive and legally binding international agreements on climate change.[76]

Multiple sectorial initiatives have also been taken, most remarkably in relation to energy governance. The International Renewable Energy Agency (IRENA) was established in 2009 in order to accelerate the development and deployment of renewable energies; as of early 2018, it counted 154 members, with 26 more States having started the formal process of becoming members. The Statute of IRENA emphasizes "the major role that renewable energy can play in reducing greenhouse gas concentrations in the atmosphere, thereby contributing to the stabilisation of the climate system, and allowing for a sustainable, secure and gentle transit to a low carbon economy."[77] Likewise, the International Energy Agency (IEA) has sought to play a greater role in advancing renewable energies and climate change mitigation more generally, while the International Atomic Energy Agency has promoted the idea that nuclear energy could be a relevant strategy in diverting from fossil fuels and mitigating climate change.

Such sectorial initiatives were also taken in other sectors. Frustrated by the stalling of negotiations on international cooperation on forestry, 50 States briefly joined a non-binding Agreement on Financing and Quick-Start Measures to Protect Rainforests concluded in 2010.[78] They established the REDD+ Partnership as "an interim platform for its partner countries to scale up actions and finance for initiatives to reduce emissions from deforestation and forest degradation (REDD+) in developing countries."[79] This partnership was discontinued several years later as a comprehensive framework for REDD+ was adopted by the parties to the UNFCCC at COP18.[80]

Likewise, the Climate and Clean Air Coalition to Reduce Short-Lived Climate Pollutants was established in 2012 by seven States; it now involves 53 country partners, 17 intergovernmental partners and 45 non-governmental organizations.[81] This initiative is dedicated to promoting a focus on GhGs other than carbon dioxide, including methane, black carbon and HFCs, including by orchestrating voluntary actions by non-State actors.

---

[76] See e.g. Sebastian Oberthür, "Global climate governance after Cancún: options for EU leadership" (2011) 46:1 International Spectator 5; Lau Øfjord Blaxekjær and Tobias Dan Nielsen, "Mapping the narrative position of new political groups under the UNFCCC" (2015) 15:6 Climate Policy 751.

[77] Statute of the International Renewable Energy Agency (IRENA), January 26, 2009, IRENA document IRENA/FC/ Statute, recital 3.

[78] Government of Norway, "Agreement on financing and quick-start measures to protect rainforest" (May 27, 2010), www.regjeringen.no/en/aktuelt/Agreement-on-financing-and-quick-start-measures-to-protect-rainforest/id605756 (accessed January 2, 2018). See also F. Seymour, "REDD reckoning: a review of research on a rapidly moving target" (2012) Plant Sciences Reviews 147.

[79] REDD+ Partnership, cited by Chris Lang, "Is the REDD+ Partnership closing down? And should we care?" REDD Monitor (July 18, 2014), www.redd-monitor.org/2014/07/18/is-the-redd-partnership-closing-down-and-should-we-care (accessed January 2, 2018).

[80] See Chapter 8. See also Ernesto Roessing Neto and Joyeeta Gupta, "REDD+ and multilevel governance beyond the climate negotiations" in Christina Voigt (ed.), Research Handbook on REDD-Plus and International Law (Edward Elgar, 2016) 289.

[81] See the website of the Clean Air Coalition to reduce short-lived climate pollutants, www.ccacoalition.org/en/partners (accessed January 2, 2018).

While the Climate and Clean Air Coalition is a State-centered initiative involving some non-governmental organizations, other transnational initiatives were established directly by non-State actors, including local governments. For instance, the World Mayors Council on Climate Change, which comprises more than 80 members, advocates "for enhanced engagement of local governments as governmental stakeholders in multilateral efforts addressing climate change and related issues of global sustainability."[82] The Western Climate Initiative is a transnational emissions trading scheme applicable in California and the Canadian provinces of Québec, Ontario, Manitoba and British Columbia.[83] Such initiatives by non-State actors, which play a growing role in the transnational efforts to promote climate change mitigation, are further discussed in Chapter 15.

IV. MAINSTREAMING ADAPTATION CONCERNS IN RELEVANT INTERNATIONAL REGIMES

The impacts of climate change affect many international regimes. New debates have arisen in recent years about the potential need for specific action to promote climate change adaptation. This section explores the debates that have arisen under the World Heritage Convention and the Convention on Biological Diversity. Similar developments are taking place in other regimes protecting human rights, addressing migration or advancing human health, among others.

A. The World Heritage Convention

The 1972 World Heritage Convention, adopted under the auspices of the United Nations Educational, Scientific and Cultural Organization (UNESCO), seeks to protect cultural and natural heritage from threats of destruction by natural and non-natural causes of damage and destruction.[84] It sets up an institutional framework to monitor a list of properties forming part of the cultural heritage which State Parties commit to protect. In particular, a list of "World Heritage in Danger" is maintained to call attention on problematic situations requiring international assistance.[85]

It was formally brought to the attention of the World Heritage Committee in 2005 that several world heritage properties could be placed at risk due to the impacts of climate change.[86] Without denying the problem, members of the Committee noted that "climate change was different from other problems the Committee was dealing with, as its potential

---

[82] "About," page on the website of the World Mayors Council on Climate Change, www.worldmayorscouncil.org/about.html (accessed January 2, 2018).

[83] Website of the Western Climate Initiative, www.wci-inc.org (accessed January 2, 2018).

[84] Convention for the Protection of the World Cultural and Natural Heritage, November 16, 1972, 1037 *UNTS* 151.

[85] *Ibid.*, art. 11.

[86] See UNESCO World Heritage Committee, 29th Session, decision 29 COM 7B.a (Durban, July 10–17, 2005), in doc. WHC-05/29.COM/22, 36; UNESCO World Heritage Committee, 29th Session, *State of Conservation Reports of Properties Inscribed on the World Heritage List*, in doc. WHC-05/29.COM/7B.Rev (June 15, 2005), paras. 14–21. The impacts of climate change had been raised previously within the World Heritage Committee, for instance, by Australia: see UNESCO World Heritage Committee, Report of the 25th Session, doc. WHC-01/CONF.208/24 (February 8, 2002), para. IX.12.

impacts were global and indirect, not local and direct."[87] They acknowledged that "the problem of climate change could not be solved in the framework of the World Heritage Convention."[88] They were nevertheless willing to consider possible action within their mandate. A workshop organized the following year suggested that, in addition to calling States to cooperate in promoting preventive actions (climate change mitigation), the World Heritage Committee could promote corrective actions (adaptation) and sharing knowledge on such actions.[89]

In 2007, the General Assembly of the States Parties to the World Heritage Convention adopted a policy statement on the impacts of climate change on world heritage. This document highlighted in particular the duty of the parties:

> to ensure that they are doing all that they can to address the causes and impacts of climate change, in relation to the potential and identified effects of climate change (and other threats) on World Heritage properties situated on their territories.[90]

In 2008, the impacts of climate change were added to the list of potential dangers for cultural and natural properties.[91] This series of decisions had a catalytic effect: from then onward, the World Heritage Committee placed much more emphasis on the impacts of climate change on the cultural and natural properties. Although there is certainly no magic bullet and it remains for States to face the challenge of protecting the world's heritage from the impacts of climate change,[92] this development has fostered greater awareness of the impacts of climate change on cultural heritage and on possible ways to protect unique natural and cultural properties.[93]

B. The Convention on Biological Diversity

The Convention on Biological Diversity was adopted on June 5, 1992 and, like the UNFCCC, it opened for signature at the Earth Summit.[94] It is almost universally ratified, with the exception

---

[87] UNESCO World Heritage Committee, 29th Session, Summary record (July 10–17, 2005), doc. WHC-05/29.COM/INF.22, at 97 (United Kingdom). See also UNESCO World Heritage Committee, 30th Session (July 8–16, 2006), Summary record, WHC-06/30.COM/INF.19, at 11 (Benin).

[88] *Ibid.*, at 98 (New Zealand representative).

[89] UNESCO World Heritage Committee, 30th Session, *Issues Related to the State of Conservation of World Heritage Properties: The Impacts of Climate Change on World Heritage Properties*, doc. WHC-06/30.COM/7.1 (June 26, 2006), at 5.

[90] UNESCO, General Assembly of States Parties to the Convention for the Protection of the World Cultural and Natural Heritage, Sixteenth Session, *Policy Document on the Impacts of Climate Change on World Heritage Properties*, doc. WHC-07/16.GA/10 (September 28, 2007), at 7.

[91] UNESCO World Heritage Committee, 32nd Session, decision 32 COM 7A.32 (Québec City, July 2–10, 2008), in doc. WHC-08/32.COM/24Rev (March 31, 2009) 40, para. 6.

[92] Stefan Gruber, "The impact of climate change on cultural heritage sites: environmental law and adaptation" (2011) 5:2 *Carbon & Climate Law Review* 209; Stefan Gruber, "Protecting China's cultural heritage sites in times of rapid change: current developments, practice and law" (2010) 10:3–4 *Asia Pacific Journal of Environmental Law* 253.

[93] See e.g. Ben Marzeion and Anders Levermann, "Loss of cultural world heritage and currently inhabited places to sea-level rise" (2014) 9 *Environmental Research Letters* 034001, Jim Perry, "World heritage hot spots: a global model identifies the 16 natural heritage properties on the World Heritage List most at risk from climate change" (2011) 17:5 *International Journal of Heritage Studies* 426; C. Sabbioni, P. Brimblecombe and M. Cassar, *The Atlas of Climate Change Impact on European Cultural Heritage: Scientific Analysis and Management Strategies* (Anthem Press, 2010).

[94] United Nations Convention on Biological Diversity, June 5, 1992, 1760 *UNTS* 79 (hereinafter CBD).

of the United States. As climate change is associated with serious risks for biological diversity,[95] there is a possibility of synergy between the biodiversity regime and the UNFCCC regime.

This potential synergy was recognized by the COP to the Conference on Biological Diversity (COP/CBD) at its third session, held in 1996, where it requested its Executive Secretary to "develop closer relationships with ... the United Nations Framework Convention on Climate Change," among other multilateral environmental agreements, "with a view to making implementation activities and institutional arrangements mutually supportive."[96] In the following years, the COP/CBD highlighted the impacts of climate change on diverse aspects of biological diversity in environments such as forests[97] or dry and sub-humid lands,[98] or issues such as coral bleaching[99] or plants conservation.[100] A joint liaison group was established between the secretariats of the UNFCCC, the Convention on Biological Diversity and the Convention to Combat Desertification.[101] Attention was particularly given to seek "opportunities to implement climate change mitigation and adaptation activities in ways that are mutually beneficial and synergistic,"[102] contributing to the achievement of the objectives of multiple aspects of sustainable development.

If climate change is clearly the cause of much concern in the biodiversity regime, as under the World Heritage Convention, there appears to be no magic bullet to address these concerns. Nevertheless, some synergies could be found concerning specific aspects of international action on climate change and on biodiversity. For instance, the COP/CBD advocated for climate action to take biodiversity into account in its efforts to protect forests.[103] Accordingly, efforts to protect and restore forests should recognize the diversity of species which inhabit these thriving ecosystems. These deliberations of the COP/CBD have more generally contributed to raising awareness of the consequences of climate change on biological diversity, thus making

---

[95] Christopher B. Field *et al.*, "Summary for policymakers" in Christopher B. Field *et al.* (eds.), *Climate Change 2014: Impacts, Adaptation, and Vulnerability. Part A: Global and Sectoral Aspects. Working Group II Contribution to the Fifth Assessment Report of the Intergovernmental Panel on Climate Change* (Cambridge University Press, 2014) 1, at 13.

[96] CBD decision III/21, "Relationship of the Convention with the Commission on Sustainable Development and biodiversity-related conventions, other international agreements, institutions and processes of relevance" (Buenos Aires, November 4–15, 1996), para. 4.

[97] CBD decision IV/7, "Forest biological diversity" (Bratislava, May 4–15, 1998), para. 9.

[98] CBD decision VIII/2, "Biological diversity of dry and sub-humid lands" (Curitiba, March 20–31, 2006), recital 3 and para. 10.

[99] CBD decision V/3, "Progress report on the implementation of the programme of work on marine and coastal biological diversity (implementation of decision IV/5)" (Nairobi, May 15–26, 2000), para. 5.

[100] "Global Strategy for Plant Conservation," in the annex of CBD decision VI/9 (The Hague, April 7–19, 2002), para. 6.

[101] CBD decision VI/20, "Cooperation with other organizations, initiatives and conventions" (The Hague, April 7–19, 2002), para. 12. See also CBD, *supra* note 94; and United Nations Convention to Combat Desertification in those Countries Experiencing Serious Drought and/or Desertification, Particularly in Africa, October 14, 1994, 1954 *UNTS* 3 (hereinafter Convention against Desertification).

[102] CBD decision VII/15, "Biodiversity and climate change" (Kuala Lumpur, February 9–20, 2004), para. 7.

[103] See CBD decision XI/19, "Biodiversity and climate change-related issues" (Hyderabad, October 8–19, 2012). Extending synergies beyond climate change adaptation, the parties to the CBD had also decided to adopt a moratorium on "all geo-engineering activities ... that may affect biodiversity." See CBD decision X/33, "Biodiversity and climate change" (Nagoya, October 18–29, 2010), para. 8(w). See also further discussions in Chapter 9, section III.B.2.

it more likely that conservation and restoration projects give due consideration to the evolving climatic conditions in which they take place.[104]

V. CONCLUSION

General climate change agreements are not the only source of international rules guiding international action on climate change. This chapter has reviewed some of the most important developments that have been taking place within other regimes, in particular the regime on the protection of the ozone layer or under the aegis of the IMO and the ICAO. Pioneering initiatives have been taken by pre-constituted or ad hoc coalitions of States, sometimes along with non-State actors. Action on climate change adaptation has also been promoted under other multilateral agreements such as the World Heritage Convention and the Convention on Biological Diversity. Growing frustration with the perceived inefficiency of international negotiations on climate change and an increased awareness of the urgent need for action to tackle climate change and its impacts have thus provided an impetus for initiatives taking place outside the UNFCCC regime, some of which have spurred further developments within the UNFCCC regime. This diversity of forums is most likely to foster new initiatives, although it could also come at a risk of duplications or inconsistencies in international efforts to tackle climate change.

---

[104] See e.g. Mark C. Urban, "Accelerating extinction risk from climate change" (2015) 348:6234 *Science* 571; Camille Parmesan *et al.*, "Beyond climate change attribution in conservation and ecological research" (2013) 16(1) *Ecology Letters* 58.

5
Relevant Norms of General International Law

Across legal systems and traditions, a distinction is often made between general and special rules. In criminal law courses throughout the world, for instance, a general part defining concepts such as criminal offense and punishment often precedes more special parts defining the rules applicable to particular offenses. A similar distinction between *general* norms and *special* rules can be made in international law.[1] General international law is composed of norms that apply in most circumstances, although their application can typically be excluded by special rules applicable to particular issue-areas.

These general norms often reflect a broad understanding, shared among nations, of principles of justice that governments ought to respect.[2] According to Christian Tomuschat, general international law comprises three sorts of norms:

1. "Axiomatic premises of the international legal order," such as the principle of sovereign equality of States, which are the foundation of the contemporary international legal order.
2. Systematic features that derive almost automatically from these premises, such as the law of treaties and the law of State responsibility.
3. Widely accepted values, such as the right to life.[3]

This chapter argues that some of these norms – in particular the no-harm principle and remedial obligations – require States to take particular steps with respect to climate change. In particular, under the no-harm principle, States must prevent activities within their territory or control from causing serious transboundary harm, including through excessive GhG emissions. They must act consistently with principles of international environmental law such as the principle of sustainable development and the principle of cooperation. They must respect and promote human rights protection through their climate action. If they breach any of the obligations above, they must pay adequate compensation to remedy the resulting injury.

---

[1] For an overview of the recognition of the concept of "general international law" in the work of the international law commission, see ILC Memorandum by the Secretariat, *Formation and Evidence of Customary International Law: Elements in the Previous Work of the International Law Commission That Could Be Particularly Relevant to the Topic* (March 14, 2013), doc. A/CN.4/659, at 35–37.

[2] I use "norm" here to connote a rule which is not only based on conventional grounds (such as driving on the left or on the right side of the street), but is also anchored on common moral principles.

[3] See Christian Tomuschat, "What is General International Law?" (Audiovisual Library of International Law), http://legal.un.org/avl/ls/Tomuschat.html# (accessed December 30, 2017).

It is argued in the following sections that international agreements on climate change do not substitute these norms of general international law.[4] But even if one were to consider that treaty regimes precluded the application of general international law, the discussions in this chapter should lead one to question the discrepancies between general norms and special rules. If international climate agreements derogated to norms of general international law on a systematic basis, they would likely fall short of the legitimate expectations of some relevant actors; being considered unfair, they may fail to gather sufficient political support in all countries to ensure their effective implementation, and could affect the legitimacy of the international law system as a whole as a promise of international justice. Thus, a better understanding of how general international law applies to climate change appears essential to a successful development of international cooperation on climate change.

I. THE NO-HARM PRINCIPLE

A. Recognition in General International Law

The principle of States' sovereign equality is an axiomatic premise of the current international legal order.[5] Even though States are very unequal in terms of geopolitical or economic power, they hold equal rights and bear equal obligations. Equal sovereignty promotes the independence of each State from any other State by implying a right for every State to be free from unjustified interference in its domestic affairs, and a corollary obligation for every State to refrain from such interference in the affairs of others.

In some circumstances, the conduct of a State is obviously incompatible with the principle of equal sovereignty. Thus, the UN Charter explicitly prohibits the threat or use of force by one State against another State without being allowed by the UN Security Council, except for what falls within States' "inherent right of individual or collective self-defense."[6] As Tomuschat noted: "Within an international system where going to war is considered lawful conduct, sovereign equality of States [would] pertain ... more to the realm of legal fiction than to a living reality."[7] The prohibition of the use of force in international relations is almost a necessary implication of the principle of sovereign equality in an international society of States with competing interests – an indispensable guarantee against domination by force. This prohibition of the use of force not only implies an obligation for a State to refrain from attacking another State; it has also been interpreted as involving a due diligence obligation to prevent actors within its territory or under its overall control from carrying out attacks against other States.[8]

Likewise, sovereign equality implies the prohibition against any State from causing serious environmental damages to another State, for instance, through pollution affecting the latter's territory or population, even if (unlike the circumstance of the use of force) this damage is

---

[4] See below, section VI.

[5] See Charter of the United Nations, June 26, 1945, 1 *UNTS* XVI, art. 2.1.

[6] See *ibid.*, art. 2.4 and art. 51.

[7] Christian Tomuschat, "International law: ensuring the survival of mankind on the eve of a new century: general course on public international law" (1999) 281 *Collected Courses of the Hague Academy of International Law* 1, at 206.

[8] See e.g. ICJ, *Corfu Channel (United Kingdom* v. *Albania)*, judgment of April 9, 1949, at 22; ICJ, *Military and Paramilitary Activities in and against Nicaragua (Nicaragua* v. *United States)*, judgment of June 27, 1986, para. 115; and, for a specific application in relation to international terrorism, UN Security Council Resolution 1373 (September 28, 2001).

not caused with the specific intent of harming the other State or its nationals. Some scholars make a distinction between two principles: the no-harm principle, which establishes a negative obligation for States to refrain from causing serious transboundary harm; and the preventive principle, a due diligence obligation to prevent activities that would cause serious transboundary harm from being carried out under their jurisdiction.[9] By commodity, "no-harm principle" refers, in this book, to the general obligation on a State to ensure that neither its own activities nor any activities conducted by other actors under its jurisdiction cause such transboundary harm.

The no-harm principle was first recognized in an arbitral sentence of 1941 in a case concerning a smelter situated in the Canadian city of Trail, in British Columbia, only a few kilometers from the US border. The Trail smelter was emitting considerable amounts of fumes which were causing significant environmental damage to its surroundings, including across the border in the United States. Following unsuccessful diplomatic consultations, the United States initiated arbitral proceedings against Canada. In its final award, the arbitral tribunal ruled out in favor of the United States on the ground that:

> Under the principles of international law ... no state has the right to use or permit the use of territory in such a manner as to cause injury by fumes in or to the territory of another of the properties or persons therein, when the case is of serious consequence and the injury is established by clear and convincing evidence.[10]

Since 1941, this principle was reaffirmed multiple times by international courts and tribunals.[11] A statement of the no-harm principle was also included in diverse international documents, notably in Principle 21 of the 1972 Stockholm Declaration on the Human Environment and Principle 2 of the 1992 Rio Declaration on Environment and Development. The latter reads as follows:

> States have, in accordance with the Charter of the United Nations and the principles of international law, the sovereign right to exploit their own resources pursuant to their own environmental and developmental policies, and the responsibility to ensure that activities within their jurisdiction or control do not cause damage to the environment of other States or of areas beyond the limits of national jurisdiction.[12]

---

9 This distinction is made for instance in Philippe Sands and Jacqueline Peel, *Principles of International Environmental Law*, 3rd edn (Cambridge University Press, 2012) 200. See generally S. Jayakumar *et al.* (eds.), *Transboundary Pollution: Evolving Issues of International Law and Policy* (Edward Elgar, 2015); Nicolas Bremer, "Post-Environmental Impact Assessment monitoring of measures of activities with significant transboundary impact: an assessment of customary international law" (2017) 26(1) *Review of European, Comparative & International Environmental Law* 80.

10 *Trail Smelter (U.S. v. Canada)*, Arbitral Award of March 11, 1941 (1949) III *UNRIAA* 1938, at 1965.

11 See e.g. ICJ, *Pulp Mills on the River Uruguay (Argentina v. Uruguay)*, judgment of April 20, 2010, para. 101; arbitral award of May 24, 2005 in the case of the *Iron Rhine Railway (Belgium v. Netherlands)* (2005) XXVII *UNRIAA* 35, para. 222; ICJ, *Gabčíkovo-Nagymaros Project (Hungary v. Slovakia)*, judgment of September 25, 1997, para. 140. See generally Sands and Peel, *supra* note 9, at 195–203.

12 UNCED, Rio Declaration on Environment and Development (June 3–14, 1992), available in (1992) 31 *ILM* 874 (hereinafter Rio Declaration), principle 2. See also UNCHE, Stockholm Declaration on the Human Environment, available in (1972) 11 *ILM* 1416 (June 5–16, 1972) (hereinafter Stockholm Declaration), principle 21.

The Preamble to the UNFCCC emphasized the "pertinent provisions" of the Rio Declaration on Environmental and Development before further "recalling" the no-harm principle.[13]

While some scholars noted a lack of general State practice,[14] the International Court of Justice clearly identified the no-harm principle as customary international law in its Advisory Opinion on *The Legality of the Threat or Use of Nuclear Weapons*.[15] Philippe Sands and Jacqueline Peel called the no-harm principle the "cornerstone of international environmental law."[16] Just like the prohibition of the use of force, the no-harm principle derives almost automatically from the principle of sovereign equality. Grave environmental harms, which could possibly render the whole of a State's territory uninhabitable, can interfere as severely with sovereign rights as an armed attack.[17] A prohibition of serious environmental harms is necessary if the principle of sovereign equality is to be taken seriously in an age of man-made global environmental changes.[18]

While the existence of the no-harm principle is well established in general international law, its modalities remain ill-defined. Due to its customary nature, there is no unique authoritative statement of the no-harm principle, resulting in a certain lack of precision. The Stockholm Declaration on the Human Environment and the Rio Declaration on Environment and Development are indications of these principles, but, as they are not directly binding, they do not constitute an authoritative statement of these principles. An important question regards the *de minimis* threshold – the limit between minimal and hence tolerable transboundary harm, and more serious harm prohibited under the no-harm principle. Another difficulty regards the definition of the standard of due diligence which defines the intensity of the efforts that a State should make to prevent activities under its jurisdiction from causing such transboundary harm. These questions on the modalities of application are an important source of difficulties and confusion, but they are not of a nature to exclude the application of the principle. The application of a norm cannot be excluded simply because the norm is vaguely defined.

B. Relevance to Climate Change

Climate change is certainly the greatest environmental damage ever caused by humankind. Its impacts not only create localized harms to individuals, societies, economies and ecosystems, but also extend to planetary systems, threatening our very existence – as civilizations, if not as a species – through the looming risk of a global civilizational collapse. The predictable impacts of climate change in the coming decades include the disappearance of all of the natural land territory of small-island developing States under a rising sea level, as well as great impacts

---

[13] United Nations Framework Convention on Climate Change, May 9, 1992, 1771 *UNTS* 107 (hereinafter UNFCCC), recitals 8 and 9.

[14] John Knox, "The myth and reality of transboundary environmental impact assessment" (2002) 96(2) *American Journal of International Law* 291 at 293.

[15] ICJ, *The Legality of the Threat or Use of Nuclear Weapons*, Advisory Opinion of July 8, 1996, para. 29.

[16] Sands and Peel, *supra* note 9, at 191. See also, regarding the preventive principle more specifically, ILC, *Draft Articles on Prevention of Transboundary Harm from Hazardous Activities*, in (2001) *Yearbook of the International Law Commission*, vol. II, part two, art. 3.

[17] See, for instance, Josha W. Busby, "Who cares about the weather? Climate change and U.S. national security" (2008) 17:3 *Security Studies* 468.

[18] See P.J. Crutzen, "The 'Anthropocene'" in Eckart Ehlers and Thomas Krafft (eds.), *Earth System Science in the Anthropocene* (Springer, 2006) 13; Will Steffen *et al.*, "The Anthropocene: conceptual and historical perspectives" (2011) 369:1938 *Philosophical Transactions of the Royal Society, A: Mathematical, Physical & Engineering Sciences* 842.

affecting the territories of many other States, especially low-lying, arid or tropical developing States. Climate change will unavoidably affect the prosperity, development and enjoyment of human rights of many in all nations. The conduct of States which causes such harms either through their direct action or through their failure to prevent activities causing excessive GhG emissions under their jurisdiction is certainly incompatible with the principle of equal sovereignty and, more specifically, with the no-harm principle. To paraphrase Tomuschat, the sovereign equality of States would pertain more to the realm of legal fiction than to a living reality within an international system where some States can cause such tremendous harm to others.

Yet, it is undeniable that climate change differs from typical transboundary harm in some significant ways. In classical cases such as the *Trail Smelter* case, activities within one State's territory release pollutants which directly affect another State's territory. In such classical cases, damage is directly caused by substances which cross an international border and, more often than not, harms are confined to a relatively small border area. By contrast, excessive GhG emissions affect the global climate system rather than any specific State or region. Whereas climate change has adverse impacts in many places throughout the world, none of these impacts is the direct consequence of the emissions of GhGs at a particular place or at a given time. Rather, it is the cumulative effect of GhG emissions in multiple places, year after year, which causes serious environmental impacts throughout the world as the chemistry of our atmosphere is gradually altered. Harm affects the territory of every State, as well as the global commons, in much less direct ways.

Alexander Zahar argued that the no-harm principle, which was generally recognized when damages were directly caused by pollutants crossing an international border, would not necessarily apply to cases where damages result from the progressive accumulation of pollutants in the atmosphere.[19] The distinction, however, does not appear to be material to the application of the no-harm principle. The International Court of Justice had the opportunity to consider the application of the no-harm principle to environmental damages of such a cumulative nature in its advisory proceedings regarding *The Legality of the Threat or Use of Nuclear Weapons*. Beyond the local consequences of the detonation of nuclear weapons and possible nuclear fallout in neighboring regions (direct damage), submissions before the Court invoked the diffuse harm of such activities on the global environment through risks of a nuclear winter or an interference with the Earth's electromagnetic field (cumulative damages).[20] The Court, however, made no distinction in its Advisory Opinion between the two types of environmental damage.[21] Two dissenting opinions suggested that a distinction should be made and that a different treatment should apply to damage affecting the global environment, but they did not exclude the application of the no-harm principle to cumulative damages; on the contrary, they argued that a *more* stringent application of this principle would be necessary in such circumstances.[22] Indeed, the prohibition of environmental damage affecting another State implies *a fortiori* the prohibition of environmental damage affecting all other States.[23]

---

[19] See Alexander Zahar, "Mediated versus cumulative environmental damage and the International Law Association's legal principles on climate change" (2014) 4:3–4 *Climate Law* 217.

[20] See e.g. Written Statement by the Government of Mexico (June 13, 1995), para. 65; Written Statement of the Government of Egypt (June 20, 1995), para. 32; Letter from the General Director for Multilateral Organizations at the Ministry of Foreign Affairs of Ecuador (June 20, 1995), para. D.

[21] See ICJ, *The Legality of the Threat or Use of Nuclear Weapons, supra* note 15, paras. 29 and 35.

[22] See *ibid.*, Dissenting Opinion of Judge Koroma and Dissenting Opinion of Judge Weeramantry, at 456–458.

[23] See ILA Resolution 2/2014, "Declaration of Legal Principles relating to Climate Change" (Washington, April 7–11, 2014), art. 7A; Benoit Mayer, "The Applicability of the Principle of Prevention to Climate Change: A Response to

Another possible argument for an exclusion of the no-harm principle in relation to climate change could relate more specifically to the multiple sources of GhGs. But here again, climate change does not differ in any material way from the scenario of a cataclysmic atomic war leading to a nuclear winter discussed before the International Court of Justice. Whereas only a few States possess atomic weapons, most GhG emissions also stem from only a handful of nations: China, the United States and the European Union represent almost half of all global GhG emissions.[24] Accordingly, there appears to be no convincing ground for excluding excessive GhG emissions from the scope of the no-harm principle. The case of climate change is certainly more complex than classical bilateral disputes on transboundary harm, but complexity alone is not a reason for excluding the application of a principle.

However, many questions arise regarding the modalities of application of the no-harm principle to climate change. One such question relates to the standard of due diligence applicable to the obligation of a State to prevent harmful activities from taking place, as mentioned above. If only negligence or gross negligence is prohibited, States would not be responsible for the consequences of GhG emissions which occurred before clear scientific evidence of climate change and its impact arose, most likely before the mid-1980s. Alternatively, a regime of liability would possibly take into account all historical GhG emissions, starting from the beginning of the Industrial Revolution in Great Britain of the late eighteenth century. The growing recognition of a precautionary principle, according to which "lack of full scientific certainly shall not be used as a reason for postponing cost-effective measures," suggests that some sovereign obligations arise as soon as credible "threats of serious or irreversible damage" are perceived.[25]

Another important issue relates to the geographic scope of the no-harm principle with regard to the activity from which harm arises. The *Trail Smelter* arbitral award defined this principle with regard to activities conducted within the "territory" of the responsible State.[26] By contrast, the Stockholm Declaration on the Human Environment and the Rio Declaration on Environment and Development extend the geographic scope to a State's "jurisdiction or control,"[27] while the International Court of Justice referred to "jurisdiction *and* control" in its Advisory Opinion on *The Legality of the Threat or Use of Nuclear Weapons*.[28] These alternatives are of great significance to circumstances where GhG emissions arise from export-oriented activities, such as some of the industrial production of China, or from the activities of transnational corporations in the developing world.[29] While the territorially competent State certainly bears some responsibility, it is arguable that the importing State or the investing State

---

Zahar" (2015) 5:1 *Climate Law* 1; ILC, *Second Report on the Protection of the Atmosphere by Special Rapporteur Shinya Murase*, doc. A/CN.4/681 (March 2, 2015), para. 51 and *Third Report on the Protection of the Atmosphere by Special Rapporteur Shinya Murase,* doc. A/CN.4/692 (February 25, 2016), para. 36.

[24] WRI, CAIT Climate Data Explorer, "Total GHG emissions excluding land-use change and forestry," according to which these three countries or regional entities represented 49 percent of global GhG emissions in 2013 (last year available).

[25] See Rio Declaration on Environment and Development, *supra* note 12, principle 15; UNFCCC, *supra* note 13, art. 3.3.

[26] *Trail Smelter, supra* note 10, at 1965.

[27] Stockholm Declaration, *supra* note 12, principle 21; Rio Declaration, *supra* note 12, principle 2.

[28] ICJ, *The Legality of the Threat or Use of Nuclear Weapons, supra* note 15, para. 29. See also *Third Report on the Protection of the Atmosphere by Special Rapporteur Shinya Murase, supra* note 23, para. 33.

[29] Questions of attribution of GhG emissions in a globalized economy are further discussed in Chapter 6, section I.A.1.

could also bear some responsibility inasmuch as they exercise a certain degree of control over the activities in question.

II. OTHER PRINCIPLES OF INTERNATIONAL ENVIRONMENTAL LAW

While the no-harm principle is certainly also the "cornerstone" of the international law on climate change,[30] several other principles of international environmental law may also be of relevance to climate change, including the concept of sustainable development, the precautionary approach, the polluter-pays principle and the principle of cooperation.

A. The Concept of Sustainable Development

Sustainable development was defined by the 1987 "Brundtland" Report as a form of development that "meets the needs of the present without compromising the ability of future generations to meet their own needs."[31] The concept of sustainable development was at the center of the attempt of reconciling environmental protection with social and economic development in the Earth Summit; it was mentioned eight times in the 27 principles of the Rio Declaration on Environment and Development.[32] Consistently, the UNFCCC recognizes the rights and obligations of its parties to promote sustainable development.[33] It also separately highlighted a collective determination "to protect the climate system for present and future generations."[34] More recently, the principle of "sustainable development" appeared central in the Paris Agreement, being mentioned no less than 12 times. Such repeated incantation of this concept, however, did little to clarify its meaning.[35]

Indeed, the legal implications of the concept of sustainable development have remained largely undefined. In the case of *Gabčíkovo-Nagymaros*, the International Court of Justice recognized its legal value as an expression of the "need to reconcile economic development with protection of the environment," thus allowing the revision of a bilateral treaty on a transboundary project between Slovakia and Hungary.[36] There is perhaps a procedural obligation for States to give consideration, formally at least, to environmental protection and to the rights of future generations in decision processes likely to affect them, for instance, through an environmental impact assessment. Beyond this, the concept of sustainable development may have some implications on the substantial obligations of States, at the very least by prohibiting reckless conduct harming the environment or future generations. However, it remains difficult to achieve the right balance between social and economic development and environmental protection, or between the interests of the present and future generations. An international jurisdiction called upon to review such assessments is most likely to give great deference to national governments, unless the conduct causes harm beyond national borders.

---

[30] See *supra* note 16.

[31] WCED, *Our Common Future* (Oxford University Press, 1987) (hereinafter Brundtland Report), at Chapter II, para. 1.

[32] See Rio Declaration, *supra* note 12, principles 1, 4, 7, 8, 9, 12, 22 and 24.

[33] UNFCCC, *supra* note 13, art. 3.4. See also decision 1/CP.8, "Delhi Ministerial Declaration on Climate Change and Sustainable Development" (November 1, 2002).

[34] UNFCCC, *supra* note 13, recital 25. See also *ibid.*, art. 3.1.

[35] See Jorge E. Viñuales, "The rise and fall of sustainable development" (2013) 22:1 *Review of European Community & International Environmental Law* 3; John C. Dernbach and Federico Cheever, "Sustainable development and its discontents" (2015) 4:2 *Transnational Environmental Law* 247.

[36] ICJ, *Gabčíkovo-Nagymaros Project, supra* note 11, para. 140.

In principle, the concept of sustainable development is clearly relevant to responses to climate change, touching on some of the most central issues – the need to reconcile economic and human development with environmental protection and to promote intergenerational equity. Yet, its normative implications are sometimes only confirming those of the no-harm principle; at other times, these normative implications are so vague that the concept adds little to the assessment of a State's rights and obligations. At least, the concept of sustainable development advocates for a more holistic vision of global environmental protection, including the need for coordination among multilateral environmental regimes[37] and for due consideration for developing States' need for assistance.[38]

B. The Precautionary Approach

International environmental law promotes a so-called "precautionary approach" to scientific uncertainty. As defined in the Rio Declaration on Environment and Development, this approach dictates that:

[w]here there are threats of serious or irreversible damage, lack of full scientific certainty shall not be used as a reason for postponing cost-effective measures to prevent environmental degradation.[39]

The modalities and legal force of this approach remain generally ill-defined.[40]

The UNFCCC highlights the relevance of the precautionary approach among the "principles" which should guide State actions. It provides that the parties "should take precautionary measures to anticipate, prevent or minimize the causes of climate change and mitigate its effects."[41] This provision was inserted at a time where, although it was clear that anthropogenic GhG emissions would cause adverse impacts, science remained inconclusive as to the magnitude and gravity of these impacts.[42] As further research ascertained prior findings, the precautionary approach has progressively lost some of its relevance, although it was invoked again, episodically, in relation to the approaches to address loss and damage.[43]

The precautionary approach remains relevant, to date, in assessing the relevance of the cataclysmic risk of a climate runaway scenario. It could also be relevant in assessing the historical responsibility of industrial nations for failing to take appropriate measures when scientific evidence was already accumulating to suggest the impact of GhG emissions. Yet, the precautionary approach was not well established under international law until at least the mid-1980s, when strong evidence of climate change started to accumulate and placed the occurrence of climate

---

[37] See e.g. decision 22/CP.5, "Institutional linkage of the Convention secretariat to the United Nations" (October 25, 1999). See also Chapter 4, section IV.B and generally Chapter 16, section I.

[38] See e.g. decision 1/CP.8, *supra* note 33.

[39] Rio Declaration, *supra* note 12, principle 15. See generally in Antônio Augusto Cançado Trindade, "Principle 15: precaution" in Jorge E. Viñuales (ed.), *The Rio Declaration on Environment and Development: A Commentary* (Oxford University Press, 2015) 403.

[40] See generally Sands and Peel, *supra* note 9, at 217–228.

[41] UNFCCC, *supra* note 13, art. 3.3.

[42] This is reflected in *ibid.*, recital 6.

[43] See e.g. decision 3/CP.18, "Approaches to address loss and damage associated with climate change impacts in developing countries that are particularly vulnerable to the adverse effects of climate change to enhance adaptive capacity" (December 8, 2012), recital 8.

change beyond uncertainty. Last but not least, as geoengineering options are being considered to counter the impact of climate change, including through solar radiation management, the precautionary approach acquires a new relevance in relation to planetary-wide projects whose impact would be serious and irreversible.[44]

C. The Polluter-Pays Principle

The polluter-pays principle suggests that the person causing some adverse environmental impact should bear the adverse consequences thereof. Assumedly, this would encourage polluters to reduce their pollution by any means available to them. When enforcement is not carefully monitored, however, there is a risk that making polluters pay creates an incentive for unlawful activities such as smuggling in controlled substances.[45] This principle is rather well-recognized as an aspirational principle within the OECD,[46] but much less beyond. Negotiations led, for instance, to a rather circumvoluted statement in the Rio Declaration on Environment and Development:

> National authorities should endeavour to promote the internalization of environmental costs and the use of economic instruments, taking into account the approach that the polluter should, in principle, bear the cost of pollution, with due regard to the public interest and without distorting international trade and investment.[47]

The polluter-pays principle was invoked before an arbitral tribunal in the *Rhine Chlorides* case. While noting the recognition of this principle in some treaties, the Tribunal rejected the view that it had become part of general international law.[48] As Philippe Sands and Jacqueline Peel noted, State practice simply "does not support the view that all the costs of pollution should be borne by the polluter, particularly in inter-state relations,"[49] at least not as a matter of law.

However, rather than an established legal norm, the polluter-pays principle ought to have some influence on the development of the international law on climate change as an idea. The idea that a person causing harm to another ought to bear the costs is after all a fundamental, ethical and a common legal concept. This does not necessarily require the polluter to pay *all* the costs in every circumstance, but it supports the view that polluters should be charged some fee that would dissuade them from polluting whenever this is likely to disincentivize pollution. Through the ad hoc principle of common but differentiated responsibility or the notion of climate justice, some States and advocates have been trying to bring back essentially the same idea on the international plane – making States responsible for the largest share of GhG emissions responsible for the

---

[44] See generally the discussion of geoengineering options in Chapter 9.

[45] See e.g. Lorraine Elliott, "Smuggling networks and the black market in ozone depleting substances" in Tanya Wyatt (ed.), *Detecting and Preventing Green Crimes* (Springer, 2016) 45; Graham Donnelly Welch, "HFC smuggling: preventing the illicit (and lucrative) sale of greenhouse gases" (2017) 44:2 *Boston College Environmental Affairs Law Review* 525.

[46] See e.g. OECD, *Recommendation of the Council on Guiding Principles Concerning the International Economic Aspects of Environmental Policies*, doc. C(72)128 (May 26, 1972), reproduced in (1975) 14 *ILM* 236.

[47] Rio Declaration, *supra* note 12, principle 16.

[48] *Case Concerning the Audit of Accounts in Application of the Protocol of 25 September 1991 Additional to the Convention for the Protection of the Rhine from Pollution by Chlorides of 3 December 1976 (Netherlands v. France)*, Arbitral Award of the March 12 2004, (2004) XXV *UNRIAA* 267, para. 103.

[49] Sands and Peel, *supra* note 9, at 232. See also Priscilla Schwartz, "Principle 16: the polluter-pays principle" in Viñuales (ed.), *supra note 39, at* 429.

impacts of climate change.[50] Similar conclusions can also be arrived at on the basis of established norms of general international law on the responsibility of States for internationally wrongful acts when States have breached their obligations under the no-harm principle.[51]

D. The Principle of Cooperation

Last but not least, cooperation to address global environmental issues has also, arguably, become a matter of legal principle. The Stockholm Declaration on the Human Environment highlighted that "[c]ooperation through multilateral or bilateral arrangements or other appropriate means is essential to effectively control, prevent, reduce and eliminate adverse environmental effects."[52] Similarly, the Rio Declaration on Environment and Development reflected on the need for States and people to "cooperate in good faith and in a spirit of partnership for the fulfilment of the principles embodied in this declaration."[53] The International Tribunal on the Law of the Sea recognized a duty to cooperate as "a fundamental principle in the prevention of pollution of the marine environment" both under the UN Convention on the Law of the Sea and under general international law.[54]

The principle of cooperation is particularly relevant to climate change, a cooperation issue *par excellence*. Climate change is caused by the cumulative effect of GhG emissions from activities taking place in multiple States; it could not be addressed by a State alone. Accordingly, the Preamble to the UNFCCC acknowledges:

that the global nature of climate change calls for the widest possible cooperation by all countries and their participation in an effective and appropriate international response, in accordance with their common but differentiated responsibilities and respective capabilities and their social and economic conditions.[55]

The principle of cooperation could be interpreted to suggest a rather vague obligation for States to pursue negotiations on climate change in good faith. Free-riding on other States' cooperation could appear inconsistent with this principle.[56] Concretely, however, it may be difficult to make a case that any particular State is breaching its obligation under this principle except in the most blatant rejections of international cooperation. The withdrawal of the United States from the Paris Agreement or its previous decision not to ratify the Kyoto Protocol would appear to be the two most likely instances of such a refusal to cooperate at all if these decisions are understood as little more than an attempt at free-riding efforts carried out by other States.

---

[50] See Teresa M. Thorp, *Climate Justice: A Voice for the Future* (Palgrave Macmillan, 2014); Simon Caney, "Cosmopolitan justice, responsibility, and global climate change" (2005) 18:4 *Leiden Journal of International Law* 747.

[51] See respectively sections I and IV of this chapter.

[52] Stockholm Declaration, *supra* note 12, principle 24.

[53] Rio Declaration, *supra* note 12, principle 27. See generally Peter H. Sand, "Principle 27: cooperation in a spirit of global partnership" in Viñuales (ed.), *supra note 39, at* 617.

[54] ITLOS, *The MOX Plant Case (Ireland* v. *United Kingdom)*, Order for Provisional Measures of December 3, 2001, para. 82. See also ITLOS, *Case Concerning Land Reclamation by Singapore in and around the Straits of Johor (Malaysia* v. *Singapore)*, Order for Provisional Measures of October 8, 2003.

[55] UNFCCC, *supra* note 13, recital 7.

[56] See Eric A. Posner and David A. Weisbach, *Climate Change Justice* (Princeton University Press, 1010) 178.

III. INTERNATIONAL HUMAN RIGHTS LAW

For a long time, international law only dealt with relations between sovereign nations. Yet, the twentieth century saw a progressive trend toward international norms applicable to internal matters. This trend is not foreign to environmental law, but it has nowhere been as discernible as with regard to the protection of human rights. The philosophy of human rights came to nuance an absolutist vision of State sovereignty by advancing the idea that national governments must restrain their use of power in order to respect human dignity within their jurisdiction, or else be held accountable for failing to do so.

This idea was not new; the existence of natural rights had been discussed in some form or shape for seemingly as long as human beings have constituted organized communities.[57] However, the roots of the modern idea of human rights stem from the work of eighteenth-century philosophers, during the Age of Enlightenment. Starting with the United States' Declaration of Independence of 1776 and France's Declaration of the Rights of Man and of the Citizen of 1789, these ideas have progressively percolated into the constitutional and legal doctrines of most nations before entering debates at the international level.

A movement for the recognition of human rights at the international level gained momentum in reaction to the atrocities committed during the Second World War. On December 10, 1948, the General Assembly of the United Nations adopted the Universal Declaration on Human Rights.[58] To translate this declaration into treaty provisions with clear legal force, the General Assembly further adopted two comprehensive covenants and a series of more specific conventions.[59] These treaties define the obligation of States to respect the rights of everyone within their jurisdiction, to protect these rights from infringement by others under their jurisdiction, and to fulfill these rights through appropriate laws, policies, programmes, projects and decisions, as necessary. Furthermore, these treaties established international mechanisms through which individual complaints or national policies could be reviewed. Some regions of the world concluded treaties containing more advanced provisions,[60] sometimes including the creation of regional human rights courts whose decisions States committed to implement.[61]

Additional steps were taken for the international protection of human rights following the end of the Cold War. The establishment of the International Criminal Court reflects a more consistent approach of punishing acts of genocide, crimes against humanity and war crimes.[62]

---

[57] The Cyrus Cylinder, now in the British Museum, is an early example of some human rights considerations in a declaration written on behalf of Achaemenid Emperor Cyrus the Great around 539 BCE.

[58] UN General Assembly Resolution 217 A, "Universal Declaration of Human Rights" (December 10, 1948).

[59] See International Covenant on Economic, Social and Cultural Rights, December 16, 1966, 993 *UNTS* 3, and International Covenant on Civil and Political Rights, December 16, 1966, 999 *UNTS* 171. See also e.g. International Convention on the Elimination of All Forms of Racial Discrimination, December 21, 1965, 660 *UNTS* 195; Convention on the Elimination of All Forms of Discrimination against Women, December 18, 1979, 1249 *UNTS* 13; Convention against Torture and Other Cruel, Inhuman or Degrading Treatment or Punishment, December 10, 1984, 1465 *UNTS* 85; International Convention on the Rights of the Child, November 20, 1989, 1577 *UNTS* 3; International Convention on the Rights of All Migrant Workers and Members of Their Families, December 18, 1990, 2220 *UNTS* 3; International Convention on the Rights of Persons with Disabilities, December 13, 2006, 2515 *UNTS* 3; International Convention for the Protection of All Persons from Enforced Disappearance, December 20, 2006, 2716 *UNTS* 3.

[60] See e.g. Convention for the Protection of Human Rights and Fundamental Freedoms, November 9, 1950, 213 *UNTS* 222; American Convention on Human Rights (hereinafter Pact of San Jose), November 22, 1969, 1144 *UNTS* 123; African Charter on Human and People's Rights (hereinafter "Banjul Charter"), June 27, 1981, 1520 *UNTS* 217.

[61] See generally Dinah Shelton (ed.), *The Oxford Handbook of International Human Rights Law* (Oxford University Press, 2013).

[62] Rome Statute of the International Criminal Court, July 17, 1998, 2187 *UNTS* 90.

When such crimes are committed, the concept of a responsibility to protect has come to provide an additional, systematic justification for an intervention of the international community.[63] Sovereignty was progressively reconceived not just as a right of national government, but also as a responsibility for the protection of everyone within their jurisdiction.[64] Through the Millennium Development Goals and then the Sustainable Development Goals, States pledged to cooperate more closely in promoting economic and human development across the world.[65]

International human rights law is relevant to climate change, albeit rather indirectly, in three different ways.[66] Firstly, the impacts of climate change are a challenge to the enjoyment of human rights. Accordingly, some consideration for climate change is needed in the development of human rights policies. Climate change may affect the way in which individuals enjoy their right to life, to health, to food, to a decent standard of living, to property and to cultural flourishing, among other things. Efforts to promote climate change adaptation can sometimes be understood as a way for States to protect the human rights of individuals within their jurisdiction when they are faced with the adverse impacts of climate change.

Secondly, the obligation of States to respect and protect human rights may also be taken to imply an obligation to restrain GhG emissions within their jurisdiction – a justification for action on climate change mitigation.[67] This, however, requires a significant stretching of the conceptual framework on human rights protection because most of the impacts of GhG emissions take place beyond the territory where these emissions occur. The obligation of a State to respect and protect human rights has generally been understood to extend to any situation under its effective control,[68] and it is uncertain whether a State's control over GhG emissions taking place within its territory or under its jurisdiction is sufficient to establish control over the remote consequences of these GhG emissions. Although international human rights law may come in support of the need to take effective measures to mitigate climate change, existing human rights institutions are not adequately designed to address issues which stem from activities scattered throughout the world and result in widespread impacts.

Thirdly, human rights are more directly relevant to assessing the impacts of response measures. Projects which aim to mitigate climate change or adapt to its impact may have unintended adverse consequences.[69] A hydroelectric dam, for instance, could provide a clean source of energy and improve water management, but it also often involves the resettlement

---

[63] See UN General Assembly Resolution 60/1, "2005 World Summit Outcome" (October 24, 2005), paras. 138–140; UN Secretary General, *Implementing the Responsibility to Protect*, doc. A/63/677 (January 12, 2009), para. 11.

[64] See Francis Deng *et al.*, *Sovereignty as Responsibility: Conflict Management in Africa* (Brookings Institution, 1996).

[65] See UN General Assembly Resolution 55/2, "United Nations Millennium Declaration" (September 8, 2000); UN General Assembly Resolution 70/1, "Transforming our world: the 2030 Agenda for Sustainable Development" (September 25, 2015).

[66] See generally Stephen Humphreys (ed.), *Human Rights and Climate Change* (Cambridge University Press, 2010); John Knox, "Human rights principles and climate change" in John S Dryzek, Richard B. Norgaard and David Schlosberg (eds.), *The Oxford Handbook of Climate Change and Society* (Cambridge University Press, 2011) 213. See also further discussions in Chapter 16, section I.

[67] See UN General Assembly Resolution 70/1, *supra* note 65, goal 13. See also Chapter 2, section III.B. See also CESCR, *Concluding Observations on the Fifth Periodic Report of Australia* (June 23, 2017), doc. E/C.12/AUS/CO/5, para. 12.

[68] See Marko Milanovic, *Extraterritorial Application of Human Rights Treaties: Law, Principles, and Policy* (Oxford University Press, 2011).

[69] See Emily Boyd, "Governing the Clean Development Mechanism: global rhetoric versus local realities in carbon sequestration projects" (2009) 41:10 *Environment & Planning A* 2380.

of populations living in what is to become a reservoir. It is essential that responses to climate change do not come at the expense of the human rights of affected populations.

IV. RESPONSIBILITY OF STATES FOR INTERNATIONALLY WRONGFUL ACTS

For a legal system to be effective, the breach of an obligation needs to be sanctioned. Some sanctions take place outside of the legal system, for instance, through social reprobation and shame. Like other normative systems (e.g. moral or religious), legal systems largely rely on such social pressure for compliance. In addition, however, legal systems seek to impose formal sanctions, such as punishment and remedial obligations, on those acting in breach of their obligations. It is thus well established as a matter of general international law, as the International Law Commission (ILC) recognized in its Articles on the Responsibility of States for Internationally Wrongful Acts, that "[e]very internationally wrongful act of a State entails the international responsibility of that State."[70] These Articles on State Responsibility define an internationally wrongful act as an action or an omission which "(a) [i]s attributable to the state under international law and (b) [c]onstitutes a breach of an international obligation of the state."[71]

A. Conditions to Invoke the Responsibility of States for Internationally Wrongful Acts

With regard to climate change, State responsibility could be invoked on different grounds. For instance, the breach of a specific obligation established in the UNFCCC regime or through relevant developments in other regimes could entail the international responsibility of a State. Thus, the international responsibility of a developed State Party could arise from its failure to "adopt national policies and take corresponding measures on the mitigation of climate change,"[72] as it committed to do when ratifying the UNFCCC, or to ensure that its GhG emissions do not exceed its assigned amount as defined by Article 3 and Annex B of the Kyoto Protocol.[73] Likewise, developing States could be held responsible for omitting to "formulate, implement, publish and regularly update ... programmes containing measures to mitigate climate change"[74] as they committed to do when ratifying the UNFCCC. Furthermore, the failure of a State to comply with its obligations under the Montreal Protocol on Substances that Deplete the Ozone Layer or under Annex VI of MARPOL, for instance, could entail the international responsibility of a State. The obligation of a State to "protect and preserve the marine environment" under the UN Convention on the Law of the Sea could provide yet another ground for responsibility when States fail to take action to curb GhG emissions under their

---

[70] ILC, *Draft Articles on Responsibility of States for Internationally Wrongful Acts with Commentaries*, in (2001) *Yearbook of the International Law Commission*, vol. II, part two (hereinafter *Articles on State Responsibility*), art. 1. See generally James Crawford, *State Responsibility: The General Part* (Cambridge University Press, 2013); James Crawford, Alain Pellet and Simon Olleson (eds.), *The Law of International Responsibility* (Oxford University Press, 2010).

[71] *Articles on State Responsibility*, *supra* note 70, art. 2.

[72] UNFCCC, *supra* note 13, art. 4.2(a).

[73] Kyoto Protocol to the United Nations Framework Convention on Climate Change, December 11, 1997, 2303 *UNTS* 162 (hereinafter Kyoto Protocol), art. 3.1.

[74] UNFCCC, *supra* note 13, art. 4.1(b).

jurisdiction, despite the inevitable impact of such pollution on the marine environment (e.g. acidification or sea-level rise).[75]

More general arguments on the responsibility of industrial States for excessive GhG emissions could invoke a breach of their obligations under general international law, most convincingly under the no-harm principle, or alternatively under other principles of international environmental law or under international human rights law. With regard to the no-harm principle, the responsibility of a State could be invoked for its own action – State-owned enterprises, in particular, often play a direct role in the extraction and distribution of fossil fuels[76] – or alternatively on the ground of its failure to regulate private actors within its territory or under its jurisdiction. As mentioned in section I above, the existence of a due diligence obligation of States to prevent activities causing excessive GhG emissions from taking place within their territory or under their jurisdiction is well established in international law.

However, certainly not all GhG emissions entail a breach of the no-harm principle. We all emit some small quantity of carbon dioxide by breathing, but it would be absurd – or criminal – to expect any government to prevent its population from breathing. On the one hand, the existence of a *de minimis* threshold was mentioned in section I – it is generally understood that the no-harm principle only prohibits transboundary harm that is of sufficiently significant or "serious" consequence. On the other hand, the law of State responsibility recognizes the existence of circumstances which can preclude wrongfulness. Necessity, in particular, can be invoked when an act "is the only way for the State to safeguard an essential interest against a grave and imminent peril" and "does not seriously impair an essential interest of the State or States towards which the obligation exists, or of the international community as a whole."[77]

Beyond the evident case of human breathing, some GhG emissions could be considered necessary for improving our living conditions to at least a minimal level of human development. In this sense, Henry Shue suggested a distinction between (excusable) "subsistence emissions" and (inexcusable) "luxury emissions."[78] Rather than an illusory dichotomy between what is or is not necessary, there may be a need to compare the benefits that particular GhG emissions bring with the harms that they cause – a complex exercise requiring a progressive interpretation of general international law.

B. Secondary Obligations

The responsibility of States for excessive GhG emissions entails two types of obligation: cessation and reparation. Firstly, the ILC's Articles on State Responsibility recognized that a State responsible for an internationally wrongful act is under an obligation "to cease that act,

---

[75] See United Nations Convention on the Law of the Sea, December 10, 1982, 1833 *UNTS* 3, art. 192. See also Meinhard Doelle, "Climate change and the use of the dispute settlement regime of the Law of the Sea Convention" (2006) 37:3-4 *Ocean Development & International Law* 319; William C.G. Burns, "Potential causes of action for climate change damages in international fora: the Law of the Sea Convention" (2006) 2:1 *McGill International Journal of Sustainable Development Law & Policy* 27; Roda Verheyen, *Climate Change Damage and International Law: Prevention Duties and State Responsibility* (Brill, 2005) at 193ff.

[76] Benoit Mayer and Mikko Rajavuori, "National fossil fuel companies and climate change mitigation under international law" (2016) 44:1 *Syracuse Journal of International Law & Commerce* 55.

[77] *Articles on State Responsibility, supra* note 70, art. 25.1. See also Alan O. Sykes, "Economic 'necessity' in international law" (2015) 109:2 *American Journal of International Law* 296.

[78] Henry Shue, "Subsistence emissions and luxury emissions" (1993) 15:1 *Law & Policy* 39.

if it is continuing" and "to offer appropriate assurances and guarantees of non-repetition, if circumstances require."[79] As the failure of many States to prevent excessive GhG emissions within their jurisdiction appears to be continuing, cessation is more relevant than appropriate assurances and guarantees of non-repetition. Accordingly, industrial States are under an obligation to take immediate measures to control their activities, or activities carried out within their jurisdiction, from causing excessive GhG emissions. Yet, the ambit of this obligation is uncertain, depending in particular on how much GhG emissions can be considered as justifiable. The word "cease," at least, seems to suggest much more stringent efforts than are being considered through "mitigation" policies. In other words, the law of State responsibility is not satisfied by an incremental reduction, but it requires drastic measures for States to comply with their obligations under the no-harm principle immediately or as soon as possible. Cessation is the most pressing secondary obligation of States responsible for a continuing internationally wrongful act, including a breach of the no-harm principle.

Secondly, a State responsible for an internationally wrongful act must provide adequate reparation for the injury resulting from that act.[80] A difficulty here relates to the fact that excessive GhG emissions do not affect any particular State directly. Instead, the damage is directly caused to the global commons, and less directly to any particular State. Yet, inasmuch as States are bound to be harmed in one way or another as a consequence of excessive GhG emissions, nothing seems to preclude an interpretation of these States as either directly injured or otherwise as having an interest in the payment of some form of reparation. States would appear as the most natural institution on which to rely to distribute reparation for the loss and damage suffered by individuals or corporations within their jurisdiction.[81] Even then, assessing the injury suffered by any given State would be extremely challenging.[82]

Yet, assessing what constitutes an "adequate" reparation is not straightforward, as even the principles are not clearly defined. The ILC's Articles on State Responsibility stated that "[t]he responsible State is under an obligation to make *full* reparation for the injury caused by the internationally wrongful act."[83] As ILC Special Rapporteur James Crawford then noted, full reparation – where the quantum of reparation is defined solely based on a valuation of the injury – appeared to be implemented in most cases that had been decided by international courts and tribunals.[84] Beyond these cases, however, full reparation was not always claimed, let alone granted, in disputes settled through diplomatic means, or even in those settled by specialized international courts and tribunals in disputes concerning international trade law,[85]

---

[79] *Articles on State Responsibility, supra* note 70, art. 30.

[80] See e.g. PCIJ, *Factory at Chorzów (Germany v. Poland)*, judgment on jurisdiction of July 26, 1927, in Series A, No. 9, at 21.

[81] See ILC, *Draft Articles on Diplomatic Protection, with Commentaries*, in (2006) *Yearbook of the International Law Commission*, vol. II, part two; PCIJ, *Mavrommatis Palestine Concessions (Greece v. United Kingdom)*, judgment on jurisdiction of August 30, 1924, in Series A, No. 2, at 12.

[82] On the valuation of the harm caused by climate change, see generally Chapter 2, section II. Additional difficulties regard the distribution of this harm among States. See also Micahel Bowman and Alan Boyle (eds.), *Environmental Damage in International and Comparative Law: Problems or Definition and Valuation* (Oxford University Press, 2002).

[83] *Articles on State Responsibility, supra* note 70, art. 31 (emphasis added).

[84] ILC, *Third Report on State Responsibility by Special Rapporteur James Crawford*, doc. A/CN.4/507 (March 15, 2000), para. 42.

[85] WTO DSB, Minutes of Meeting on February 11, 2000, doc. WT/DSB/M/75, at 5.

international investment law,[86] or regarding war reparations or reparations for other mass atrocities.[87] The ILC itself was divided as to whether reparation should be qualified as "full" or simply "as complete as possible" in the particular circumstances of each case.[88] Some members of the ILC argued that full reparation "was often impossible and even undesirable."[89]

Unlike the cases typically decided by international courts and tribunals, climate change involves extensive and serious damage occurring throughout the world and spreading over time. Reparations could impose a great burden on responsible States, especially if reparation were to be paid at once for the consequences of GhG emissions that have been and will be faced over a long period of time. There is a reasonable argument for reparation to be defined not only in relation to the injury (as per the requirement of a full reparation), but also in relation to the respective capacity to pay of responsible States and to the need to sanction the continuation of a wrongful act.[90]

Some reparation would nevertheless be important, not only in supporting the States most affected by the adverse impacts of climate change, but also in incentivizing compliance with the international obligations breached. A parallel can be drawn with transitional justice, whereby limited remedies often play a great role in fostering reconciliation and triggering social processes of reconciliation. Because States are more likely to support an agreement that they consider as fair, international cooperation to reduce GhG emissions throughout the world could greatly benefit from a formal recognition of past wrongdoing and from at least some symbolic forms of atonement on the part of developed States.

Thus, just like the no-harm principle, the law of State responsibility defines some general norms which must guide States' conduct, for instance, by requiring them to take measures to cease excessive GhG emissions and to provide remedies in some shape or form. These norms are not purely conventional rules; they are almost necessarily implied by the axiomatic premises of the international legal order, in particular the principle of equal sovereignty, and they are firmly established in the practice that States have long accepted as law. Yet, the modalities of application of these general norms remain largely indeterminate. What level of GhG emissions constitutes a breach of the no-harm principle? How fast should excessive GhG emissions cease? What quantum of reparation should be paid to the States most affected by climate change? Such questions cannot be answered on the basis of existing law alone, but they would need to

---

[86] Muthucumaraswamy Sornarajah, *The International Law on Foreign Investment* (Cambridge University Press, 2010) 417.

[87] See, for instance, Eritrea-Ethiopia Claims Commission (EECC), decision of August 17, 2009, Final Award on Ethiopia's Damages Claims, in (2009) XXVI *UNRIAA* 631, paras. 18–22.

[88] See ILC, *Summary Record of the 2615th Meeting* (May 4, 2000), doc. A/CN.4/SR.2615, paras. 52 (R. Goco) and 55 (P.S. Rao).

[89] ILC, *Summary Record of the 2392nd Meeting* (May 31, 1995), doc. A/CN.4/SR.2392, para. 31 (Igor Lukashuk). See also *ibid.*, para. 37 (C Tomuschat); *Summary Record of the 2314th Meeting* (June 30, 1993), doc. A/CN.4/SR.2314, para. 84 (A. Mahiou); *Summary Record of the 2454th Meeting* (July 5, 1996), doc. A/CN.4/SR.2454, para. 19 ("some members of the Drafting Committee").

[90] See Elena Kosolapova, *Interstate Liability for Climate Change-Related Damage* (Eleven, 2013); Christina Voigt, "State Responsibility for Climate Change Damages" (2008) 77:1 *Nordic Journal of International Law* 1; Benoit Mayer, "State responsibility and climate change governance: a light through the storm" (2014) 13:3 *Chinese Journal of International Law* 539; Benoit Mayer, "Climate change reparations and the law and practice of state responsibility" (2016) 7:1 *Asian Journal of International Law* 185; Benoit Mayer, "Less-than-full reparation in international law" (2016) 56:3–4 *Indian Journal of International Law* 465.

be answered to clarify States' obligations under general international law in relation to climate change; they require a progressive interpretation of general international law.

V. LIABILITY OF STATES FOR TRANSBOUNDARY HARM ARISING OUT OF HAZARDOUS ACTIVITIES

During its lengthy study of the law of State responsibility, the ILC distinguished this theme from the question regarding liability for possible injurious consequences arising out of the performance of certain lawful activities.[91] A parallel work program on the latter question led to the adoption of two separate documents: the 2001 Articles on the Prevention of Transboundary Harm from Hazardous Activities[92] and the 2006 Principles on the Allocation of Loss in the Case of Transboundary Harm Arising out of Hazardous Activities.[93] This separate work program has not been as influential on the development of international law as the work program on State responsibility has been; reception of the ILC's study of State liability has sometimes been rather skeptical.[94] These two documents apply to activities creating a risk of significant transboundary harm, including "a high probability of causing significant transboundary harm" or "a low probability of causing disastrous transboundary harm."[95] Massive GhG emissions certainly create a very high probability – indeed, a virtual certainty – of significant transboundary harm, and a low yet non-negligible probability of a cataclysmic consequences caused by a runaway climate change scenario. Accordingly, a State's liability could arise in relation to massive GhG emissions taking place within its territory or under its jurisdiction.[96]

Unlike the law of State responsibility, State liability may arise without any internationally wrongful act. Thus, whereas State responsibility refers to a fault-based regime of reparations, State liability defines preventive and remedial obligations on the sole basis of risk. The two regimes do not appear to be mutually exclusive. The ILC analyzed cases of transboundary harm, such as the *Trail Smelter* case, as an illustration of both regimes,[97] thus suggesting that some cases could at the same time fall under the law of State responsibility and the law of

---

[91] See ILC, *Report of the Twenty-Fifth Session* (May 7–July 13, 1973), doc. A/9010/Rev.1, at 169, para. 39; ILC, *Report of the Forty-Fourth Session* (May 4–July 24, 1992), doc. A/CN.4/SER.A/1992/Add.l (Part 2), at 51, paras. 344–348.

[92] ILC, *Draft Articles on Prevention of Transboundary Harm, supra* note 16.

[93] ILC, *Draft Principles on the Allocation of Loss in the Case of Transboundary Harm Arising out of Hazardous Activities*, in (2006) *Yearbook of the International Law Commission*, vol. II, part two.

[94] See generally Alan Boyle, "State responsibility and international liability for injurious consequences of acts not prohibited by international law: a necessary distinction?" (1990) 31:1 *International & Comparative Law Quarterly* 1; Jutta Brunnée, "Of sense and sensibility: reflections on international liability regimes as tools for environmental protection" (2004) 53:2 *International & Comparative Law Quarterly* 351; Alan Boyle, "Liability for injurious consequences of acts not prohibited by international law" in James Crawford, Alain Pellet and Simon Olleson (eds.), *The Law of International Responsibility* (Oxford University Press, 2010) 95.

[95] ILC, *Draft Articles on Prevention of Transboundary Harm, supra* note 16, art. 2(a). See also *Draft Principles on the Allocation of Loss, supra* note 93, Commentary under principle 2, para. 24.

[96] Michael G. Faure and André Nollkaemper, "International liability as an instrument to prevent and compensate for climate change" (2007) 43 *Stanford Journal of International Law* 123.

[97] The *Trail Smelter* award, for instance, was discussed as alternatively an illustration of State responsibility for an internationally wrongful act or as a case of State liability for transboundary harm arising out of hazardous activities. See *Articles on State Responsibility, supra* note 70, Commentary under art. 14, para. 14; ILC, *Draft Articles on Prevention of Transboundary Harm, supra* note 16, Commentary under art. 2, para. 6; and *Draft Principles on the Allocation of Loss, supra* note 93, Commentary under principle 2, para. 1.

State liability. Likewise, both regimes are relevant to climate change. The law of State responsibility applies inasmuch as excessive GhG emissions constitute a breach of the no-harm principle, whereas the law of State liability also applies as long as massive GhG emissions can be considered as hazardous activities causing significant transboundary harm.

Some provisions of the 2001 Articles on the Prevention of Transboundary Harm from Hazardous Activities appear particularly relevant to the context of climate change. Some of these provisions are redundant with the obligations of States deriving from the no-harm principle. In particular, States accordingly have an obligation to "take all appropriate measures to prevent significant transboundary harm or at any event to minimize the risk thereof."[98] Yet, the 2001 Articles on the Prevention of Transboundary Harm from Hazardous Activities contain more specific provisions with regard to situations where risks of transboundary harm can be foreseen. In such a situation, States that are so requested must accept to "enter into consultations" and seek to reach an agreement on appropriate preventive measures based on "an equitable balance of interests."[99]

Furthermore, the 2006 Principles on the Allocation of Loss in the Case of Transboundary Harm Arising out of Hazardous Activities call for remedial measures intended "to ensure prompt and adequate compensation to victims of transboundary harm and to preserve and protect the environment."[100] These remedial measures "should include the imposition of liability on the operator or, where appropriate, other person or entity."[101] While remedial obligations under the law of State responsibility consist in full or appropriate reparation for the injury suffered by States (either exercising diplomatic protection or on their own right, such as in cases of loss of territory or environmental damages), the 2006 Principles consider "prompt and adequate compensation"[102] that the operator should pay to individual or corporate victims. Beyond the law of State responsibility, however, the 2006 Principles require the liable State, "with the appropriate involvement of the operator," to "ensure that appropriate response measures are taken" to avoid or mitigate harms.[103] Lastly, the 2006 Principles also call for the development of specific international regimes "[w]here, in respect of particular categories of hazardous activities, specific global, regional or bilateral agreements would provide effective arrangements concerning compensation, response measure and international and domestic remedies."[104]

The law of State liability completes rather than replaces the law of State responsibility. Its emphasis on the prevention of damages arising from hazardous activities is too narrow in circumstances where colossal GhG emissions have already wreaked havoc around the world and their continuation is certain to exacerbate these issues. Its emphasis on the civil liability of the operator is also interesting, although perhaps more directly relevant to cases relating to damages caused by an industrial disaster originating from a particular facility at a particular point in time rather than to GhG emissions spread over time and space. With regard to climate change, a focus on individual or corporate activities alone could create a useful economic incentive for less GhG emissions, but it would not be sufficient. States have a leading role to play – and their share of

---

[98] ILC, *Draft Articles on Prevention of Transboundary Harm*, *supra* note 16, art. 3.
[99] *Ibid.*, arts. 9.1 and 9.2.
[100] *Draft Principles on the Allocation of Loss*, *supra* note 93, principle 3.
[101] *Ibid.*, principle 4.2.
[102] *Ibid.*, principle 3.
[103] *Ibid.*, principle 5(b).
[104] *Ibid.*, principle 7.1.

responsibilities to bear – in making important collective decisions, for instance, relating to transport infrastructures and to the promotion of particular forms of development.

VI. THE APPLICATION OF GENERAL INTERNATIONAL LAW

The previous sections have exposed general norms of international law which are relevant to climate change, in particular the no-harm principle and the law of State responsibility. This last section seeks to articulate these general norms with the special rules discussed in the previous two chapters on the UNFCCC regime and on relevant developments in other regimes. In contrast to the norms of general international law analyzed in this chapter, the rules adopted through the UNFCCC and other regimes are based on treaties addressing particular issues or areas of international law. Yet, this does not necessarily mean that the latter rules ought to prevail over the former norms.

It is widely accepted that a special rule may derogate from a general norm (*lex specialis derogat lege generali*). The ILC's Articles on State Responsibility, for instance, clearly recognize that "special rules of international law" may exclude the application of norms of general international law.[105] The World Trade Organization Dispute Settlement Understanding and the European Convention on Human Rights are among the special rules of international law which define different remedial obligations and thus exclude the application of norms of general international law on State responsibility.[106] Yet, special rules do not seek to exclude the application of general international law. Some special rules may, for instance, be adopted to supplement pre-existing norms by specifying their modalities of application or facilitating their implementation under particular circumstances.[107] In other words, the *lex specialis* principle does not provide that *every* rule addressing a specific issue excludes the application of general norms to the same subject matter.

Assessing whether a special rule supplements or derogates from a general norm requires consideration of their nature as well as their object and purpose. As the ILC suggested in its Commentary on the Draft Articles on State Responsibility:

> For the *lex specialis* principle to apply it is not enough that the same subject matter is dealt with by two provisions; there must be *some actual inconsistency between them*, or else a *discernible intention that one provision is to exclude the other.*[108]

The work of the ILC on the fragmentation of international law, led by Martti Koskenniemi, calls for a narrow interpretation of the *lex specialis* principle. Accordingly, the principle only

---

[105] *Articles on State Responsibility, supra* note 70, art. 55.

[106] *Ibid.*, Commentary under art. 55, para. 3.

[107] See ILC, *Fragmentation of International Law: Difficulties Arising from the Diversification and Expansion of International Law* (April 13, 2006), doc. A/CN.4/L.682, paras. 56–57, 88. Some general rules cannot be overruled by special rules: those are called "peremptory rules of general international law" (or "jus cogens"). See Vienna Convention on the Law of Treaties, May 23, 1969, 1155 *UNTS* 331, art. 53.

[108] *Articles on State Responsibility, supra* note 70, Commentary under art. 55, para. 4 (emphasis added). See also ILC, *Fragmentation of International Law, supra* note 107, paras. 56ff; Sir Gerald Fitzmaurice, "The law and procedure of the International Court of Justice 1951–4: treaty interpretation and other treaty points" (1957) 33 *British Yearbook of International Law* 203, at 236–238.

applies when there is a "relationship ... of conflict" between two norms, but not when the relationship is only one of "interpretation."[109] The ILC highlighted the "principle of harmonization" according to which "when several norms bear on a single issue they should, to the extent possible, be interpreted so as to give rise to a single set of compatible obligations."[110] Absent some discernible intention of the parties to exclude general international law or some actual inconsistency with it, specific rules on climate change are not of such a nature as to exclude the application of norms of general international law.

An examination of the treaty rules related to climate change provide little indication of the intention of the parties either to exclude or to supplement norms of general international law. In these treaties, references to general international law are systematically confined to vague preambular provisions.[111] In particular, the preamble to the UNFCCC recalls "the pertinent provisions" of the Stockholm Declaration and, in particular, the no-harm principle:

> *Recalling also* that States have, in accordance with the Charter of the United Nations and the principles of international law, the sovereign right to exploit their own resources pursuant to their own environmental and developmental policies, and the responsibility to ensure that activities within their jurisdiction or control do not cause damage to the environment of other States or of areas beyond the limits of national jurisdiction.[112]

A mere "recalling" of a legal principle says little as to whether the object and purpose of the treaty is to supplement this principle or to derogate from it. Meanwhile, neither the UNFCCC nor any subsequent international climate change treaty mentions the law of State responsibility (or liability) at all.

As reflected by the statements of the parties during and in relation to the negotiations, this ambivalence is indicative of the inability of negotiating parties to achieve a more far-reaching agreement. On the one hand, the United States and other industrial States have long shown great hostility to any mention of principles within the UNFCCC regime.[113] On the other hand, several developing States particularly exposed to climate change impacts, such as low-lying small island developing States, have clearly and repeatedly registered their "understanding" that the UNFCCC, the Kyoto Protocol and the Paris Agreement do not

> constitute a renunciation of any rights under international law concerning state responsibility for the adverse effects of climate change, and that no provisions in the Convention can be interpreted as derogating from the principles of general international law.[114]

---

[109] ILC, *Fragmentation of International Law*, *supra* note 107, para. 2.

[110] *Ibid.*, para. 4.

[111] The preamble to a treaty can be used to assert its object and purpose. See Vienna Convention on the Law of Treaties, *supra* note 107, art. 31.2.

[112] UNFCCC, *supra* note 13, recital 9.

[113] See Chapter 3, section I.B.

[114] See, for instance, the declarations of Kiribati, Fiji, Nauru and Tuvalu upon signature of the UNFCCC (1992), 1771 *UNTS* 317–318.

Thus, there was clearly no common understanding among the parties that the UNFCCC regime would exclude the application of general norms of international law.

Nor does any actual inconsistency between special rules and general norms justify the consideration of the rules as derogatory. What appears closest to an actual inconsistency is perhaps the provisions of international climate agreements, in particular Annex B of the Kyoto Protocol, which have sometimes been understood as "emissions quotas" or as a "right" to emit.[115] A treaty defining a right for a State to emit a particular amount of GhG would be inconsistent with a more demanding obligation under general international law. Yet, nothing justifies such an interpretation of international climate agreements at odds with their object and purpose of reducing GhG emissions. Through their successive national commitments, States assume an *obligation* to take particular steps and sometimes achieve particular outcomes, but these commitments do not involve any *right* to emit any particular amounts of GhGs.[116] Two sets of obligations can coexist, one being more demanding than the other, without any actual inconsistency: the more liberal obligation does not necessarily indicate a right to violate the stricter one. Thus, a State which complies with its obligations under international climate agreements may still be in breach of its obligations under general international law.

Absent any common intention to exclude the application of general international law or any actual inconsistency between special rules and general norms, it appears that the special rules on climate change contained in the UNFCCC and other regimes do not derogate from general international law. Rather, special rules remain limited in terms of their scope and ambit. The UNFCCC regime, in particular, does little to promote any form of adequate reparation for the injury caused by excessive GhG emissions, and it does not do enough to ensure that this breach of the no-harm principle is promptly discontinued. Treaty-based rules have been revealed to be extremely difficult to negotiate where substantial national interests are involved. Lack of consensus has frustrated the understanding of many negotiators that much more needs to be done. While the UNFCCC regime has spurred some steps toward compliance with general international law, it has not yet entirely fulfilled this objective.[117]

The political opposition of industrial States to a recognition of general international law has extended beyond negotiations on international climate agreements. The development of an authoritative interpretation of general international law in the context of climate change has repeatedly been obstructed by efforts made by representatives of developed States.

Thus, the ILC's project of a study program on the protection of the atmosphere has met with great political hostility.[118] Western diplomats repeatedly warned the ILC against "interfer[ing] with relevant political negotiations,"[119] while also questioning its competence to deal with

---

[115] Alexander Zahar, "Methodological issues in climate law" (2015) 5:1 *Climate Law* 25, at 32.

[116] See Charlotte Streck and Moritz von Unger, "Creating, regulating and allocating rights to offset and pollute: carbon rights in practice" (2016) 13:3 *Carbon & Climate Law Review* 178, at 181 (noting that the Kyoto Protocol "arguably has not created a right to emit"); and decisions 15/CP.7, "Principles, nature and scope of the mechanisms pursuant to Articles 6, 23 and 17 of the Kyoto Protocol" (November 10, 2011), recital 6; and 2/CMP.1, "Principles, nature and scope of the mechanisms pursuant to Articles 6, 12 and 17 of the Kyoto Protocol" (November 30, 2005), recital 6.

[117] See Chapter 13.

[118] See generally Peter H. Sand and Jonathan B. Wiener, "Towards a new international law of the atmosphere?" (2016) 7(2) *Göttingen Journal of International Law* 195.

[119] UN General Assembly, Sixty-Eighth Session, *Summary Record of the 18th Meeting* (October 29, 2013), in doc. A/C.6/68/SR.18, para. 102, Mr. Válek (Czech Republic).

"[t]he scientific and technical aspects of that topic."[120] Shortly after the resounding failure of the Copenhagen Summit and despite all evidence that the Cancún pledges were insufficient to meet the collective ambition and to prevent great damage throughout the world, Western diplomats did not shy away from claiming that political negotiations on climate change were after all "relatively effective,"[121] that they had "provided sufficient general guidance to States"[122] and that the topic "was already well-served by established legal arrangements."[123] If the ILC could eventually initiate the study of the protection of the atmosphere in 2013, the scope of the study had to be significantly constrained. It was in particular agreed that the study would not deal with:

> questions such as the liability of States and their nationals, the polluter-pays-principle, the precautionary principle, common but differentiated responsibilities, and the transfer of funds and technology to developing countries, including intellectual property rights.[124]

As an alternative to the ILC, international courts and tribunals could also help develop an authoritative interpretation of the relevant norms of general international law.[125] However, the attempts of some of the world's least-influential States have faced the fierce opposition of some of the most powerful ones. The uphill campaign of the small island State of Palau for the UN General Assembly to request an Advisory Opinion from the International Court of Justice, for instance, was discontinued in 2014 after the United States threatened to interrupt the provision of development aid to Palau.[126] Likewise, Tuvalu, another small island developing State highly dependent on international aid, has never carried out its threat of seeking the responsibility of Australia or the United States for excessive GhG emissions before an international jurisdiction. While the ILC's study on the protection of the atmosphere and other initiatives are ongoing, the diplomatic efforts of industrial nations to obstruct such developments will continue to hinder the determination of States' respective rights and obligations in relation to climate change.

VII. CONCLUSION

Some norms of general international law are applicable to climate change. On the one hand, the no-harm principle requires States not to cause transboundary harm through excessive GhG

---

[120] UN General Assembly, Sixty-Seventh Session, *Summary Record of the 19th Meeting* (November 2, 2012), in doc. A/C.6/67/SR.19, para. 91, Ms. Belliard (France).

[121] UN General Assembly, Sixty-Sixth Session, *Summary Record of the 20th Meeting* (October 26, 2011), in doc. A/C.6/66/SR.20, para. 15, Mr. Simonoff (United States).

[122] UN General Assembly, Sixty-Seventh Session, *Summary Record of the 19th Meeting* (November 2, 2012), in doc. A/C.6/67/SR.19, para. 118, Mr. Buchwald (United States).

[123] UN General Assembly, Sixty-Eighth Session, *Summary Record of the 18th Meeting* (October 29, 2013), in doc. A/C.6/68/SR.18, para. 21, Mr. Macleod (United Kingdom).

[124] ILC, *First Report on the Protection of the Atmosphere by Special Rapporteur Shinya Murase*, doc. A/CN.4/667 (February 14, 2014), para. 5(a). See also ILA Resolution 2/2014, *supra* note 23, Oslo Principles on Global Obligations to Reduce Climate Change, adopted on March 1, 2015 by a group of experts in international law, human rights and environmental law.

[125] See further discussions in Chapter 14, section I.

[126] Stuart Beck and Elizabeth Burleson, "Inside the system, outside the box: Palau's pursuit of climate justice and security at the United Nations" (2014) 3:1 *Transnational Environmental Law* 17, at 26.

emissions and to prevent activities under their jurisdiction which would cause such damage. On the other hand, the law of State responsibility requires them to cease the continuing breach of these obligations and to pay adequate reparations for the injury which has been caused. These norms apply despite the development of a UNFCCC regime and despite developments in other regimes because such special rules do not display any actual inconsistency or discernible intention to exclude their application. Yet, many of the modalities of these norms of general international law remain indeterminate, in part because of developed States' efforts to obstruct the development of any authoritative interpretation.

This chapter and the two previous chapters have discussed the three building blocks of the international law on climate change: a series of specific international agreements constituting the UNFCCC regime; some relevant developments scattered across different regimes; and relevant norms of general international law. The following chapters will look at particular substantive issues, starting with the question of differentiation among States.

6
Differentiation

Many multilateral treaties assume that every State, being equally sovereign, should hold the same rights and bear the same obligations. When national circumstances differ significantly, however, formal equality could result in extremely unfair conclusions. In such circumstances, differentiation is necessary in order to pursue substantive equality.

Differentiation is not exclusive to the international law on climate change. It has sometimes been used in a colonial context of domination, to which the UN Charter[1] and the ICJ Statute[2] continue to bear witness. Differentiation has also been used as a tool of emancipation, in particular as a way of promoting substantive equality between decolonized States and their former colonial powers, when formal equality would have caused an unjust disadvantage for the newly created States with weak political and economic structures. It has thus long been accepted that "divergences in levels of economic development or unequal capacities to tackle a given problem"[3] may justify differentiation among the parties to an international agreement. The International Covenant on Economic, Social and Cultural Rights provides that developing countries "may determine to what extent they would guarantee the economic rights recognized in the present Covenant to non-nationals,"[4] without recognizing the same option to developed States. Likewise, some provisions in WTO law aim to protect the economic development of developing States.[5] Other fields in international environmental law have also recognized different national circumstances through different obligations.[6]

---

[1] See Charter of the United Nations, June 26, 1946, 1 *UNTS* XVI, Chapters XI "regarding non-self-governing territories" and XII on an "international trusteeship system."

[2] See Statute of the International Court of Justice, June 26, 1945, 3 *Bevans* 1179, art. 38.1(c), referring to "general principles of law recognized by civilized nations."

[3] Philippe Cullet, "Differential treatment in international law: towards a new paradigm of inter-state relations" (1999) 10:3 *European Journal of International Law* 549.

[4] International Covenant on Economic, Social and Cultural Rights, December 16, 1966, 993 *UNTS* 3, art. 2.3.

[5] WTO Committee on Trade and Development, *Implementation of Special and Differential Treatment Provisions in WTO Agreements and Decisions* (October 25, 2000), doc. WT/COMTD/W/77.

[6] See generally Christina Voigt, "Equity in the 2015 Climate Agreement: lessons from differential treatment in multilateral environmental agreements" (2014) 4:1–2 *Climate Law* 50; Tuula Honkonen, "The development of the principle of common but differentiated responsibilities and its place in international environmental regimes" in Thomas Kuokkanen *et al.* (eds.), *International Environmental Law-Making and Diplomacy: Insights and Overviews* (Routledge, 2016) 160; Philippe Cullet, "Differential treatment in environmental law: addressing critiques and conceptualizing the next steps" (2016) 5:2 *Transnational Environmental Law* 305; Lavanya Rajamani, *Differential Treatment in International Environmental Law* (Oxford University Press, 2006); Philippe Cullet, *Differential Treatment in International Environmental Law* (Routledge, 2003).

Yet, few, if any, fields of international law have put as much emphasis on differentiation as the international law on climate change. Within the UNFCCC regime, differentiation has become the norm rather than the exception. Provisions on differentiation are also to be found in other relevant treaty-based rules, in particular under the Montreal Protocol on Substances that Deplete the Ozone Layer.[7] Differentiation in the international law on climate change reflects the fact that States have different capacities to reconcile environmental priorities with economic and social priorities within a vision of sustainable development. In addition, differentiation also reflects the respective contributions of States to global GhG emissions and their responsibility to promote climate change mitigation and adaptation to the impacts of climate change.

However, there is no unique and consensual formula for differentiation in the international law on climate change. Potential grounds for differentiation are multiple, ranging from past or present GhG emissions to financial capacity, population, geographic circumstances or exposure to climate impacts, among others. These grounds for differentiation unfold differently with respect to particular fields of international cooperation on climate change, for instance, mitigation, adaptation or international support. Disagreements as to the relevant grounds for differentiation and their respective weight have been the greatest obstacle to international agreements on climate change over the last quarter of a century.[8] While such disputes are exacerbated by conflicting interests, they are also deeply rooted in different moral assumptions, for instance, regarding the responsibilities that a State may bear for the conduct of previous generations.

This chapter discusses the main arguments for differentiation in international responses to climate change. It then turns to exploring how these arguments have shaped international climate agreements, in particular through the principle of common but differentiated responsibilities and respective capabilities.

I. GROUNDS FOR DIFFERENTIATION

Reasons to differentiate among States in their efforts to tackle climate change abound. Some States are emitting colossal amounts of GhG, have done so over decades or will be doing so in the foreseeable future. Others are severely affected by the impacts of climate change. Some States have the financial or technological capacity to make an effective contribution to global efforts, which others cannot afford to do so without jeopardizing the basic needs of their populations. Some States have large populations which may excuse greater GhG emissions; others may need higher levels of GhG emissions due to the need to develop a vast territory or to adapt to harsh climatic conditions.

In the following, a distinction is made between grounds for differentiation which relate to States' respective *responsibilities* and differentiation based on their respective *capacities*. Responsibility-based grounds for differentiation relate to the current, past and possibly future levels of GhG emissions attributable to a State, on the ground that a State which contributed the most to causing climate change should also contribute the most to addressing it. By

---

[7] Montreal Protocol on Substances that Deplete the Ozone Layer, September 16, 1987, 1522 *UNTS* 3, art. 5.

[8] Some have also argued for differentiation and support within States. See in particular Daniel A. Farber, "Beyond the North-South dichotomy in international climate law: the distinctive adaptation responsibilities of the emerging economies" (2013) 22:1 *Review of European Community & International Environmental Law* 42.

contrast, capacity-based grounds for differentiation relate to the ability of a State to reduce its net GhG emissions, to promote climate change adaptation and to provide international support for climate action. These factors include population, development level, financial capacity and geographic circumstances.

A. Responsibility-Based Grounds for Differentiation

1. Current GhG Emissions

The GhG emissions of some States are of a larger order of magnitude than those of others. According to data compiled by the World Resources Institute, a think tank, China has contributed to 26 percent of global GhG emissions in 2013, followed by the United States (14 percent), the European Union (9 percent), India (6 percent) and Russia (5 percent). The five permanent members of the UN Security Council – China, France, Russia, the United Kingdom and the United States – account for almost half of all global GhG emissions. One hundred countries, taken together, account for less than 3 percent.[9] The case for a breach of the no-harm principle and for remedial obligations under the law of State responsibility could be made in relation to some of the industrial nations which contribute the largest part of current GhG emissions.

Yet, even though current GhG emissions can appear a relatively straightforward ground for differentiation, several difficulties appear when attempting to measure national GhG emissions.

Firstly, accounting for GhG emissions can be challenging. The current system relies heavily on States themselves reporting their data.[10] States are supposed to collect data in compliance with guidelines developed by the IPCC and endorsed by the COP to the UNFCCC.[11] National data are verified by expert review teams (ERTs).[12] The complexity of the process of reporting and verifying GhG accountancy means that data are typically available with a delay of around three to four years. Despite such efforts, data are not always reliable, especially in developing countries. In 2015, for instance, China revised its estimated coal consumption from 2000 to 2013 with a staggering 14 percent increase.[13]

In particular, accounting for GhG emissions and sinks from Land Use, Land-Use Change and Forestry (LULUCF) presents formidable methodological challenges, especially so in countries which do not keep a clear track of the state of their forests. LULUCF emissions are often excluded when national GhG emissions are compared because they are too unreliable. However, LULUCF contributes a non-negligible share of global GhG emissions[14] and is a very

---

[9] WRI, CAIT Climate Data Explorer, "Total GHG emissions excluding land-use change and forestry" (2013).

[10] See United Nations Framework Convention on Climate Change, May 9, 1992, 1771 *UNTS* 107 (hereinafter UNFCCC), arts. 4.1(a) and 12.1(a); Kyoto Protocol to the United Nations Framework Convention on Climate Change, December 11, 1997, 2303 *UNTS* 162 (hereinafter Kyoto Protocol), arts. 5 and 7; decision 1/CP.16, "The Cancún Agreements: Outcome of the work of the Ad Hoc Working Group on Long-Term Cooperative Action under the Convention" (December 10–11, 2010), paras. 40(a) and 60(c); Paris Agreement, December 12, 2015, in the annex of decision 1/CP.21, "Adoption of the Paris Agreement" (December 12, 2015), art. 13.7(a).

[11] See e.g. Simon Eggleston *et al.*, *2006 IPCC Guidelines for National Greenhouse Gas Inventories* (IGES, 2006).

[12] See Chapter 13, section II.B.3.

[13] International Energy Agency, "Special data release with revisions for People's Republic of China" (2015), www.iea.org/publications/freepublications/publication/SpecialdatareleasewithrevisionsforPeoplesRepublicofChina04.11.2015.pdf (accessed January 3, 2018).

[14] The IPCC estimated that carbon dioxide emissions from forestry and other land use represented 11 percent of global GhG emissions in 2010. See R.K. Pachauri *et al.*, *Climate Change 2014: Synthesis Report. Contribution of Working Groups I, II and III to the Fifth Assessment Report of the Intergovernmental Panel on Climate Change* (IPCC, 2015) 5.

important consideration for assessing the contribution of certain countries to climate change. Data from the World Resource Institute suggest that LULUCF activities represent more than three-quarters of national GhG emissions in Zambia, Botswana, the Democratic Republic of the Congo, Guyana, Paraguay, the Solomon Islands and Papua New Guinea, and about two-thirds of Indonesia's GhG emissions.[15] On the other hand, in some countries which have engaged in large-scale reforestation programs such as Gabon, Bhutan, Romania and Chile, removal of GhGs from LULUCF does more than balancing emissions from all other sectors.[16]

Secondly, another rather technical difficulty appears when comparing emissions of different GhGs. Once released in the atmosphere, carbon dioxide can stay there for many centuries before being removed by natural processes. By contrast, methane – a more powerful GhG – degrades quickly into different molecules, in particular when it reaches the higher atmosphere; its average atmospheric lifetime is 12 years.[17] Therefore, methane emissions have a strong but short-lived impact on the global climate, whereas the same mass of carbon dioxide emissions has a slower but more prolonged impact. Aggregate GhG emissions are usually expressed in terms of carbon dioxide equivalence, but this equivalence is only possible in a particular timeframe. As a report of the IPCC concluded, "[n]o single metric can accurately compare all consequences of different emissions."[18] In practice, GhG inventories generally use the respective global warming potential of GhGs on a 100-year period.[19] This somewhat arbitrary technical choice has some implications when comparing the emission of countries with different baskets of GhGs. Methane is often emitted from the degradation of organic waste, in particular in developing countries which lack proper waste-treatment facilities. Accordingly, developing countries could incline to compare emissions in different GhGs across a longer timeframe in order to put less emphasis on methane.

Thirdly and most importantly, transnational production processes make it problematic to attribute GhG emissions to a particular State when, actually, economic activities commonly span international borders. Mostly for convenience, GhG emissions are usually attributed to the State on whose territory they occur, for instance the place where fossil fuel is burnt. Yet, in a globalized economy, the production process may be operated by a foreign company and the production may target consumers in yet another country. Rough estimates suggest that up to one-third of GhG emissions occurring in China are export-oriented, often consisting in industrial commodities produced for Western markets.[20] Glen Peters and colleagues estimated that in 2008, 7.8 $GtCO_2$ emissions representing 26 percent of global carbon dioxide emissions were emitted in one country for consumption in another country.[21] Furthermore, empirical evidence suggests that mitigation efforts in

---

[15] WRI, CAIT Climate Data Explorer, "Total GHG emissions excluding land-use change and forestry" and "Total GHG emissions including land-use change and forestry" (2013).

[16] *Ibid.*

[17] See Stefanie Kirschke *et al.*, "Three decades of global methane sources and sinks" (2013) 6:10 *Nature Geoscience* 813. Likewise, some HFCs are very potent GhGs, but have an atmospheric lifetime of a few weeks only. See generally G. Myhre *et al.*, "Anthropogenic and natural radiative forcing" in T.F. Stocker *et al.* (eds.), *Climate Change 2013: The Physical Science Basis. Contribution of Working Group I to the Fifth Assessment Report of the Intergovernmental Panel on Climate Change* (Cambridge University Press, 2013) 659, at 731, Appendix A.

[18] L.V. Alexander *et al.*, "Summary for policymakers" in Stocker *et al.* (eds.), *supra* note 17, 3, at 17.

[19] See Thomas F. Stocker *et al.*, "Technical summary" in Stocker *et al.* (eds.), *supra* note 17, 33, at 58–59.

[20] Ben Block, "*Exports account for one-third of China's emissions*" (Worldwatch Institute, n.d.), www.worldwatch.org/node/5846 (accessed January 3, 2018).

[21] See Glen P. Peters *et al.*, "Growth in emission transfers via international trade from 1990 to 2008" (2011) 108:21 *Proceedings of the National Academy of Sciences* 8903.

Western countries have displaced polluting activities in emerging economies, a phenomenon often called "carbon leakage."[22] It is arguably unfair to put all the responsibility on a State for GhG emissions when the real benefit is for foreign consumers and sometimes foreign investors – all the more so when these activities often coincide with the emissions of other pollutants which cause public health issues in the producing country. More pragmatically, this way of accounting for GhG emissions leads to the neglect of consumption-side mitigation policies, which could seek, for instance, to discourage the purchase or use of goods whose production requires large GhG emissions.[23] Other countries could be under a legal obligation to reduce GhG emissions from transnational production processes if one is to interpret the no-harm principle as imposing extra-territorial obligations, for instance, through control over international trade and investments.

As a consequence of these technical difficulties, there are multiple ways of comparing current GhG emissions in different States. Accounting for LULUCF puts more blame on countries where agriculture is extending spatially, like in Indonesia or Brazil, rather than those whose forests are long gone; this approach benefits countries which are able to undergo large reforestation programs like China and the Member States of the European Union. Likewise, the timeframe for comparing GhG emissions has different implications for States with contrasting GhG emissions profiles. But the most impactful methodological choice regards the attribution of GhGs on a territorial basis rather than, for instance, on a benefit basis.

2. Historical and Future Emissions

The impacts of climate change observed today are caused by cumulative GhG emissions, which started with the beginning of the Industrial Revolution in 1750s England. Countries responsible for large shares of cumulative GhG emissions arguably have a historical responsibility. Having already had a great impact on the global climate in order to foster their transition to a modern economy, these countries should take extra steps to reduce their GhG emissions. Having caused changes that now impact many throughout the world, they also have particular moral duties or legal obligations to make adequate reparation.[24]

Until the second half of the twentieth century, fossil fuels were being used almost exclusively in a few Western States, and even China was not a major GhG emitter until the rapid expansion of its economy in the 1990s. Accordingly, consideration for cumulative historical rather than current GhG emissions suggests that Western countries should bear a greater share of the responsibility. According to data from the World Resources Institute, the United States represented more than a quarter of global carbon dioxide emissions during the twentieth century – twice as much as China.[25]

---

[22] See e.g. Boqiang Lin and Chuanwang Sun, "Evaluating carbon dioxide emissions in international trade of China" (2010) 38:1 *Energy Policy* 613; Mustafa H. Babiker, "Climate change policy, market structure, and carbon leakage" (2005) 65:2 *Journal of International Economics* 421.

[23] Paul G. Harris and Jonathan Symons, "Norm conflict in climate governance: greenhouse gas accounting and the problem of consumption" (2013) 13:1 *Global Environmental Politics* 9.

[24] See generally Lukas H. Meyer and Pranay Sanklecha (eds.), *Climate Justice and Historical Emissions* (Cambridge University Press, 2017).

[25] Respectively 26.7 and 11.8 percent of carbon dioxide emissions from 1902 to 2013. Data from WRI, CAIT Climate Data Explorer, "Total CO2 emissions excluding land-use change and forestry." No comprehensive and reliable data on century-long historical emissions of other GhGs could be found.

The responsibility of developed States for their historical emissions depends on whether these emissions were, at the time, an internationally wrongful act – the breach of an international obligation attributable to the State.[26] Such has certainly been the case since at least the mid- to late 1980s, when a scientific consensus emerged on the existence of a change in our climate system caused by anthropogenic GhG emissions. But are States also responsible for failing to prevent excessive GhG emissions at the time when scientists did not know or were not sure of their impact? Such an argument would require us to interpret the no-harm principle as necessitating a particularly strong standard of care or even as a regime of strict liability, where knowledge is not a necessary element of the breach of an international obligation. This is probably not the case. Nor is it possible to rely on the precautionary approach before this principle even started to emerge.[27]

Besides historical GhG emissions, one could also argue that predictable future GhG emissions could be relevant to determining respective obligations. Under virtually any predictable circumstances, global GhG emissions will continue to grow in the years to come. States' development strategies imply such emissions – coal plants, highways and airports are being built throughout the world. The INDCs that most States submitted in the run-up to the 2015 Paris Summit, if fully implemented, would only reduce the rate of increase in global GhG emissions.[28] GhG emissions are expected to increase at a fast rate in most newly industrialized economies.[29]

As a matter of principle, States cannot be held responsible for future conduct: responsibility arises from acts, not from mere intention. It is only inasmuch as a State has had a chance to change its course of action that it could be held responsible for failing to do so. However, States have already been making arrangements which imply future GhG emissions in the coming years, if not by pure necessity, then at least by making it much more difficult for the State to adapt a different development pathway. By building coal mines, highways and airports, a State commits to a particular development pathway which implies certain levels of GhG emissions. Arguments for future-oriented State responsibility could be based on such past or present acts of States which imply future GhG emissions.

However, arguments for responsibility for future GhG emissions are highly unlikely to be successful. In the case of the Gabčíkovo-Nagymaros Project, the ICJ appeared justly reluctant to consider preparatory action as constitutive of an internationally wrongful act as long as that act could still have been avoided.[30] Admittedly, there might be circumstances where a State's preparatory action would go as far as making it virtually impossible for that State to avoid future GhG emissions, thus possibly entailing its responsibility.

---

[26] ILC, *Draft Articles on Responsibility of States for Internationally Wrongful Acts with Commentaries*, in (2001) *Yearbook of the International Law Commission*, vol. II, part two (hereinafter *Articles on State Responsibility*), arts. 1 and 2.

[27] See Antônio Augusto Cançado Trindade, "Principle 15: precaution" in Jorge E. Viñuales, *The Rio Declaration on Environment and Development: A Commentary* (Oxford University Press, 2015) 403, at 404, retracing the recognition of the precautionary approach at the international level to the mid-1980s.

[28] See UNFCCC Secretariat, *Aggregate Effect of the Intended Nationally Determined Contributions: An Update*, doc. FCCC/CP/2016/2 (May 2, 2016), para. 34, estimating that global GhG emissions would reach 55 $GtCO_2eq$ by 2025 and 56 $GtCO_2eq$ by 2030.

[29] See OECD, *OECD Environmental Outlook to 2050: The Consequences of Inaction* (OECD Publishing, 2012).

[30] ICJ, *Gabčíkovo-Nagymaros Project (Hungary v. Slovakia)*, judgment of September 25, 1997, para. 79.

B. Capacity-Based Grounds for Differentiation

Capacity-based grounds for differentiation reflect the understanding that not every State has the same capacity to reduce its GhG emissions or to adapt to the impacts of climate change. A given State may need to emit more GhGs in order to satisfy the need of its population if this population is large. Developing States may need to increase their GhG emissions for a transitory phase in order to satisfy the basic needs of their populations. Financial capacity could be considered as a ground for States to assist others. Geographic circumstances may also imply particular needs for GhG emissions, for instance, to develop a vast territory, or its ability to exploit sources of renewable energy.

1. Population

While States are equal sovereigns, their circumstances are very different. China's population of 1.4 billion contrasts sharply with Tuvalu's 10,000 citizens. Such differences in population should arguably be considered as one of the elements justifying a certain level of GhG emissions as necessary, especially if such necessity is defined in relation to the basic needs or the fundamental rights of individuals under the State's jurisdiction.[31] If a State needs to permit some GhG emissions in order to offer a decent life to its citizens, then a State with more citizens needs to permit more GhG emissions. Rather than comparing the absolute levels of GhG emissions in different States as if all States were identical, it is fairer to compare per capita GhG emissions, that is, the quotient of the GhG emissions of a State or another entity according to the size of its population.

A comparison of national per capita emissions suggests a totally different attribution of responsibilities. On average, each person on Earth causes 6.3 tCO_2eq of GhG emissions, but there are vast inequalities across countries. The average US citizen emits 19.9 tCO_2eq/year, compared with 8.7 tCO_2eq/year for the average Chinese citizen, 8.4 tCO_2eq/year for the average EU citizen, 2.3 tCO_2eq/year for the average Indian citizen, 1 tCO_2eq/year for the average Bangladeshi citizen and 0.5 tCO_2eq/year for the average Congolese citizen.[32] Fossil fuel-producing countries, where oil is cheap, are generally characterized by high per capita emissions,[33] while other developing States and *a fortiori* least developed States have generally much lower per capita emissions. Whereas China as a whole is the world's largest single GhG emitter, its per capita emissions are comparable to those of the European Union as a whole, but lower than Germany and far below those of the United States.[34] Like the national GhG emissions data on which they are based, however, these figures do not properly account for the fact that large amounts of GhG emissions take place in emerging economies for the benefit of consumers overseas.

While population is certainly an element to consider, population policies could also be a tool for addressing climate change and other environmental issues. Family planning policies could be an important component in a policy toolkit to limit demographic growth and thus

---

[31] See Chapter 5, section III.

[32] WRI, CAIT Climate Data Explorer, "Total GHG emissions excluding land-use change and forestry" (2013) and "Population" (2013).

[33] For instance, Kuwait (54.5 tCO_2eq/year), Brunei (46.2 tCO_2eq/year), Qatar (39.4 tCO_2eq/year), Belize (28.1 tCO_2eq/year) and Oman (26.1 tCO_2eq/year) have the highest per capita rates of GhG emissions.

[34] Germany's per capita emissions are 10.9 tCO_2eq/year.

the environmental impact of societies – arguably a necessary step toward sustainable development.[35] Although population policies have sometimes been associated with human rights abuses, they can also be rights-based and empowering to women.[36]

2. Development Levels

Besides population, GhG emissions should also, arguably, be excused if and inasmuch as they seek to fulfill our most basic needs. Developed States may be able to take measures to reduce their GhG emissions without jeopardizing the basic needs of their population; in developing States, limitations on the growth in GhG emissions may hinder policies to eradicate poverty. The distinction that Henry Shue suggested between "subsistence emissions" and "luxury emissions"[37] reflects the idea that the first units of GhG emissions are the most important to fulfill our basic needs, while the last units of emissions are the most wasteful. While development has progressively been recognized as a human right,[38] it remains difficult to conceive of human development without at least some minimal level of GhG emissions.

Yet, if development is a necessity, it is not an excuse for any level of GhG emissions. The idea of sustainable development promotes a quest for the right balance between economic development, social equity and environmental protection together.[39] In this sense, international climate change agreements have repeatedly affirmed that "responses to climate change should be coordinated with social and economic development in an integrated manner with a view to avoiding adverse impacts on the latter," especially in developing States.[40]

The GhG intensity of a national economy is calculated by dividing a State's GhG emissions by its gross domestic product (GDP). Some States prefer to express their mitigation target in terms of GhG intensity rather than in absolute terms.[41] When a State's GhG emissions increase in absolute terms, its GhG intensity may remain stable or may even decrease due to economic growth. Thus, GhG intensity is a more favorable metric for countries with a high economic growth as it may "conceal" a growth in absolute levels of GhG emissions.[42] This metric is particularly favorable when a country is transitioning from an industrial development to a "cleaner" service-based economy, a context in which the GhG intensity of the overall economy on a production basis can decrease rapidly. For instance, China intends to lower its carbon dioxide intensity by 60–65 percent between 2005 and 2030 without necessarily reducing its

---

[35] See e.g. UNCHE, Stockholm Declaration on the Human Environment, available in (1972) 11 *ILM* 1416 (June 5–16, 1972) (hereinafter Stockholm Declaration), principle 16; UNCED, Rio Declaration on Environment and Development (June 3–14, 1992), available in (1992) 31 *ILM* 874 (hereinafter Rio Declaration), principle 8.

[36] See e.g. Clare Heyward, "A growing problem? Dealing with population increases in climate justice" (2012) 19:4 *Ethical Perspectives* 703, at 704; Christina Voigt, "Principle 8: sustainable patterns of production and consumption and demographic policies" in Viñuales (ed.), *supra note 27*, 245, at 258. The real benefit of population policies on climate change mitigation is uncertain in the short or medium term. While reducing population over time, such policy favors a more rapid development, which often comes along with higher rates of GhG emissions per capita, at least during a transition period. Yet, a more rapid development rate could also imply a quicker transition toward a more advanced, less GhG-intensive economic system and hence a lesser overall impact on the climate system.

[37] Henry Shue, "Subsistence emissions and luxury emissions" (1993) 15:1 *Law & Policy* 39.

[38] See in particular UN General Assembly Resolution 41/128, "Declaration on the Right to Development" (December 4, 1986); and Rio Declaration, *supra note 35*, principle 3. See also Paris Agreement, *supra note 10*, 12th recital, for a rare mention of the right to development within a multilateral treaty.

[39] See WCED, *Our Common Future* (Oxford University Press, 1987) (hereinafter Brundtland Report).

[40] UNFCCC, *supra note 10*, recital 25.

[41] See UNFCCC Secretariat, *supra note 28*, at para. 9(c).

[42] See Timothy Herzog *et al.*, *Target: Intensity. An Analysis of Greenhouse Gas Intensity Targets* (WRI, 2006) 8–10.

overall amount of GhG emissions.[43] Likewise, India's commitment to reduce its emission intensity by 33–35 percent during the same period does not preclude a significant growth of its GhG emissions.[44]

Beyond a favorable framing of mitigation commitments, expressing mitigation targets in terms of GhG intensity reflects an understanding that climate change mitigation should not come at the cost of economic growth. A low GhG intensity reflects the ability to achieve a certain level of development with less interference to the climate system. The States in which little industrial development has taken place – least developed States in particular – typically have moderate GhG intensity.[45] By contrast, newly industrialized economies and economies which rely on the production of oil and gas typically have among the highest GhG intensities.[46] The shift of Western countries to a service-based economy has led to a lowering of their GhG intensity,[47] although this largely came at the cost of outsourcing industrial activities in emerging economies.

3. Financial Capacity

A distinct ground for differentiation relates to States' financial capacities. Generally speaking, measures to mitigate climate change or adapt to its impact are costly. While some reports highlight that renewable energies can be exploited at overall costs which are comparable with fossil fuels, the transition requires costly investments in new infrastructure for the production and distribution of electricity.[48] Not every country has the financial resources necessary for such investments. In this context, support to developing States, for instance through investments and the transfer of technology, may enable them to implement action on climate change mitigation. Some economists suggest that appropriate support would enable developing States to "leapfrog" by turning directly to "more sustainable, low-carbon development pathways and avoid the more emissions-intensive stages of development that were previously experienced by industrialized nations."[49]

Financial capacity is also relevant to assessing the duty of a State to provide assistance to other States and their populations when they are affected by adverse impacts of climate change. Arguments on a duty of assistance generally suggest a combination of targeted development aid and humanitarian assistance to the populations of the States most affected by climate change. Yet, unlike arguments based on a State's excessive GhG emissions, arguments

---

[43] See China, *INDC* (June 30, 2015), at 5. The same document announces that China's carbon dioxide emissions will peak in or before 2030.

[44] India, *INDC* (October 1, 2015), at 29. See also India's country profile on Climate Action Tracker, www.climateactiontracker.org/countries/india (accessed January 3, 2018).

[45] For instance, Nigeria has a GhG intensity of 715 tCO2eq/million USD, according to data from WRI, CAIT Climate Data Explorer, "Total GHG emissions excluding land-use change and forestry" (2013) and "GDP-USD" (2010).

[46] China's GhG intensity is 1,511 tCO2eq/million USD, compared with 397 tCO2eq/million USD in the United States, according to data from *ibid*.

[47] Thus, Switzerland (85 tCO2eq/million USD), Sweden (100 tCO2eq/million USD), Norway (104 tCO2eq/million USD) and France (162 tCO2eq/million USD) have the lowest GhG intensities, according to data from *ibid*.

[48] See IRENA, *Renewable Power Generation Costs in 2014* (January 2015), at 14; REN21, *Renewables 2016: Global Status Report* (2016). While the running costs of exploiting renewable energies are typically very low, these projects require higher upfront investments than coal plants. This might be more difficult in emerging economies where interest rates are typically higher than in Western countries.

[49] S. Agrawala *et al.*, "Regional development and cooperation" in O. Edenhofer *et al.* (eds.), *Climate Change 2014: Mitigation of Climate Change. Contribution of Working Group III to the Fifth Assessment Report of the Intergovernmental Panel on Climate Change* (Cambridge University Press, 2014) 1083, at 1106.

based on financial capacity tend to be framed with more nuance. If States have a duty to assist other States when the latter are unable to protect their population, this obligation is certainly not specific to the context of climate change.

4. Geographic Circumstances

A last potential ground for differentiation regards the geographic circumstances of a State. Some geographic circumstances represent constraints on States' mitigation efforts. A higher level of GhG emissions may be necessary to satisfy the basic needs of a population when this population is scattered over a vast territory, thus requiring more GhG emissions from transportation. A close-knit electrical railway network can promote low-carbon intercity transportation in densely populated centers such as eastern China or Western Europe, but the airplane is likely to remain the main mode of transportation between cities and towns scattered in Australia, in the American Midwest, or *a fortiori* between isolated islands across the world's oceans. Climatic conditions can also entail greater energy needs, whether for heating or for air-conditioning (the latter also being associated with emissions of carbon compounds with a high global warming potential).[50]

On the other hand, geographic circumstances may also provide opportunities for climate change mitigation projects. Large mountain rivers or regions exposed to sunshine and wind make it possible for a State to exploit sources of renewable energy as an alternative to fossil fuels. Access to geothermal energy in addition to hydroelectricity enabled Iceland to become a leader in the use of renewable resources, with 85 percent of its primary energy use coming from renewables.[51] By contrast, a city state such as Singapore has little space or natural resources to use in order to exploit renewable energies, unless technological development allows for the exploitation of wave and tidal energy.[52]

II. DIFFERENTIATION IN INTERNATIONAL NEGOTIATIONS

While all States have accepted differentiation as a matter of principle throughout the international negotiations on climate change, they have adopted different positions on the reason to differentiate and on what needs to be differentiated.[53] Some have viewed differentiation as a matter of responsibility and remedial obligations, while others have viewed it as a matter of needs and capacities. Despite the difficulties that it brings in already-complex international negotiations, there remains general agreement that no fair and equitable international climate agreement can be reached without differentiation and that no international climate agreement can be broadly implemented unless it is largely seen as relatively fair and equitable.[54] The

---

[50] Such arguments were, for instance, put forward by Canada in its INDC (May 15, 2015), at 1.

[51] Orkustofnun (National Energy Authority), "Iceland: a leader in the use of Renewable Resources" (n.d.), http://os.is/gogn/Frettir/Iceland_Leader_RenewableEnergy.pdf (accessed January 3, 2018).

[52] See D. Arvizu *et al.*, "Technical summary" in O. Edenhofer *et al.* (eds.), *IPCC Special Report on Renewable Energy Sources and Climate Change Mitigation* (Cambridge University Press, 2011) 33, at 87.

[53] See e.g. Thomas Deleuil and Tuula Honkonen, "Vertical, horizontal, concentric: the mechanics of differential treatment in the climate regime" (2015) 5:1 *Climate Law* 82; Harald Winkler and Lavanya Rajamani, "CBDR&RC in a regime applicable to all" (2014) 14:1 *Climate Policy* 102; Lavanya Rajamani, "Ambition and differentiation in the 2015 Paris Agreement: interpretative possibilities and underlying politics" (2016) 65:2 *International & Comparative Law Quarterly* 493; Jutta Brunnée and Charlotte Streck, "The UNFCCC as a negotiation forum: towards common but more differentiated responsibilities" (2013) 13:5 *Climate Policy* 589.

[54] Cullet, "Differential treatment," *supra* note 6.

following retraces some aspects of the many developments relating to differentiation in the international law on climate change. It looks in turn at differentiation in the regime on the protection of the ozone layer, in the UNFCCC regime and finally in civil aviation and maritime transportation.

A. Differentiation in the Regime on the Protection of the Ozone Layer

When negotiations on the protection of the ozone layer were initiated in the mid-1980s, the vast majority of ozone-depleting substances were produced and consumed in developed States. Eighty-five percent of CFCs were then emitted by industrialized States representing 25 percent of the world's population, compared with 2 percent from China and India, which by then represented a third of the world's population.[55] Although concerns were raised regarding the projected growth in consumption of such substances in developing countries, most efforts clearly had to be focused on developed States. Access to refrigeration in developing countries was considered an important condition for their development, among others, amid concerns for food safety. In 1972, the Stockholm Declaration had provided some recognition to the particular circumstances of developing States.[56] In this context, the 1985 Vienna Convention for the Protection of the Ozone Layer recognized clearly that international cooperation should "tak[e] into account in particular the needs of the developing countries."[57] Similarly, the Preamble to the Montreal Protocol acknowledged "that special provision [was] required to meet the needs of developing countries for these substances."[58]

Differentiation was implemented through more specific provisions inserted into the Montreal Protocol. In particular, developing States were allowed to delay compliance with the obligation to reduce emissions and consumption of ozone-depleting substances by ten years.[59] They could also benefit from financial assistance and, later, from the transfer of technology.[60] The benefit of this provision was limited to developing countries "whose annual calculated level of consumption of ... controlled substances ... is less than 0.3 kilograms per capita,"[61] on the condition that they should never exceed this level of consumption.[62]

"Developing countries" are not defined in the Protocol. In practice, the distinction was generally facilitated by ample differences between developed and developing States' per capita emissions. A list of countries were recognized as developing countries for the purpose of the Protocol in a decision adopted by the First Meeting of the Parties.[63] The Ozone Secretariat was then requested to verify which, among these countries, satisfied the condition of a level

---

[55] See James T.B. Trip, "The UNEP Montreal Protocol: industrialized and developing countries sharing the responsibility for protecting the stratospheric ozone layer" (1987–1988) 20:3 *New York University Journal of International Law & Politics* 733, at 743–744.

[56] Stockholm Declaration, *supra* note 35, principles 12 and 23.

[57] Vienna Convention for the Protection of the Ozone Layer, March 22, 1985, 1513 *UNTS* 293, art. 4.2. See also *ibid.*, recital 4.

[58] Montreal Protocol, *supra* note 7, recital 7.

[59] *Ibid.*, art. 5.1. More specific rules have been adopted on particular substances. For a synthesis, see Ozone Secretariat, *Handbook for the Montreal Protocol on Substances that Deplete the Ozone Layer*, 11th edn (UNEP, 2017) 33–42.

[60] Montreal Protocol, *supra* note 7, arts. 10 and 10A. See also Stephen O. Andersen, K. Madhava Sarma and Kristen N. Taddonio (eds.), *Technology Transfer for the Ozone Layer: Lessons for Climate Change* (Earthscan, 2007).

[61] Montreal Protocol, *supra* note 7, art. 5.1.

[62] *Ibid.*, art. 5.2.

[63] Meeting of the Parties to the Montreal Protocol decision I/12E, "Clarification of terms and definitions: developing countries" (May 2–5, 1989).

of consumption below 0.3 kilograms of controlled substances per capita and per year.[64] Only four developing States originally had a higher level of consumption.[65] Accepting the States' requests for classification as developing countries on an individual basis,[66] the Meeting of the Parties to the Montreal Protocol deemed it unnecessary to adopt any particular criteria.[67]

B. The Principle of Differentiation in the UNFCCC Regime

Like the depletion of the ozone layer, climate change is caused by activities largely concentrated in a handful of developed States. A clear divide existed in the early 1990s between a relatively small but powerful group of developed, industrial States and a larger but less powerful group of developing States, many of which recently freed themselves from colonial domination. In 1989, a meeting of the Group of 77 (G77), a loose coalition of developing States, defined the following negotiating position:

> Since developed countries account for the bulk of the production and consumption of environmentally damaging substances, they should bear the main responsibility in the search for long-term remedies for global environmental protection and should make the major contribution to international efforts to reduce consumption of such substances.[68]

This common position of developing States had a significant influence on the pursuit of international negotiations which culminated with the definition of the principle of common but differentiated responsibilities at the 1992 Earth Summit. Principle 7 of the Rio Declaration on Environment and Development provides the following:

> In view of the different contributions to global environmental degradation, States have common but differentiated responsibilities. The developed countries acknowledge the responsibility that they bear in the international pursuit of sustainable development in view of the pressures their societies place on the global environment and of the technologies and financial resources they command.[69]

While this provision suggested that developed States should bear the blame, it also allowed for another interpretation. When the Rio Declaration was adopted, the United States registered a written statement to reserve its interpretation of this principle as "highlight[ing] the special leadership role of the developed countries, based on [their] industrial development, [their] experience with environmental protection policies and actions, and [their] wealth, technical

---

[64] Meeting of the Parties to the Montreal Protocol decision II/10, "Data of developing countries" (June 27–29, 1990).

[65] Meeting of the Parties to the Montreal Protocol decision III/3, "Implementation Committee" (June 19–21, 1991), para. (d) (Bahrain, Malta, Singapore and the United Arab Emirates).

[66] Meeting of the Parties to the Montreal Protocol decision III/5, "Definition of developing countries" (June 19–21, 1991).

[67] Meeting of the Parties to the Montreal Protocol decision IV/7, "Definition of developing countries" (November 23–25, 1992).

[68] Caracas Declaration of the Ministers of Foreign Affairs of the Group of 77 on the Occasion of the Twenty-Fifth Anniversary of the Group (June 21–23, 1989), reproduced in doc. A/44/361, para. II–34.

[69] Rio Declaration, *supra* note 35, principle 7.

expertise and capabilities." The United States also stated that it did "not accept any interpretation of principle 7 that would imply a recognition or acceptance by the United States or any international obligations or liabilities, or any diminution in the responsibilities of developing countries."[70]

The negotiations within the Intergovernmental Negotiating Committee for a Framework Convention on Climate Change (INC/FCCC) led to a slightly different statement of this principle. In the UNFCCC, the mention of common but differentiated responsibilities is accompanied by a reference to "respective capabilities." The Preamble to the UNFCCC mentions the need for the "widest possible cooperation by all countries" and their participation "in accordance with their common but differentiated responsibilities and respective capabilities and their social and economic conditions."[71] Article 3 lays down the principle in general terms:

> The Parties should protect the climate system for the benefit of present and future generations of humankind, on the basis of equity and in accordance with their common but differentiated responsibilities and respective capabilities. Accordingly, the developed country Parties should take the lead in combating climate change and the adverse effects thereof.[72]

Lastly, Article 4.1 defines the obligations that all parties shall assume "taking into account their common but differentiated responsibilities and their specific national and regional development priorities, objectives and circumstances."[73] The commitments of developed States are generally more demanding than those of developing States, although all commitments remain rather general. Commitments to international support for climate action make a clear distinction between developed States, which are to provide support, and developing States, which are to receive it.[74]

If the principle of common but differentiated responsibilities was thus endorsed in a universally ratified agreement, this was based on a constructive ambiguity opened to alternative interpretations. There was – and still is – no consensus on what is "common" or "differentiated," or on the basis for or extent of differentiation. For some, "responsibility" connotes the responsibility of polluting States for a breach of their obligations under general international law; for others, it should be construed in a weaker sense, as reflecting a duty of assistance of developed States, based on their capacity and on the needs of developing States for support for effective climate action. Thus, while States agreed that some of their obligations under the UNFCCC regime could differ, they also largely differed in their understanding of this principle – and, indeed, on whether this is a principle at all.[75]

The UNFCCC contains two lists of developed countries. Annex I is a list of originally 36 parties which commit to take the lead in climate change mitigation and adaptation action

---

[70] Written statement of the United States on Principle 7 of the Rio Declaration, in *Report of the United Nations Conference on Environment and Development*, UN document A/CONF.151/26 (vol. IV) (September 28, 1992), para. 16.

[71] UNFCCC, *supra* note 10, recital 7.

[72] *Ibid.*, art. 3.1.

[73] *Ibid.*, art. 4.1.

[74] See *ibid.*, arts. 4.3–5.

[75] While the heading of *ibid.*, art. 3 announces "Principles," a footnote on the heading of art. 1 states that "Titles of articles are included solely to assist the reader."

generally.[76] This list was based on the membership to the OECD. Annex II is identical to Annex I, except for the exclusion of countries from the former Soviet Bloc which were "undergoing the process of transition to a market economy." The 25 parties originally listed in Annex II commit to provide assistance to others in tackling climate change and its effects.[77]

Inserting these lists into an Annex to a treaty introduced some rigidity. As some developing countries became richer, more industrialized and thus contributed more to global GhG emissions, their changing circumstances could not easily be recognized in the UNFCCC regime. From a fifth of global economic production in 1992, non-Annex I countries grew to comprise more than a third in 2010[78] and further since; likewise, their share in global GhG emissions grew from 42 percent in 1992 to 63 percent in 2013.[79] As discussed above, emerging economies reached levels of per capita GhG emissions comparable to some Western countries.[80]

Yet, only a few, relatively anecdotal revisions have been made to the annexes of the UNFCCC. A first revision was made in 1997 to take account of the dissolution of Yugoslavia and Czechoslovakia and to add Croatia, the Czech Republic, Liechtenstein, Monaco, Slovakia and Slovenia to Annex I.[81] Two subsequent revisions, in 2009 and 2011, added Malta and Cyprus into Annex I.[82] In addition to such formal revisions of Annex I, Article 4.2(g) allows any non-Annex I party to notify the Depository (i.e. the UN Secretary General) of its intention to implement the more stringent commitments contained in Articles 4.2(a) and (b).[83] Kazakhstan made a declaration in this sense on March 23, 2000; thus, although Kazakhstan is still not formally an Annex I party, it is bound by the same national commitments under the UNFCCC as Annex I parties.[84] Turkey, which had requested to have its name removed from Annexes I and II on account of national circumstances, was removed from Annex II (but not from Annex I) by a decision of the parties adopted at COP7, paving the way for Turkey's accession to the UNFCCC.[85]

The architecture of the Kyoto Protocol relied on this binary approach to differentiation. The Berlin Mandate made it clear that the Protocol would "not introduce any new commitments for Parties not included in Annex I."[86] Accordingly, Annex B of the Kyoto Protocol defines Quantified Emission Limitation or Reduction Commitments (QELRCs) that Annex I parties must implement between 2008 and 2012,[87] while developing States were subjected to no new

---

[76] See *ibid.*, Annex I. See also *ibid.*, art. 4.2.

[77] See in particular *ibid.*, arts. 4.3, 4.4 and 4.5.

[78] WRI, CAIT Climate Data Explorer, "GDP-USD."

[79] WRI, CAIT Climate Data Explorer, "Global greenhouse gas emissions excluding LULUCF."

[80] See above, section I.B.1.

[81] See decision 4/CP.3, "Amendments to the list in Annex I to the Convention under Article 4.2(f) of the Convention" (December 11, 1997).

[82] Decisions 3/CP.15, "Amendment to Annex I to the Convention" (December 18–19, 2009) and 10/CP.17, "Amendment to Annex I to the Convention" (December 11, 2011).

[83] See UNFCCC, *supra* note 10, art. 4.2(g).

[84] Although Kazakhstan is not an Annex I party to the UNFCCC, it is accordingly considered as an Annex I party for the purpose of the application of the Kyoto Protocol. For the purpose of the Kyoto Protocol, *supra* note 10, as defined in art. 1.7 thereof, "a 'Party included in Annex I' means a Party included in Annex I to the Convention, as may be amended, or a Party which has made a notification under Article 4, paragraph 2(g), of the Convention."

[85] Decision 26/CP.7, "Amendment to the list in Annex II to the Convention" (November 9, 2001). See generally Farhana Yamin and Joanna Depledge, *The International Climate Change Regime: A Guide to Rules, Institutions and Procedures* (Cambridge University Press, 2004) 106.

[86] Decision 1/CP.1, "Berlin Mandate" (April 7, 1995), para. 2.

[87] Kyoto Protocol, *supra* note 10, art. 3.1 and Annex B.

mitigation obligations.[88] This approach has proved increasingly divisive. On July 25, 1997, just a few months before the adoption of the Kyoto Protocol, the US Senate adopted the Byrd-Hagel Resolution by a vote of 95–0 (with five abstentions) to express the "sense of the Senate" that the United States should not ratify an agreement imposing mitigation obligations only on developed States.[89] Consistently, the US Senate never ratified the Kyoto Protocol.

By the time the Kyoto Protocol entered into force, GhG emissions had significantly increased in emerging economies and the developing world generally. In 2004, half of the world's GhG emissions originated from the developing world; the following year, China replaced the United States as the world's largest GhG emitter.[90] Without the ratification of the United States and given the withdrawal of Canada, the Kyoto Protocol only imposed mitigation commitments on parties representing 23 percent of global GhG emissions between 2008 and 2012.[91] Nevertheless, per capita GhG emissions remained significantly higher in developed countries. Different conceptions of the principle of common but differentiated responsibilities gave rise to incompatible expectations, thus hindering international climate change negotiations.

Differentiation is not confined to North–South relations. To define respective QELRCs, the negotiators of the Kyoto Protocol had to compare the circumstances of developed States. The negotiations first focused on the criteria which should be taken into consideration. Several developing countries, seeking to set a precedent for further negotiations, suggested that differentiation among Annex B parties should be based on the principle of common but differentiated responsibilities, in particular on each party's current and historical GhG emissions.[92] Instead, developed States agreed among themselves that they would focus on capacity-related criteria such as:

> the differences in starting points and approaches, economic structures and resource bases, the need to maintain strong and sustainable growth, available technologies and other individual circumstances, as well as the need for equitable and appropriate contributions by each of these Parties to the global effort.[93]

Likewise, negotiations among the Member States of the European Union distinguished national circumstances based on capacity rather than historical responsibility.[94]

---

[88] *Ibid.*, art. 10.

[89] US Senate Resolution 98, 105th Cong., 143 Cong. Rec. S8138–39 (July 25, 1997) (hereinafter Byrd-Hagel Resolution).

[90] WRI, CAIT Climate Data Explorer, "Total GHG emissions excluding land-use change and forestry."

[91] Data from *ibid.*

[92] See UNFCCC, Ad Hoc Group on the Berlin Mandate, *Framework Compilation of Proposals from Parties for the Elements of a Protocol or Another Legal Instrument, Note by the Chairman* (February 3, 1997), doc. FCCC/AGBM/ 1997/2, at 42 (Iran); UNFCCC, Ad Hoc Group on the Berlin Mandate, *Implementation of the Berlin Mandate Proposals from Parties* (February 26, 1997), doc. FCCC/AGBM/1997/Misc.1/Add.1, at 33 (Venezuela, Iran, Saudi Arabia and the United Arab Emirates); and UNFCCC, Ad Hoc Group on the Berlin Mandate, *Implementation of the Berlin Mandate: Additional Proposals from Parties* (May 30, 1997), doc. FCCC/AGBM/1997/Misc.1/Add.3, at 7 (Brazil). See generally Lasse RingiusAsbjørn Torvanger and Arild Underdal, "Burden sharing and fairness principles in international climate policy" (2002) 2:1 *International Environmental Agreements* 1.

[93] UNFCCC, COP3, *Adoption of a Protocol or Another Legal Instrument: Fulfilment of the Berlin Mandate: Revised Text under Negotiation* (November 12, 1997), doc. FCCC/CP/1997/2, at 31.

[94] EU Council Decision 2002/358/EC concerning the approval, on behalf of the European Community, of the Kyoto Protocol to the United Nations Framework Convention on Climate Change and the joint fulfilment of commitments thereunder (April 25, 2002), doc. 32002D0358.

C. Self-Differentiation as an Alternative Approach to Differentiation under the UNFCCC Regime

Unlike the Montreal Protocol, the UNFCCC regime was not able to foster the technological changes in developed States that could have allowed developing States to "leapfrog," that is to say, to turn directly to a low-carbon development model. GhG emissions in developed States did not significantly decrease, while emissions in developing States continued to increase rapidly. The simplistic dichotomy between developed and developing States became less and less relevant. Although the UNFCCC had provided for a review and possible amendments to Annexes I and II by the end of 1998,[95] any significant changes to the Annexes faced a strong political resistance. International negotiations on what would follow the Kyoto Protocol's first commitment period stalled on the challenging question of defining an alternative approach to differentiation, in negotiations led on a consensual basis.

The implementation of the 2007 Bali Action Plan, including the failure to reach agreement at the 2009 Copenhagen Summit, was characterized by a fierce opposition between developed and developing States, and more specifically between the United States and China. Highlighting the increasing share of developing States in global GhG emissions, some developed States advanced the idea that mitigation efforts could not be successful without the participation of emerging economies. By contrast, developing States continued to support a binary differentiation between developed and developing States, arguing that their development process should remain unconstrained in order to eradicate poverty.[96] Representatives of developed States suggested that differentiation should only be based on capacity-related elements such as GDP per capita, demographic or economic growth, mitigation potential and stage of economic development.[97] From the perspective of developing States, however, differentiation was more about justice and responsibility. It was largely felt, as Lavanya Rajamani put it, that "[a]ny erosion of differentiation would blur the lines of responsibility, shift a disproportionate (to their contribution) burden of mitigation on to developing countries, and thereby limit their development prospects."[98]

In order to go beyond a binary approach of differentiation between developed and developing parties, negotiations have progressively turned toward an approach of self-differentiation, where each party essentially assesses what, on its own account, would be its fair contribution. The Copenhagen Accord and the Cancún Agreements were the first steps in this transition, where all States were invited to define their own efforts and ambitions, although a clear distinction was maintained between developed States' "quantified economy-wide emissions targets"[99] and developing States' "nationally appropriate mitigation actions."[100] The need to reconsider differentiation was reflected in the mandate of the Durban Platform for Enhanced Action to adopt an agreement "applicable to all Parties"[101] – a

---

[95] UNFCCC, *supra* note 10, art. 4.2(f).

[96] See e.g. Submission by Philippines on Behalf of the G-77/China (December 10, 2008), reproduced in doc. FCCC/AWGLCA/2008/MISC.5/Add.2, pt. II.

[97] See Lavanya Rajamani, "Differentiation in the emerging climate regime" (2013) 14:1 *Theoretical Inquiries in Law* 151, at 158.

[98] *Ibid.*, at 159.

[99] Copenhagen Accord, in the annex of decision 2/CP.15 (December 18–19, 2009), para. 4; decision 1/CP.16, *supra* note 10, para. 37.

[100] Copenhagen Accord, *supra* note 99, para. 5; decision 1/CP.16, *supra* note 10, para. 50.

[101] Decision 1/CP.17, "Establishment of an Ad Hoc Working Group on the Durban Platform for Enhanced Action" (December 11, 2011), para. 2. See also Rajamani, *supra* note 97, at 164.

decision which did not contain any direct reference to the principle of common but differentiated responsibilities.

Achieving this gradual transition, the Paris Agreement clearly renounces a bilateral approach to differentiation, preferring an approach of self-differentiation. Its Preamble notes that the parties are "guided by [the] principles [of the UN Framework Convention on Climate Change], including the principle of equity and common but differentiated responsibilities and respective capabilities, in the light of different national circumstances."[102] Article 2 adds that the Agreement will "be implemented to reflect equity and the principle of common but differentiated responsibilities and respective capabilities, in the light of different national circumstances."[103] What these provisions mean, however, is largely for the parties to decide for themselves when determining their national commitment.[104] According to Article 4, NDCs should reflect the party's "highest possible ambition, reflecting its common but differentiated responsibilities and respective capabilities, in the light of different national circumstances."[105] Although "national circumstances" are mentioned five times as a ground for differentiation, there is no indication of what kind of circumstances are relevant, although emphasis seems to be generally on capacity rather than responsibility.

Nevertheless, some provisions of the Paris Agreement suggest a distinction between developed and developing parties. In particular, under Article 4.4, developed parties are to "continue taking the lead by undertaking economy-wide absolute emission reduction targets," whereas developing parties "should continue enhancing their mitigation effort, and are encouraged to move over time towards economy-wide emission reduction or limitation targets."[106] Unlike the UNFCCC and the Kyoto Protocol, these provisions do not establish a clear-cut distinction between developed and developing parties. Unlike previous agreements, the Paris Agreement contains no reference to the list of developed parties contained in Annex I of the UNFCCC or to any other definition of "developed" and "developing" parties. Differentiation in the Paris Agreement accordingly appears as much more progressive and flexible than in previous agreements.

The approach of the Paris Agreement is a daring bet. There is a risk that differentiation will, over time, come at the expense of the countries willing to cooperate in good faith, while others will be able to invoke their national circumstances to free-ride. The success of the approach of self-differentiation in the Paris Agreement will depend on the ability of socio-political processes to incentivize bona fide cooperation. Eventually, it will be for other States and civil society organizations to judge whether each State's nationally determined contribution is sufficiently ambitious in the light of its national circumstances.[107]

D. Differentiation in Civil Aviation and Maritime Transportation

In the negotiations on climate change mitigation in international transportation, differentiation has been limited by the need to avoid flags of convenience or market distortion. In

[102] Paris Agreement, *supra* note 10, recital 4.
[103] *Ibid.*, art. 2.2.
[104] *Ibid.*, art. 3.
[105] *Ibid.*, art. 4.3. See generally Christina Voigt and Felipe Ferreira, "'Dynamic differentiation': the principles of CBDR-RC, progression and highest possible ambition in the Paris Agreement" (2016) 5:2 *Transnational Environmental Law* 285.
[106] *Ibid.*, art. 4.4.
[107] See, for instance, on Climate Action Tracker, www.climateactiontracker.org/countries/india (accessed January 3, 2018). See generally Chapter 13.

international maritime law, consideration for the specific circumstances of developing States are generally confined to the provision of technical assistance to facilitate compliance with a unique set of standards.[108] The UN Convention on the Law of the Sea contains a few special arrangement for developing States, recognizing "the special interests and the needs of developing countries"[109] and calling for international cooperation against pollution of the marine environment from land-based sources "taking into account ... the economic capacity of developing States and their need for economic development."[110] Likewise, Annex VI on the Prevention of Air Pollution from Ships of MARPOL, as amended in 2011, includes an article promoting technical cooperation and transfer of technology to the benefit of "States, especially developing States, that request technical assistance."[111]

Likewise, differentiation is not deeply anchored in international aviation law. On the contrary, one of the objectives of the ICAO, as defined by the 1944 Chicago Convention on International Civil Aviation, is to "avoid discrimination between contracting States."[112] Non-discrimination and equal treatment among States has progressively been recognized as a general principle in international aviation law, being viewed as instrumental to the realization of ICAO's objective of promoting the development of international civil aviation.[113] Nevertheless, the ICAO Assembly formally acknowledged the "principle of common but differentiated responsibilities and respective capabilities, in light of different national circumstances"[114] in its statement of policy on climate change. Accordingly, it was agreed in principle that the Carbon Offsetting and Reduction Scheme for International Aviation (CORSIA) would be implemented gradually in order "to accommodate the special circumstances and respective capabilities of States, in particular developing States."[115] Differentiation in the implementation of CORSIA is first self-determined, with the pilot phase (2021–2023) and the first phase (2024–2026) being limited to States which volunteer to participate. States with a limited contribution to international air traffic as well as least developed countries, small island developing States and landlocked developing States are not required to participate in the second phase.[116]

Any attempt to differentiate among States in transnational activities such as maritime transportation and international civil aviation faces significant obstacles. Attributing GhG emissions to States may be challenging with regard to such transnational activities. In principle, GhG

---

[108] See Convention on the International Maritime Organization, March 6, 1948, 289 *UNTS* 48, arts. 2(e) and 15(k).

[109] United Nations Convention on the Law of the Sea, December 10, 1982, 1833 *UNTS* 3, recital 6.

[110] *Ibid.*, art. 207.4.

[111] MARPOL resolution MEPC.203(62), "Amendments to the Annex of the Protocol of 1997 to Amend the International Convention for the Prevention of Pollution from Ships, 1973, as Modified by the Protocol of 1978 Relating thereto" (July 15, 2011), in doc. MEPC 62/24/Add.1, Regulation 23, para. 1. See also MARPOL resolution MEPC.229(65), "Promotion of technical cooperation and transfer of technology relating to the improvement of energy efficiency of ships" (May 17, 2013), in doc. MEPC 65/22.

[112] Convention on International Civil Aviation, December 7, 1944, 15 *UNTS* 295 (hereinafter Chicago Convention), art. 44(g). See Beatriz Martinez Romera and Harro van Asselt, "The international regulation of aviation emissions: putting differential treatment into practice" (2015) 27:2 *Journal of Environmental Law* 259.

[113] Alejandro Piera Valdés, *Greenhouse Gas Emissions from International Aviation: Legal and Policy Challenges* (Eleven, 2015) 46.

[114] ICAO Assembly Resolution A39-2, "Consolidated Statement of Continuing ICAO Policies and Practices Related to Environmental Protection – Climate Change" (September 27–October 7, 2016), recital 11.

[115] ICAO Assembly Resolution A39-3, "Consolidated statement of continuing ICAO policies and practices related to environmental protection – Global Market-Based Measure (MBM) scheme" (September 27–October 7, 2016), para. 9.

[116] *Ibid.*, para. 9(e).

emissions from international aviation or international maritime transportation could alternatively be attributed on the basis of the registration of the vessel or of its operator, of the port of origin and destination, or of the nationality of the passengers or the final destination of the cargo, among others. However, the design of CORSIA suggests that a combination of a *de minimis* threshold, consideration to the least developed States and voluntary participation in initial phases can accommodate the constraint of differentiation in international transportation regimes.

III. CONCLUSION

The international law on climate change determines rights and obligations which depend on the specific circumstances of each State. Obligations arising from the no-harm principle arguably relate, among other things, to the capacity of each State to limit GhG emissions without relinquishing its essential interests, in particular in fulfilling the basic needs of its population.[117] The need for differentiation has generally been recognized within the UNFCCC regime through the ambiguous principle of common but differentiated responsibilities (and respective capabilities), but also in efforts carried out in other regimes, for instance, under the Montreal Protocol on Substances that Deplete the Ozone Layer and under the ICAO.

Differentiation is an important characteristic of the international law on climate change. Yet, differentiation has also appeared as a considerable challenge to international climate change negotiations, from the outset of negotiations held within the Intergovernmental Negotiation Committee for a Framework Convention on Climate Change in the early 1990s to the tumultuous Copenhagen Summit. It will remain a contentious aspect of the negotiations which, under the Paris Agreement, should largely be channeled through regular stock-taking exercises.[118] There is surely a risk that, absent any sort of consensus on the grounds for differentiation, taking national circumstances into account could initiate "a movement towards the lowest common denominator."[119] This risk, however, is to be managed rather than avoided, because renouncing differentiation would result in a regime that is unfair, and hence unacceptable, to many developing States.

[117] See in particular Chapter 5, section III.
[118] See Chapter 13, section III.C.
[119] Christopher D. Stone, "Common but differentiated responsibilities in international law" (2004) 98:2 *American Journal of International Law* 276.

7

International Action on Climate Change Mitigation

The main object of the international law on climate change is to mitigate climate change. Immediate action against the rapid increase in GhG concentrations in the atmosphere is essential to ensuring that our civilizations can flourish durably and reducing the risk of a global civilizational collapse. Climate change mitigation encompasses efforts to counter the increase in GhG concentrations in the atmosphere. The relevant working group of the IPCC defines climate change mitigation as "[a] human intervention to reduce the sources or enhance the sinks of greenhouse gases."[1]

In contrast to the urgency of its object, the notion of "mitigation" suggests a rather incremental action. The obligation of States under general international law is arguably more demanding that what this term suggests. Accordingly, States have the obligation to prevent activities under their jurisdiction which cause harms to the global environment, including through excessive GhG emissions. Under the law of State responsibility, inasmuch as their breach of this international obligation is continuing, States are under an obligation to take all relevant measures to ensure the *cessation* of such breaches.[2] "Cessation" suggests more pressing action than "mitigation." Harmful anthropogenic interference with our climate system ought not just to be *mitigated*, but to have been *avoided* altogether, and excessive GhG emissions ought not just to be *limited* or *reduced*, but *stopped* entirely. If this book uses the concept of "mitigation" to refer to the obligation of States to stop excessive GhG emissions, it is by convenience (given the widespread use of this term) rather than as an endorsement.

This chapter analyzes the obligations of States relating to climate change mitigation generally. The next chapter focuses on a more particular aspect of climate change mitigation: so-called "flexibility mechanisms," whereby one State may implement its mitigation objectives through support to mitigation action in other States. Finally, Chapter 9 reviews more specific questions raised in current debates on geoengineering, some forms of which might be considered to be climate change mitigation.

---

[1] JM Allwood *et al.*, "Annex I: glossary, acronyms and chemical symbols" in O. Edenhofer *et al.* (eds.), *Climate Change 2014: Mitigation of Climate Change. Contribution of Working Group III to the Fifth Assessment Report of the Intergovernmental Panel on Climate Change* (Cambridge University Press, 2014) 1249, at 1266.

[2] ILC, *Draft Articles on Responsibility of States for Internationally Wrongful Acts with Commentaries*, in (2001) *Yearbook of the International Law Commission*, vol. II, part two (hereinafter *Articles on State Responsibility*), art. 30(a).

The present chapter starts by delineating the obligations of States under general international law relevant to climate change mitigation. Then, it discusses the developments which have taken place under the UNFCCC regime by analyzing the collective mitigation objectives and the national commitments made under successive international climate agreements. A similar architecture, including collective objectives and national commitments, is retraced though developments taking place in other regimes. Finally, the last section describes some of the typical measures of implementation on climate change mitigation, including classical command and control regulation, the use of economic incentives and ways for national authorities to exercise leadership.

I. STATE OBLIGATIONS UNDER GENERAL INTERNATIONAL LAW

The conduct of a State causing serious harm to another State cannot be tolerated in a system where States are equal and sovereign. This conduct is a breach of the State's obligation under general international law. This holds true whether or not the harm results from activities directly conducted by the State or on its behalf (action) or indirectly from its failure to prevent persons under its jurisdiction from causing such harms (omission). As discussed in Chapter 5, the no-harm principle involves a negative obligation for States to refrain from causing transboundary environmental damages, but it also implies a due diligence obligation to prevent activities which would cause transboundary environmental harm (which is sometimes considered as a distinct principle, the "preventive principle").[3] There is no convincing reason to exclude the application of the no-harm principle from circumstances such as climate change, despite the extra layer of complexity stemming from the multiple sources of the harm and the global nature of the damage caused by excessive GhG emissions.[4] As discussed in Chapter 6, some elements such as population or unmet development needs could justify some GhG emissions, yet certainly not all.[5]

It is therefore possible to conclude that every State has, in application of the no-harm principle, the obligation not to cause GhG emissions and to prevent such emissions from activities under their jurisdiction, when such emissions would interfere with the climate system and cause significant harm to areas beyond their jurisdiction in a way that is clearly disproportionate to the benefit drawn from these emissions. In addition, in application to general international law on State responsibility, States whose territory is currently used for activities causing massive GhG emissions have an obligation not just to limit or marginally reduce these GhG emissions; their obligation is to take all possible immediate measures to ensure the cessation of excessive GhG emissions.[6]

Yet, there remains considerable uncertainty regarding the modalities of this obligation. As a customary norm without a unique, authoritative written statement, the no-harm principle remains rather ill-defined. Even in archetypical cases of harms affecting a neighboring

---

[3] Philippe Sands and Jacqueline Peel, *Principles of International Environmental Law*, 3rd edn (Cambridge University Press, 2012). See generally *Trail Smelter (U.S. v. Canada)*, Arbitral Award of March 11, 1941, (1949) III *UNRIAA* 1938, at 1965; ICJ, *Legality of the Threat or Use of Nuclear Weapons*, Advisory Opinion of July 8, 1996, para. 29. See Chapter 5, section I.

[4] See Chapter 5, section I.B.

[5] See *ibid*.

[6] *Articles on State Responsibility*, *supra* note 2, art. 30(a).

country, vexing questions such as the existence of a *de minimis* threshold remain unanswered in the present state of development of international law.[7] The standard of due diligence applicable to States' obligation under the no-harm principle remains unclear, causing uncertainty as to the weight which should be given to national circumstances as a justification for some level of GhG emissions. In other words, it remains extremely difficult to define the nature of the obligation of States to take "immediate measures" or to determine what constitutes "excessive" GhG emissions.

Furthermore, applying the no-harm principle to climate change raises a number of thorny issues. The harmful impact of GhG emissions on our climate is very slow, but long and widespread. This differs sharply from most cases previously decided by international courts and tribunals, which typically involved a unique State alleged to be in breach of its obligation and a unique injured State regarding transboundary damages directly caused by a pollutant. By contrast, climate change is caused by the action of many States and affects all. A creative interpretation of the no-harm principle is needed to determine the respective obligations of multiple States. Such a consensual or otherwise authoritative interpretation is still lacking.[8]

Some additional questions arise with regard to the intergenerational dimension of climate change. Future generations will greatly suffer from the GhG emissions caused by current generations – their very existence is even at risk. Here again, however, a simple abstract legal principle is insufficient to determine a particular threshold of tolerable intergenerational harm or a sufficient level of precaution.

In this context, it is tempting to approach international climate agreements as successive attempts to defining the modalities of the no-harm principle in relation to climate change. When international climate agreements reflect a common understanding of the existing law, they may help to clarify the modalities of application of the no-harm principle in customary international law.[9] For example, international climate agreements have generally recognized the particular circumstances of developing States and more specifically their development needs as a basis for assessing their respective mitigation obligations,[10] thus implying that obligations related to the no-harm principle are partly contingent on the capacity of States. Yet, many States have repeatedly denounced the shortcomings of mitigation commitments included in successive international climate agreements. In these circumstances, it seems that the UNFCCC regime so far has only achieved a very partial and selective implementation of the no-harm principle.[11]

---

[7] See Chapter 5, section I.B.

[8] See Chapter 5, section I; and Chapter 16, section II.A.

[9] On the treaties as evidence of the identification of customary international law – and hence possibly an interpretative tool – see ILC, *Third Report on Identification of Customary International Law by Special Rapporteur Michael Wood* (March 27, 2015), doc. A/CN.4/682, paras. 31–44.

[10] See e.g. United Nations Framework Convention on Climate Change, May 9, 1992, 1771 *UNTS* 107 (hereinafter UNFCCC), recital 24 and art. 3.2; Kyoto Protocol to the United Nations Framework Convention on Climate Change, December 11, 1997, 2303 *UNTS* 162 (hereinafter Kyoto Protocol), art. 3.1; Paris Agreement, December 12, 2015, in the annex of decision 1/CP.21, "Adoption of the Paris Agreement" (December 12, 2015), recital 4 and art. 4.1.

[11] This is further discussed in Chapter 13, where the UNFCCC regime is described as a gradual attempt to bridge a widespread lack of compliance with general international law.

II. ACTION ON CLIMATE CHANGE MITIGATION UNDER THE UNFCCC REGIME

Successive international climate agreements have defined collective objectives and national commitments for action on climate change mitigation. These objectives and commitments are clearer and more specific than the obligations of States under general international law. Yet, national commitments in particular appear markedly less ambitious than the obligations of States under general international law.

A. Collective Objectives

The ultimate objective of the UN Framework Convention on Climate Change is "to achieve ... stabilization of greenhouse gas concentrations in the atmosphere at a level that would prevent dangerous anthropogenic interference with the climate system."[12] This provides only a very vague definition of what international cooperation under the UNFCCC is designed to achieve. There is no unique way of assessing the threshold at which GhG concentration in the atmosphere or its evolution becomes "dangerous." More specific proposals for a collective objective had been made. In a 1990 report, for instance, the Stockholm Environment Institute defined targets for sea-level rise (by no more than 0.5 meters), mean global temperature increase (by no more than 1–2 degrees Celsius) and atmospheric GhG concentrations (reaching no more than 400–560 ppm).[13] The final formulation of an ultimate objective was simply as far a consensus as could be reached in 1992, when oil-producing States were still denying the need for virtually any action on climate change mitigation at all.[14] Despite its vagueness, or thanks to it, the "ultimate" objective of the Convention has long remained the principal touchstone for international climate agreements as a whole.[15]

However, more specific objectives for action of Annex I parties on climate change mitigation could be adopted. Article 4.2 of the UNFCCC, which defines the commitments of Annex I parties, emphasizes that "the return by the end of [the 1990s] to earlier levels of anthropogenic [greenhouse gas] emissions" would contribute to "modifying longer-term trends in anthropogenic emissions consistent with the objective of the [UNFCCC]."[16] In other words, GhG emissions by Annex I parties would reach a peak and start decreasing during the 1990s. This aspirational goal was not achieved. While the GhG emissions of Annex I parties were between 16.4 and 16.8 GtCO2eq from 1992 to 2000, they reached their historical peak, 17.5 GtCO2eq, in 2007, before decreasing slightly to 16.4 GtCO2eq by 2013.[17]

---

[12] UNFCCC, *supra* note 10, art. 2.

[13] F.R. Rijsberman and R.J. Swart, *Targets and Indicators of Climatic Change* (SEI, 1990). See generally Michael Oppenheimer and Annie Petsonk, "Article 2 of the UNFCCC: historical origins, recent interpretations" (2005) 73:3 *Climatic Change* 195.

[14] See e.g. Daniel Bodansky, "The United Nations Framework Convention on Climate Change: a commentary" (1993) 18:2 *Yale Journal of International Law* 451, at 481.

[15] See e.g. decision 1/CP.13, "Bali Action Plan" (December 14–15, 2007), recital 2; decision 1/CP.16, "The Cancún Agreements: outcome of the work of the Ad Hoc Working Group on Long-Term Cooperative Action under the Convention" (December 10–11, 2010), recital 4; decision 1/CP.21, "Adoption of the Paris Agreement" (December 12, 2015), recital 7.

[16] UNFCCC, *supra* note 10, art. 4.2(a).

[17] WRI, CAIT Climate Data Explorer, "Total GHG emissions excluding land-use change and forestry." Similar trends appear in *ibid.*, "Total GHG emissions including land-use change and forestry."

Another collective objective for mitigation action by Annex I parties was defined several years later in the Kyoto Protocol. Article 3 of the Kyoto Protocol defined the mitigation commitment of Annex I parties "with a view to reducing their overall emissions of [greenhouse gases] by at least 5 per cent below 1990 levels in the commitment period of 2008 to 2012."[18] Due to the non-participation of the United States and the withdrawal of Canada, this objective has not been achieved: the overall GhG emissions of Annex I States have generally remained stable from 1990 to 2012.[19] Nevertheless, the Doha Amendment to the Kyoto Protocol defined the further objective of reducing the overall emissions of GhGs by Annex I parties "by at least 18 per cent below 1990 levels in the commitment period 2013 to 2020."[20] As of early 2018, it is uncertain whether this objective will be achieved.

In 2007, the Bali Action Plan called for "a shared vision for long-term cooperative action" which would extend beyond developed States. This involved in particular the adoption of "a long-term global goal for emission reductions."[21] Two years later, the Copenhagen Accord "recogniz[ed]" as a scientific view that the increase in global temperature "should be below 2 degrees Celsius" and that peaking in global GhG emissions should be achieved "as soon as possible."[22] Furthermore, the Copenhagen Accord called for a reassessment of this long-term goal by 2015, "including in relation to temperature rises of 1.5 degrees Celsius."[23] The Cancún Agreements endorsed these objectives,[24] in addition calling for immediate "work towards identifying a time frame for global peaking of greenhouse gas emissions."[25]

The adoption of the Paris Agreement endorsed the objective of "holding the increase in the global average temperature to well below 2°C above pre-industrial levels." It also stressed the need for "efforts to limit the temperature increase to 1.5°C above pre-industrial levels."[26] Furthermore, the Paris Agreement reaffirmed the aim of:

> reach[ing] global peaking of greenhouse gas emissions as soon as possible, recognizing that peaking will take longer for developing country Parties, and to undertake rapid reductions thereafter ... so as to achieve a balance between anthropogenic emissions by sources and removals by sinks of greenhouse gases in the second half of this century.[27]

The Paris Agreement is thus the first climate agreement to define a long-term aspiration to carbon-neutral humankind, which, in the second half of the twenty-first century, would, at

---

[18] Kyoto Protocol, *supra* note 10, art. 3.1.

[19] See WRI, CAIT Climate Data Explorer, "Total GHG emissions excluding land-use change and forestry." Some data are missing for 1990 and 1991. Annex I parties emitted 16.8 GtCO2eq in 1992, and an average of 16.6 GtCO2eq from 2008 to 2012. Similar trends appear in *ibid.*, "Total GHG emissions including land-use change and forestry."

[20] Kyoto Protocol, *supra* note 10, as modified by the Doha Amendment to the Kyoto Protocol, December 8, 2012, in the annex of decision 1/CMP.8, "Amendment to the Kyoto Protocol pursuant to its Article 3, paragraph 9 (Doha Amendment)" (December 8, 2012) (hereinafter Doha Amendment), art. 3, para. 1 bis.

[21] Decision 1/CP.13, *supra* note 15, para. 1(a).

[22] Copenhagen Accord, in the annex of decision 2/CP.15 (December 18–19, 2009), paras. 1 and 2.

[23] *Ibid.*, para. 12.

[24] Decision 1/CP.16, *supra* note 15, para. 4.

[25] *Ibid.*, para. 6.

[26] Paris Agreement, *supra* note 10, art. 2.1(a).

[27] *Ibid.*, art. 4.1.

last, bring States back to compliance with their obligation under general international law. But while the objective may appear ambitious, it is unclear whether the UNFCCC regime contains the rules necessary for its achievement.

B. National Commitments

Collective objectives on climate change mitigation are pursued by States, with developed States taking the lead.[28] In addition to defining collective objectives, successive international agreements on climate change have called on States to commit to particular efforts to promote climate change mitigation.

The historical development of the UNFCCC regime, including national mitigation commitments, was retraced in Chapter 1. The UN Framework Convention on Climate Change outlines only some broad mitigation commitments, some being applicable to all parties,[29] others only to developed parties.[30] The Kyoto Protocol defines QELRCs applicable to developed States over an initial commitment period running from 2008 to 2012;[31] the Doha Amendment establishes a second commitment period of 2013–2020.[32] Voluntary mitigation pledges by all parties were invited by the Copenhagen Accord and the Cancún Agreements for the period 2013–2020, with a distinction between quantified economy-wide emission reduction targets to be implemented by developed parties and other nationally appropriate mitigation actions by developing parties.[33] Lastly, the Paris Agreement calls on every party to determine and communicate its national contribution to international action on mitigation.[34]

Going beyond this historical account, the following details three types of national commitments related to climate change mitigation. Firstly, the parties to successive international climate agreements committed to limiting and reducing the net GhG emissions taking place within their territory. Secondly, they have also committed to accounting for the GhG emissions taking place within their territory, then making it possible to verify compliance with emission limitation and reduction commitments. Lastly, the parties to successive international climate agreements have repeatedly committed to participation in negotiations toward additional commitments on climate change mitigation. Overall, these national commitments contribute to, but do not guarantee, the realization of collective objectives on climate change mitigation.

1. Emission Limitation and Reduction Commitments

National commitments on the limitation and reduction of GhG emissions are the centerpiece of international climate agreements. The nature of these commitments has varied in successive international climate agreements. Thus, States have alternatively pledged to pursue some

---

[28] UNFCCC, *supra* note 10, art. 3.1.
[29] See in particular *ibid.*, art. 4.1.
[30] See in particular *ibid.*, art. 4.2.
[31] Kyoto Protocol, *supra* note 10, art. 3 and Annex B.
[32] Doha Amendment, *supra* note 20.
[33] Decision 1/CP.16, *supra* note 15, paras. 36 and 49.
[34] Paris Agreement, *supra* note 10, art. 4.2.

measures, to achieve emission limitation and reduction targets, or to pursue measures intended to achieve a pre-determined contribution.

Firstly, the most rudimental commitment is that of *pursuing some measures* aimed at curbing GhG emissions within its territory. Thus, through the UNFCCC, States agreed to a relatively blurred, unspecific obligation to "[f]ormulate, implement, publish and regularly update ... measures to mitigate climate change."[35] Developed States agreed to only a little more: to "adopt ... policies and take corresponding measures" which "will demonstrate that [they] are taking the lead in modifying longer-term trends in anthropogenic emissions."[36] Elaborating on the latter commitment of developed States, the Kyoto Protocol mentions some relevant policies and measures that developed parties could take, such as enhancing energy efficiency, protecting sinks and reservoirs of GhGs (e.g. forests), promoting sustainable agricultural practices, developing and deploying new and renewable energies, phasing out fossil-fuel subsidies, improving transportation or reducing methane emissions through waste management.[37] Under the Copenhagen Accord and the Cancún Agreements, non-Annex I parties committed to implementing "nationally appropriate mitigation actions ... as communicated by them."[38] Lastly, under the Paris Agreement, all parties must "prepare, communicate and maintain successive nationally determined contributions that it intends to achieve"[39] every five years.[40] The parties to the Paris Agreement are also encouraged to "strive to formulate and communicate long-term low greenhouse gas emission development strategies."[41]

States' commitments to pursuing specific measures are not merely procedural obligations (e.g. adopting some laws or policies). Formally, at least, these commitments also involve substantive obligations: each State must make at least a nominal effort to limit its GhG emissions. Although these commitments do not initially involve any specific measure, the failure of a State to communicate any measure at all to curb its GhG emissions or its systematic failure to make reasonable efforts towards the implementation of the measures which it has communicated would constitute a breach of its treaty obligation.

Secondly, States have committed to *achieving particular emission limitation and reduction targets*. This form of commitment was introduced by the Kyoto Protocol, whereby Annex I parties pledged to

> ensure that their aggregate anthropogenic carbon dioxide equivalent emissions of the greenhouse gases listed in Annex A do not exceed their assigned amounts, calculated pursuant to their quantified emission limitation and reduction commitments inscribed in Annex B.[42]

Annex B defined QELRCs applicable to each Annex I party as a percentage of its 1990 GhG emissions applicable to the average of the five years between 2008 and 2012. These

---

[35] UNFCCC, *supra* note 10, art. 4.1(b).

[36] *Ibid.*, art. 4.2(a).

[37] Kyoto Protocol, *supra* note 10, art. 2.1(a). See also, regarding developing parties, art. 10(b)(i).

[38] Decision 1/CP.16, *supra* note 15, para. 49. See also Copenhagen Accord, *supra* note 22, para. 5.

[39] Paris Agreement, *supra* note 10, art. 4.2.

[40] *Ibid.*, art. 4.9.

[41] *Ibid.*, art. 4.19.

[42] Kyoto Protocol, *supra* note 10, art. 3.1.

commitments ranged from a reduction by 8 percent (European Union), to a limitation to 8 or 10 percent increase (Australia and Iceland, respectively). New commitments of the same kind were defined by the Doha Amendment for a second commitment period from 2013 and 2020, although some parties to the Kyoto Protocol (Japan, New Zealand and Russia) refused to commit to an emission limitation or reduction during a second commitment period.[43] Similarly, under the Copenhagen Accord and then the Cancún Agreements, developed States pledged to implement quantified economy-wide emission reduction targets for the same period.[44]

Some authors have seen in these provisions the affirmation of a "right" to emit given amounts of GhG emissions.[45] Yet, this interpretation is not required by the texts: defining "assigned amounts" of GhG emissions does not, strictly speaking, imply any right to these amounts. As Charlotte Streck and Moritz von Unger noted, the Kyoto Protocol "defined emission limitations only and arguably has not created a right to emit."[46] Nor does the object and purpose of a treaty regime seeking to "achieve ... stabilization of greenhouse gas concentrations in the atmosphere"[47] support the view that the Kyoto Protocol or subsequent agreements created any "right" to emit GhGs. The COP confirmed that "the Kyoto Protocol has not created or bestowed any right, title or entitlement to emissions of any kind on Parties included in Annex I."[48] The commitment of States to achieve particular emission limitation and reduction targets is without prejudice of their obligation under general international law to cease excessive GhG emissions.

As defined in the Kyoto Protocol and (albeit less clearly) in the Cancún Agreements, these commitments involve an obligation of result – expressed as the obligation to "ensure" or "implement" particular mitigation targets.[49] Consistently, it is only based on the verification of the actual GhG emissions that compliance with such commitments to achieving particular mitigation targets can be ascertained. Such an "obligation of result" may seem attractive in circumstances where it may be difficult for international institutions to assess the effectiveness of complex domestic measures of implementation. A State committed to "ensuring" a particular reduction in national GhG emissions is allowed to decide which measures are the most appropriate and effective.

Framing mitigation commitments as obligation of results assumes that States can easily ensure particular mitigation pathways years in advance. Yet, the evolution of GhG emissions in a country largely depends on circumstances beyond the State's control. The 2007–2008 financial crisis resulted in a significantly – although transient – contraction in the GhG emissions of developed parties, thus greatly facilitating a formal achievement of mitigation commitments without really forcing these parties to engage in a path toward a carbon-neutral economy.[50]

---

[43] Doha Amendment, *supra* note 20. Although the Doha Amendment had not entered into force as of mid-2017, parties were invited to apply it provisionally, at least regarding their QELRCs. See decision 1/CMP.8, *supra* note 20, paras. 5 and 6.

[44] Copenhagen Accord, *supra* note 22, para. 4; decision 1/CP.16, *supra* note 15, para. 36.

[45] See, for instance, Alexander Zahar, "Methodological issues in climate law" (2015) 5:1 *Climate Law* 25, at 32. On a similar reasoning on the municipal plane, see below, note 149.

[46] See Charlotte Streck and Moritz von Unger, "Creating, regulating and allocating rights to offset and pollute: carbon rights in practice" (2016) 13:3 *Carbon & Climate Law Review* 178, at 181.

[47] UNFCCC, *supra* note 10, art. 2.

[48] Decision 15/CP.7, "Principles, Nature and Scope of the Mechanisms Pursuant to Articles 6, 23 and 17 of the Kyoto Protocol" (November 10, 2011), recital 6. See also decision 2/CMP.1, "Principles, Nature and Scope of the Mechanisms Pursuant to Articles 6, 12 and 17 of the Kyoto Protocol" (November 30, 2005), recital 6.

[49] See Kyoto Protocol, *supra* note 10, art. 3.1; decision 1/CP.16, *supra* note 15, para. 36.

[50] See Igor Shishlov, Romain Morel and Valentin Bellassen, "Compliance of the parties to the Kyoto Protocol in the first commitment period" (2016) 16:6 *Climate Policy* 768.

In other circumstances, a State with strong economic growth would certainly face unexpected difficulties in complying with its obligations. Flexibility mechanisms appeared as a convenient way for signatory States to make sure that they could fulfill their respective QELRC, if necessary, through buying carbon units.[51] Yet, the strict nature of the obligation to "ensure" was also associated with relatively unambitious targets.

Furthermore, defining national mitigation commitments as obligation of result does not really avoid the difficulty of assessing whether actual measures of implementation are sufficient. A close oversight of national measures of implementation is unavoidable in order to build trust during the implementation period and to reassure each State that others are genuinely taking steps toward compliance. Nevertheless, during the implementation of the Kyoto Protocol, this approach prevented any formal *finding* of non-compliance before the completion of the implementation period. Canada notoriously exploited this flaw. With emissions roughly 25 percent above its QELRC, the government of Stephen Harper prevented a finding of non-compliance by withdrawing from the Kyoto Protocol with effect on December 15, 2012, two weeks before the completion of the implementation period.[52]

Thirdly and lastly, States may commit to *pursuing measures intended to achieve a predetermined contribution*. Thus, Article 4.2 of the Paris Agreement calls on each party, having communicated its NDC, to "pursue domestic mitigation measures, with the aim of achieving the objectives of such contributions."[53] This unique wording was in part a political compromise between some States keen to define an obligation of result – small island developing States and the European Union – and the United States, which wished to avoid provisions that would have required ratification by the Senate.[54] This commitment requires States to act in such a way that they can expect in good faith to lead to compliance with the mitigation target expressed or implied by their NDC.[55]

Unlike emission limitation and reduction commitments under the Kyoto Protocol, Article 4.2 of the Paris Agreement does not involve an obligation of result. Parties could accordingly justify their inability to achieve the objective mentioned in their NDC, conceivably based on an economic growth higher than expected or delays in the deployment of a major clean technology project. On the other hand, this formula allows for earlier reviews of compliance. It could also help overcome the reluctance of States to make ambitious commitments that they are not entirely certain can be achieved.

The INDCs communicated by the parties to the Paris Agreement are of very different forms. Developed parties are expected to communicate "economy-wide absolute emission reduction targets," whereas developing States are "encouraged to move over time towards economy-wide

---

[51] See Chapter 8.

[52] See UNFCCC, Compliance Committee, Facilitative Branch, "Report on the Meeting" (November 9, 2012), doc. CC/FB/12/2012/3, in particular the correspondence between the chairperson of the facilitative branch and the Canadian ambassador for climate change in the annex. See also Chapter 13, section II.C.

[53] Paris Agreement, *supra* note 10, art. 4.2. See generally M.J. Mace, "Mitigation commitments under the Paris Agreement and the way forward" (2016) 6:1–2 *Climate Law* 21.

[54] See Chapter 3, section IV.A.

[55] See Benoit Mayer, "Obligations of conduct in the international law on climate change: a defence" *Review of European, Comparative and International Environmental Law* (forthcoming). See also Lavanya Rajamani, "The 2015 Paris Agreement: interplay between hard, soft and non-obligations" (2016) 28:2 *Journal Environmental Law* 337, at 354.

emission reduction or limitation targets."[56] Besides an overall target, most parties provided information on the measures they intended to take, such as market-based mechanisms or economic diversification.[57] This results from the wording of Article 4.2 that the parties to the Paris Agreement are under no obligation to implement the specific measures described in their NDCs; they may decide to implement other measures of a similar effect.[58] As of early 2018, the aggregate anticipated effect of the implementation of INDCs is clearly insufficient to ensure the swift reduction in global GhG emissions necessary to achieve the collective objective of the Paris Agreement of holding the increase in global average temperature below 2 degrees Celsius above pre-industrial levels.[59]

2. Accounting Commitments

Compliance with emission limitation and reduction commitments could not be assessed without a robust system for accounting GhG emissions. Therefore, successive international climate agreements have called upon parties to account for their GhG emissions in a reliable way, making use of harmonized methodologies that allow for comparison between parties. In particular, the UNFCCC requires every party to

[d]evelop, periodically update, publish and make available to the Conference of the Parties ... national inventories of anthropogenic emissions by sources and removals by sinks of all greenhouse gases not controlled by the Montreal Protocol.[60]

This is one of the few aspects of climate change mitigation where the obligations of developed States do not substantially differ from those of developing States, although the former are required to "take into account the best available scientific knowledge" when realizing their national inventories,[61] while the latter are eligible for international support.

A more rigorous accounting regime became necessary to allow for an assessment of the compliance of Annex I parties with their QELRCs under the Kyoto Protocol. In this agreement, developed parties committed to "a national system for the estimation of anthropogenic emissions by sources and removals by sinks of all greenhouse gases not controlled by the Montreal Protocol,"[62] in accordance with guidelines adopted by the Meeting of the Parties to the Protocol.[63] Developed States must also assess their base-year emissions[64] and monitor compliance with their QELRCs.[65]

---

[56] Paris Agreement, *supra* note 10, art. 4.4.
[57] UNFCCC Secretariat, *Aggregate Effect of the Intended Nationally Determined Contributions: An Update*, doc. FCCC/CP/2016/2 (May 2, 2016), para. 8.
[58] See Paris Agreement, *supra* note 10, art. 4.2. See also Susan Biniaz and Daniel Bodansky, "*Legal issues related to the Paris Agreement*" (C2ES, May 2017).
[59] UNFCCC Secretariat, *supra* note 57, at para. 5. See also UNEP, *The Emissions Gap Report 2017: A UN Environment Synthesis Report* (November 2017).
[60] UNFCCC, *supra* note 10, art. 4.1(a). See also Kyoto Protocol, *supra* note 10, art. 10(a).
[61] UNFCCC, *supra* note 10, art. 4.2(c).
[62] Kyoto Protocol, *supra* note 10, art. 5.1.
[63] See decision 19/CMP.1 "Guidelines for national systems under Article 5, paragraph 1, of the Kyoto Protocol" (November 30, 2005); decision 20/CMP.1 "Good practice guidance and adjustments under Article 5, paragraph 2, of the Kyoto Protocol" (November 30, 2005); and decision 21/CMP.1 "Issues relating to adjustments under Article 5, paragraph 2, of the Kyoto Protocol" (November 30, 2005).
[64] Kyoto Protocol, *supra* note 10, art. 3.4.
[65] *Ibid.*, art. 7.1.

The Paris Agreement extends rigorous accounting commitments to all States. It requires every party to provide an inventory report "regularly."[66] The decision of the parties on the adoption of the Paris Agreement added that "regularly" means, "as appropriate, no less frequently than on a biennial basis," except for the least developed country parties and small island development States, which may submit their inventories at their discretion.[67] This could naturally be changed – and a higher frequency imposed – by a subsequent decision of the Meeting of the Parties to the Paris Agreement.

The characterization of these obligations has led to the development of technical guidance on measurement, reporting and verification (MRV), some of the most technically challenging aspects of international law on climate change, in an attempt at promoting reliable and comparable national data reporting. In this regard, COP decisions have largely relied on the guidance and good practices compiled in multi-volume reports of the IPCC.[68] Unfortunately, despite such efforts, national data on GhG emissions remain largely incomplete and often unreliable, especially in emerging economies and other developing States, and more particularly in relation to emissions from land use, land-use change and forestry.[69]

3. Commitments to Promoting and Facilitating Further Action

In addition to their own emission limitation and reduction commitments, the parties to successive international climate agreements have also committed to promoting and facilitating further action on climate change mitigation. In particular, they committed to promoting further mitigation action in three ways: beyond their current emission limitation and reduction commitments through further negotiations; beyond their territory by providing support to other countries; and beyond the sectors covered by international climate agreements through cooperation in other forums.

Firstly, States have committed to continuing to negotiate further commitments through ongoing climate change negotiations. Arrangements have been made in all international climate agreements for States to pursue further negotiations. The UN Framework Convention on Climate Change called for the COP, at its first session, to review the adequacy of its provisions on mitigation and to take "appropriate action, which may include the adoption of amendments to the[se] commitments."[70] This provision led to the adoption of the Kyoto Protocol, which provided for a "first"[71] commitment period and called upon the Meeting of the

---

[66] Paris Agreement, *supra* note 10, art. 13.7 (a).

[67] Decision 1/CP.21, *supra* note 10, para. 91.

[68] See e.g. decisions 24/CP.19, "Revision of the UNFCCC reporting guidelines on annual inventories for Parties included in Annex I to the Convention" (November 22, 2013); and 3/CMP.11, "Implications of the implementation of decisions 2/CMP.7 to 4/CMP.7 and 1/CMP.8 on the previous decisions on methodological issues related to the Kyoto Protocol, including those relating to Articles 5, 7 and 8 of the Kyoto Protocol, part I: implications related to accounting and reporting and other related issues" (December 10, 2015). See also Simon Eggleston *et al.*, *2006 IPCC Guidelines for National Greenhouse Gas Inventories* (IGES, 2006).

[69] See e.g. P. Ciais *et al.*, "Carbon and other biogeochemical cycles" in T.F. Stocker *et al.* (eds.), *Climate Change 2013: The Physical Science Basis. Contribution of Working Group I to the Fifth Assessment Report of the Intergovernmental Panel on Climate Change* (Cambridge University Press, 2007) 465, at 489 (section 6.3.2.1); H. Kunreuther *et al.* "Integrated risk and uncertainty assessment of climate change response policies" in Edenhofer *et al.* (eds.), *supra note 1,* 151, at 182–183 (section 2.6.4.3).

[70] Convention on the International Maritime Organization, March 6, 1948, 289 *UNTS* 48, art. 4.2(d).

[71] Kyoto Protocol, *supra* note 10, art. 3.7.

Parties to establish "commitments for subsequent periods."[72] The decision which adopted the Doha Amendment, in accordance with this provision, calls on each developed party to "revisit its quantified emission limitation and reduction commitment for the second commitment period at the latest by 2014" with a view to increasing its ambition.[73] The Cancún Agreements "urge[d] developed country Parties to increase the ambition"[74] while also "invit[ing]" mitigation commitments from developing countries.[75]

Rather than through a succession of formal agreements, the Paris Agreement encourages further developments through unilateral declarations by the parties. Each party to the Paris Agreement must communicate "successive nationally determined contributions" which "represent a progression beyond the Party's then current nationally determined contribution."[76] This provision creates a ratchet effect which will, if effective, tie States to always more ambitious mitigation commitments. Thus, it would surely go against the object and purpose of the Paris Agreement for a party to review its NDC by downgrading its ambition, something briefly contemplated by US President Trump.[77] Beyond this ratchet effect, the parties to the Paris Agreement have also promised that their successive NDCs would reflect their "highest possible ambition."[78] From these provisions arises an obligation for the parties to negotiate further emission limitation and reduction commitments in good faith. Interestingly, even when US President Trump announced his intention to pull the United States out of the Paris Agreement, he implicitly recognized a duty to negotiate an alternative "deal."[79] This pledge to further negotiations relates closely to the principle of cooperation in a spirit of global partnership, as recognized among others in the Rio Declaration on Environment and Development.[80]

Secondly, successive international climate agreements require their parties to facilitate mitigation action carried out by other parties, through financial support, transfer of technology and capacity building. In the UNFCCC, Annex II parties (developed States except for former soviet States) committed to "provide new and additional financial resources to meet the agreed full costs incurred by developing country Parties" in accounting for their GhG emissions.[81] Furthermore, Annex II parties committed to "provide such financial resources, including for the transfer of technology, needed by the developing country Parties to meet the agreed full incremental costs" of measures that these countries would adopt to limit or reduce GhG emissions.[82] Countries other than Annex II parties became increasingly committed to providing support to

---

[72] *Ibid.*, art. 3.9.

[73] Decision 1/CMP.8, *supra* note 20, para. 7. Only three States have reviewed their commitment: Japan, Kazakhstan and New Zealand. See UNFCCC, *Compilation of Economy-Wide Emission Reduction Targets to Be Implemented by Parties Included in Annex I to the Convention* (May 9, 2014), doc. FCCC/SBST A/2014/INF.6, para. 5.

[74] Decision 1/CP.16, *supra* note 15, para. 37.

[75] *Ibid.*, para. 50.

[76] Paris Agreement, *supra* note 10, art. 4.3.

[77] See Lavanya Rajamani, "The US and the Paris Agreement: in or out and at what cost?" *EJIL: Talk!* (May 10, 2017), www.ejiltalk.org/the-us-and-the-paris-agreement-in-or-out-and-at-what-cost (accessed January 3, 2018). See *contra* Biniaz and Bodansky, *supra* note 58.

[78] Paris Agreement, *supra* note 10, art. 4.3.

[79] See Mythili Sampathkumar and Alexandra Wilts, "Donald Trump confirms withdrawal from Paris Agreement on climate change in huge blow for global deal" *The Independent* (June 1, 2017).

[80] UNCED, Rio Declaration on Environment and Development (June 3–14, 1992), available in (1992) 31 *ILM* 874 (hereinafter Rio Declaration), principle 27. See also Chapter 5, section II.A.

[81] UNFCCC, *supra* note 10, art. 4.3.

[82] *Ibid.*

mitigation action, including through "South-South cooperation." In the Paris Agreement, the provision of support was no longer limited to developed States. Rather, the Paris Agreement noted, in the passive form, that "[s]upport shall be provided to developing country Parties for the implementation" of action on climate change mitigation.[83]

Thirdly, States have also committed to pursuing negotiations in third forums. In particular, by ratifying the Kyoto Protocol, Annex I parties pledged to "pursue limitation or reduction of emissions of greenhouse gases ... from aviation and marine bunker fuels, working through the International Civil Aviation Organization and the International Maritime Organization, respectively."[84] As such activities extend beyond international borders, their attribution to a particular State is problematic from the perspective of QELRC made by individual States; it simply appeared more convenient to call on developed States to seek a negotiated solution within specialized forums. Likewise, efforts were made to explore synergies between negotiations on climate change and negotiations under the Montreal Protocol on Substances that Deplete the Ozone Layer.[85] Cooperation with all relevant actors, including international organizations, infra-national authorities, non-governmental organizations and for-profit organizations, has also been encouraged as a way of galvanizing international cooperation beyond the UNFCCC regime.[86]

III. ACTION ON CLIMATE CHANGE MITIGATION UNDER OTHER TREATY REGIMES

Efforts to promote climate change mitigation have not been limited to the UNFCCC regime. From the Vienna Convention on the Protection of the Ozone Layer and the Montreal Protocol on Substances that Deplete the Ozone Layer all the way to the ICAO and the IMO, diverse legal regimes promoted international cooperation on climate change mitigation with regard either to particular gases or to a particular economic sector. The Kyoto Protocol directed cooperation under the ICAO and the IMO.[87] These and other developments were also spurred by the momentum created under the UNFCCC regime – thus, it is no coincidence that important developments took place in 2010 and 2016, in the months following the very public Copenhagen and Paris Summits.

A. Action Specific to Particular Gases

The Vienna Convention for the Protection of the Ozone Layer and the Montreal Protocol on Substances that Deplete the Ozone Layer pursue an objective distinct from climate change mitigation.[88] They seek "to protect human health and the environment against adverse effects

---

[83] Paris Agreement, *supra* note 10, art. 4.5. See also *ibid.*, art. 9.2. International support to climate action is further discussed in Chapter 12.

[84] Kyoto Protocol, *supra* note 10, art. 2.2.

[85] See decision 12/CP.8, "Relationship between efforts to protect the stratospheric ozone layer and efforts to safeguard the global climate system: issues relating to hydrofluorocarbons and perfluorocarbons" (November 1, 2002).

[86] See e.g. decision 1/CP.16, *supra* note 15, paras. 121(f), 125; decision 1/CP.21, *supra* note 10, recital 16 and para. 74(a); Paris Agreement, *supra* note 10, recital 15 and arts. 10.2, 11.3 and 12.

[87] See Kyoto Protocol, *supra* note 10, art. 2.2.

[88] See Chapter 4, section I.

resulting or likely to result from human activities which modify or are likely to modify the ozone layer."[89] Nevertheless, because many ozone-depleting substances are potent GhGs, the efforts which have been carried out under the ozone regime have had significant co-benefits for climate change mitigation and thus have significantly advanced the objective of the UNFCCC.[90] The parties to the Montreal Protocol put it on record that they were "[c]onscious of the potential climatic effects of emissions of these substances."[91]

Like the UNFCCC would later do,[92] the Vienna Convention on the Protection of the Ozone Layer called on parties to "take appropriate measures" to ensure its implementation. Building on this, the Montreal Protocol on Substances that Deplete the Ozone Layer was adopted to define more specific national commitments. Each of its parties committed to ensure that its calculated level of consumption of listed substances "does not exceed" particular amounts calculated on the basis of its historical emissions,[93] with some specific provisions having been adopted to take into consideration the special situation of developing countries with low per capita emissions.[94]

The parties to the Montreal Protocol recognized the need to adopt particular measures to address the growing emissions of HFCs, powerful GhGs which are being used in place of ozone-depleting substances, even though they do not deplete the ozone layer. In 2010, 108 of the parties to the Montreal Protocol on Substances that Deplete the Ozone Layer signed a Declaration on the Global Transition away from Hydrochlorofluorocarbons (HCFCs) and Chlorofluorocarbons (CFCs). In this declaration, they expressed a common "intent to pursue further action under the Montreal Protocol aimed at transitioning the world to environmentally sound alternatives to HCFCs and CFCs."[95] This led, six years later, to the adoption of the Kigali Amendment to the Montreal Protocol to phase out the consumption and production of HFCs.[96] Since HFCs are not ozone-depleting substances, the Kigali Amendment does not directly advance the objective of the ozone regime. Rather, this development within the ozone regime is intended to promote climate change mitigation and thus to further the objective of the UNFCCC.

National commitments under the ozone regime overlap with some, but not all, of the national commitments under the UNFCCC regime. A distinction needs to be made between the national commitments under the UNFCCC, the Kyoto Protocol and the Paris Agreement:

- Any overlap is excluded under the UNFCCC, where national commitments on climate change mitigation are confined to "greenhouse gases not controlled by the Montreal Protocol."[97] This formulation has the merit to allow a dynamic interpretation, whereby an extension of the

---

[89] Vienna Convention for the Protection of the Ozone Layer, March 22, 1985, 1513 *UNTS* 293, art. 2.1.
[90] UNFCCC, *supra* note 10, art. 2.
[91] Montreal Protocol on Substances that Deplete the Ozone Layer, September 16, 1987, 1522 *UNTS* 3 (hereinafter Montreal Protocol), recital 5.
[92] See above, section II.B.
[93] Montreal Protocol, *supra* note 91, art. 2, paras. 1, 2, 3 and 4.
[94] *Ibid.*, art. 5.
[95] Declaration on the Global Transition away from Hydrochlorofluorocarbons (HCFCs) and Chlorofluorocarbons (CFCs), adopted by 108 parties at and around the Twenty-Second Meeting of the Parties (November 8–12, 2010), para. 6.
[96] Amendment to the Montreal Protocol on Substances that Deplete the Ozone Layer, adopted in Kigali on October 15, 2016, reproduced in (2017) 56 *ILM* 196 (hereinafter Kigali Amendment).
[97] UNFCCC, *supra* note 10, arts. 4.1, 4.2 and *passim.*

scope of the Montreal Protocol through the Kigali Amendment automatically implies a limitation on the scope of the UNFCCC.[98]

- Overlaps will occur following the entry into force of the Kigali Amendment to the Montreal Protocol on January 1, 2019 and pursuant to the entry into force of the Doha Amendment of the Kyoto Protocol (or inasmuch as States decide to implement their obligations under the Doha Amendment prior to its entry into force). This is because national commitments under the Kyoto Protocol extend to a list of GhGs included in Annex A, which excluded the substances controlled by the Montreal Protocol and its earliest amendments, but included HFCs. Thus, emissions of HFCs will concurrently be controlled under two different treaty regimes. As these treaties define obligations to reduce emissions but no right to emit, no inconsistency is to be anticipated.
- Overlaps could appear between the NDCs communicated by the parties to the Paris Agreement and their obligation under the Montreal Protocol, possibly even prior to the entry into force of the Kigali Amendment. The Paris Agreement contains no list of gases and, although it aims to "enhance the implementation of the Convention,"[99] it is not clearly limited by the scope of national commitments made under the UNFCCC. The INDCs of several States cover HFCs.[100]

Such overlaps are not particularly problematic; they have been accepted and are even intentional. The parties to the Montreal Protocol clearly stated that the Kigali Amendment "is not intended to have the effect of excepting [HFCs] from the scope" of the UNFCCC regime.[101] HFCs do not represent a sufficiently large proportion of GhG emissions for such overlap to significantly distort the implementation of more general agreements reached under the UNFCCC regime.

B. Action Specific to Particular Sectors

Action on climate change mitigation has also taken place in a range of other regimes. Two of the most advanced sectorial processes of cooperation on climate change mitigation regard aspects of international transportation which were not included within the scope of the Kyoto Protocol: international aviation and maritime transportation.

The 1973 International Convention for the Prevention of Pollution from Ships (MARPOL) was adopted in order "to achieve the complete elimination of intentional pollution of the marine environment by oil and other harmful substances and the minimization of accidental discharge of such substances."[102] An amendment was adopted in 1997 to regulate emissions of air pollutants, including nitrogen oxides, some of which are GhGs. A further amendment, in 2011, introduced regulations on the intensity of GhG emissions from ships.[103] Negotiations were still ongoing as of early 2018 regarding possible further steps, including market-based measures.[104]

---

[98] This limitation on the scope of the UNFCCC is arguably contingent upon the entry into force of the relevant amendment to the Montreal Protocol and limited to the parties to this amendment.

[99] Paris Agreement, *supra* note 10, art. 2.1.

[100] UNFCCC Secretariat, *supra* note 57, para. 16.

[101] Kigali Amendment, *supra* note 96, art. III.

[102] 1973 International Convention for the Prevention of Pollution from Ships, November 2, 1973, 1340 *UNTS* 184, recital 4.

[103] MARPOL resolution MEPC.203(62), "Amendments to the Annex of the Protocol of 1997 to Amend the International Convention for the Prevention of Pollution from Ships, 1973, as Modified by the Protocol of 1978 Relating Thereto" (July 15, 2011), in IMO document MEPC 62/24/Add.1.

[104] See Chapter 4, section II.A.

The Assembly of the ICAO adopted in 2010 "a collective medium-term global aspirational goal of keeping the global net carbon emissions from international aviation from 2020 at the same level."[105] It made clear, however, that this goal was adopted "without any attribution of specific obligation to individual States."[106] Three years later, in September 2016, the Assembly adopted concrete provisions for an international market-based mechanism applicable to GhG emissions from international civil aviation – the Carbon Offsetting and Reduction Scheme for International Aviation (CORSIA).[107] In contrast to other efforts to promote climate change mitigation, this international market-based mechanism could involve a more central role for international institutions, although implementation will remain the responsibility of States. As CORSIA relies on offsetting emissions through funding mitigation projects in other sectors, it does not guarantee net mitigation outcomes in the aviation sector. Accordingly, the "aspirational global fuel efficiency improvement rate of 2 per cent per annum from 2021 to 2050"[108] adopted by the ICAO Assembly in 2010 will need to be implemented either through the modalities of CORSIA or through additional national commitments. Lastly, no plan has yet been announced to implement the commitment of institutions representative of the international air transport industry to reduce the carbon emissions of air transport by 50 percent by 2050 compared to the 2005 level.[109]

Diverse other initiatives may be noted. Some seek to complement mitigation action through other actions with mitigation co-benefits. Thus, discussions were held by the COP to the "Ramsar" Convention on Wetlands of International Importance Especially as Waterfowl Habitat of 1971 (COP/Ramsar).[110] In 2008, the COP/RAMSAR agreed "to take urgent action" to protect wetlands as greenhouse sinks and "to encourage expansion of demonstration sites on peatland restoration and wise use management in relation to climate change mitigation."[111] In 2015, the COP/RAMSAR further agreed "to consider limiting activities that ... may cause ... the emission of greenhouse gases."[112]

Other developments sought to spur action within the UNFCCC through pioneering initiatives. Such was the case of the non-binding Agreement on Financing and Quick-Start Measures to Protect Rainforests that 50 States joined in 2010 to spur development of action on forests within the UNFCCC regime.[113] Likewise, the members of the G20 committed, through the "Leaders Statement" adopted at the 2009 Pittsburgh Summit, to "phase out and rationalize over the

---

[105] ICAO Assembly Resolution A37-19, "Consolidated statement of continuing ICAO policies and practices related to environmental protection – climate change" (September 28–October 8, 2010), para. 6.

[106] ICAO Assembly Resolution A38-18, "Consolidated statement of continuing ICAO policies and practices related to environmental protection – climate change" (September 24–October 4, 2013), para. 7.

[107] ICAO Assembly Resolution A39-3, "Consolidated statement of continuing ICAO policies and practices related to environmental protection – Global Market-Based Measure (MBM) scheme" (September 27–October 7, 2016), para. 5.

[108] ICAO Assembly Resolution A37-19, *supra* note 105, para. 4.

[109] *Ibid.*, para. 22.

[110] Convention on Wetlands of International Importance Especially as Waterfowl Habitat, February 2, 1971, 996 *UNTS* 246 (hereinafter Ramsar Convention, from the name of the town where the Convention was adopted).

[111] Conference of the Parties to the Ramsar Convention, Resolution X.24, "Climate change and wetlands" (October 28–November 4, 2008), para. 32. See also Conference of the Parties to the Ramsar Convention, Resolution VIII.3, "Climate change and wetlands: impacts, adaptation, and mitigation" (November 18–26, 2002), para. 15.

[112] Conference of the Parties to the Ramsar Convention, Resolution XII.11, "Peatlands, climate change and wise use: Implications for the Ramsar Convention" (June 1–9, 2015), para. 21.

[113] Government of Norway, Agreement on Financing and Quick-Start Measures to Protect Rainforest (May 27, 2010), www.regjeringen.no/en/aktuelt/Agreement-on-financing-and-quick-start-measures-to-protect-rainforest/id605756 (accessed January 3, 2018).

medium term inefficient fossil fuel subsidies"[114] – an objective whose adoption within the UNFCCC regime continues to face significant resistance from a number of developing States.

Some other initiatives, discussed more thoroughly in Chapter 15, rely in part or in whole on non-State actors for their implementation. One example out of many, the "Zero Routine Flaring by 2030" initiative of the World Bank, aims to eliminate flaring, an important source of GhG emissions and air pollution in the process of extracting and refining oil and gas.[115] As of early 2018, the objective was endorsed by 26 national governments and one subnational government (California), 31 oil companies and 15 development agencies.[116] Non-State actors have developed multiple frameworks, including standards and verification processes, to encourage voluntary commitments on climate change mitigation and publicly monitor their effective implementation.

IV. AN OVERVIEW OF IMPLEMENTATION MEASURES

Although this textbook focuses on the international law on climate change, it may be useful to provide a general overview of the measures that States and the European Union have adopted to fulfill their international obligations regarding climate change mitigation. These measures of implementation seek either to reduce sources of GhG through gains in efficiency and changes in modes of consumption or to enhance sinks and reservoirs of GhG such as forests and soils.[117] In addition to "command and control" measures that impose standards or prohibit certain courses of action, governments have sought to provide economic incentives, for instance, through the creation of a market-based mechanism or carbon tax. States have also sought to exercise leadership through, among other things, planning, investment, ownership, research and education.

The international law on climate change mitigation defines national obligations, but there are few rules on the particular measures that a State must take to implement these obligations. The UNFCCC recognizes "the principle of sovereignty of States in international cooperation to address climate change."[118] Accordingly, it is down to each State to determine what measures of implementation are preferable, on its own analysis, in order to mitigate climate change without disproportionately affecting other national objectives such as economic development and the welfare and rights of its population. Nonetheless, the Kyoto Protocol went some way toward promoting particular ways of mitigating climate change, encouraging market-based approaches and strategies seeking to maximize economic efficiency rather than, say, social equity. This was possible because the emission limitation and reduction commitments were only imposed on Annex I parties with relatively similar circumstances (e.g. generally well-functioning governments). Extending firm mitigation commitments on all parties, the Paris Agreement encourages its parties to envisage more diverse mitigation actions, including market and non-market approaches.[119]

---

[114] G20 Leaders' Statement: The Pittsburgh Summit (September 24–25, 2009), para. 24.

[115] Flaring and fugitive emissions represent 6 percent of global GhG emissions, according to David G. Victor *et al.*, "Introductory chapter" in Edenhofer *et al.* (eds.), *supra note 1*, 111, at 123.

[116] The World Bank, "Zero routine flaring by 2030" (January 31, 2017), www.worldbank.org/en/programs/zero-routine-flaring-by-2030#1 (accessed January 3, 2018).

[117] Carbon capture and storage techniques, which consist in artificial sinks, are discussed in Chapter 9.

[118] UNFCCC, *supra* note 10, recital 10.

[119] See Paris Agreement, *supra* note 10, art. 6.

National measures of implementation can be anchored in domestic laws in multiple ways. Some national constitutions include provisions on the promotion of sustainable development,[120] the protection of the environment[121] or even the mitigation of climate change.[122] Beyond their symbolic importance, such provisions may offer a basis for litigation. Yet, more specific measures are necessary, including statutes, regulations and policy documents. Mitigation strategies typically involve virtually every sector of a modern economy, including power generation, transportation, building, industries, agriculture and human settlements. These measures are developed and implemented by governmental institutions that are often not used to working together, creating issues of coordination. Some national mitigation strategies include market-based mechanisms of considerable complexity.

A. Command and Control Regulation

The first and most obvious method for a government to implement its international obligations on climate change mitigation is to impose obligations on legal persons under its jurisdiction, if necessary, with the help of administrative and criminal sanctions. Some regulations are adopted with the aim of mitigating climate change; others, which pursue objectives such as combating air quality or ensuring energy security, incidentally contribute to climate change mitigation. Many States, especially developed States, have imposed detailed regulations on sectors that emit large amounts of GhGs and other pollutants. These regulations typically consist of emission standards, technology standards or product standards.[123]

Emission standards relate to the quantity of pollutant discharged in the environment by a given person over a given period. This quantity can be fixed or can depend on other variables, such as production levels. Performance standards are a subcategory of emission standards which are expressed in terms of emission intensity – the level of emissions divided by a unit of production, for instance, the power output of a power plant. Separate emission standards can be imposed on one or several GhGs, or one emission standard can be imposed on a basket of GhGs with weighting factors. The implementation of emission standards requires provisions for measuring, reporting and verification, which often rely on self-reporting and external verification by accredited agencies. As such, emission standards are typically limited to large, stationary sources of pollutants which can more easily be controlled.

Technology standards impose the use of specified technology in a production process. For instance, a technology standard can impose the use of specified equipment in a coal plant in order to promote better environmental performance. Technology standards may also impose specified methodologies, for instance, agricultural practices that maximize carbon intakes in soils.[124] Although enforcement may be easier than for emission standards, technology standards

---

[120] Charter for the Environment (June 24, 2004), in Constitution of France (1958), art. 6.

[121] Constitution of India (November 26, 1949); Constitution of Fiji (August 22, 2013), art. 40.

[122] Constitution of Tunisia (June 26, 2014), art. 45; Constitution of Ecuador (September 28, 2008), art. 414.

[123] See e.g. E. Somanathan *et al.*, "National and Sub-national Policies and Institutions" in Edenhofer *et al.* (eds.), *supra* note 1, 1141, at 1155; and S. Gupta *et al.*, "Policies, instruments and co-operative agreements" in B. Metz *et al.* (eds.), *Climate Change 2007: Mitigation. Contribution of Working Group III to the Fourth Assessment Report of the Intergovernmental Panel on Climate Change* (Cambridge University Press, 2007) 745, at 753–755.

[124] See Jonathan Verschuuren, "Toward a regulatory design for reducing emissions from agriculture: lessons from Australia's carbon farming initiative" (2014) 7:1 *Climate Law* 1.

are specific to the activities they regulate and they require frequent and complex updates, particularly in sectors where technology is rapidly improving. They leave comparatively little room for spontaneous innovation.

Product standards impose specific characteristics to what is produced or sold on a market. Product standards can impose performance levels, for instance, power-efficiency standards applicable to new electrical appliances, fuel-efficiency standards applicable to cars[125] or thermic insulation standards applicable to new buildings,[126] in order to reduce the quantity of energy and GhG emissions that the product will cause during its lifetime. They can also impose production methods, such as a specified proportion of biofuels (fuels produced from plants and trees rather than from the extraction of fossil fuels) in diesel fuel.[127] Product standards are more readily applied to scattered sources of GhG emissions by end-users, where it would be challenging to monitor emissions or control processes. These standards can be mandatory or optional, as in the case of energy efficiency labeling, adopted in numerous States in order to "enable … consumers to make informed decisions by providing information about the energy consumption of an appliance."[128]

B. Price-Based Mechanisms

Price-based mitigation incentives seek to put a price on the emission of GhGs rather than prohibiting it. Such mechanisms relate to the understanding of climate change as a market failure. GhG emissions are, in economic parlance, "negative externalities" – a "cost" suffered by individuals other than the producer and the consumer, and hence not properly accounted for in the rational economic decisions made by the producer and the consumer. Price-based mitigation mechanisms seek to "internalize" these "negative externalities" by requiring the producer or (directly) the consumer to pay a cost equivalent to the value of the harm because of GhG emissions, thus reflecting the "polluter pays principle."[129]

Yet, as discussed in Chapter 2, there can be no objective economic valuation of climate change, in particular because assessing the "cost" of climate change would require us to put a price tag on things which are not on the market or to assess the value of harm which is to unfold over centuries or millennia.[130] For the same reasons, it is impossible to determine the objective value of the "negative externality" associated with the emission of a unit of GhG. If a price is given to GhG emissions, this is inevitably based on value-based judgments on how much we care for, among other things, human lives and societies, biodiversity and future generations. In practice, the price that municipal laws give to GhG emissions often relates to

---

[125] See, for instance, US Code Chapter 85 (1963), Title 42, *Air Pollution Prevention and Control*, Subchapter II, "Emission Standards for Moving Sources."

[126] E.g. US Energy Policy Act, August 8, 2005, Pub. L. 109–158; US Energy Independence and Security Act, December 19, 2007, Pub. L. 110–140; India Energy Conservation Building Code (May 2007).

[127] E.g. India, *INDC* (October 1, 2015), at 16; Brazil, *INDC* (September 28, 2015), at 3.

[128] EU Directive 2010/30/EU on the indication by labelling and standard product information of the consumption of energy and other resources by energy-related products (May 19, 2010), doc. 32010L0030. See also India, *INDC* (October 1, 2015), at 11.

[129] See Chapter 5, section II.C.

[130] See Chapter 2, section II. See also Endre Tvinnereim and Michael Mehling, "Carbon pricing and the 1.5°C target: near-term decarbonisation and the importance of an instrument mix" *Carbon and Climate Law Review* (forthcoming).

the political acceptability of a given level of economic constraint rather than to a genuine attempt to value the impact of these emissions.

The most direct way to impose a price on GhG emissions is by requiring economic actors to pay a tax in proportion to their emissions. Sweden introduced a GhG tax as early as 1991 and increased its amount from 29 EUR/tCO$_2$eq to 137 EUR/tCO$_2$eq.[131] India started levying taxes on GhG emissions in 2010.[132] A tax of 25.40 AUD/tCO$_2$eq was imposed in Australia in 2012, but repealed in 2014 after a change in government.[133] Chile announced in 2014 that it would impose a tax of 5 USD/tCO$_2$eq from 2018 onward.[134] Portugal established a tax equivalent to 7 USD/tCO$_2$eq in 2015.[135] Likewise, Canada announced that a tax of 10 CAD/tCO$_2$eq would be imposed on GhG emissions in eight of its ten provinces from 2018 onward.[136]

Typically, these direct GhG taxes only apply to large emitters, namely power plants and industrial facilities, whose activity can be monitored, reported and imposed without excessive difficulty. Taxes can also be imposed on the production or purchase of fossil fuels, thus creating a burden on any economic agent using fossil fuels throughout the economic chain. Some specific taxes, such as a tax on the registration of cars (Singapore), may also impose economic disincentives to some activities causing significant GhG emissions.[137]

Taxes, however, are unpopular.[138] In political debates, economic interests are more strongly represented than the climate system, allowing much more weight to be given to the alleged economic impact of GhG taxes than to their environmental benefit. When asked to balance climate change mitigation and immediate economic growth, most voters prefer the latter. In principle, the introduction of a greenhouse tax creates a source of income for the State which would replace other sources of income, thus taking the place of other taxes.[139] A greenhouse tax can be the opportunity to discontinue subsidies to fossil fuels which, in many countries, continue to provide the wrong incentive to many economic actors. Alternatively, the income levied by a greenhouse tax can also be used to promote climate change mitigation, for instance, through the production of renewable energy, whether through direct subsidies, preferential taxation policies[140] or financial support.[141]

[131] "When it comes to emissions, Sweden has its cake and eats it too" *World Bank* (May 16, 2016), www.worldbank.org/en/news/feature/2016/05/16/when-it-comes-to-emissions-sweden-has-its-cake-and-eats-it-too (accessed January 3, 2018).
[132] India, Finance Act (2010), section 83. See generally Ipshita Chaturvedi, "The 'carbon tax package': an appraisal of its efficiency in India's clean energy future" (2016) 10:4 *Carbon & Climate Law Review* 194.
[133] See Lenore Taylor, "Australia kills off carbon tax," *The Guardian* (July 17, 2014).
[134] Kate Galbraith, "Climate change concerns push Chile to forefront of carbon tax movement" *New York Times* (October 29, 2014).
[135] World Bank, *State and Trends of Carbon Pricing 2016* (Washington, D.C., October 2016) 27.
[136] Reuters in Ottawa, "Canada sets its first national carbon price at C$10 a ton" *The Guardian* (December 10, 2016). For a review of fiscal measures promised in NDCs, see Anika Terton et al., "Fiscal instruments in INDCs: how countries are looking to fiscal policies to support INDC implementation" (IISD and Global Subsidies Initiative, December 2015).
[137] See "Tax structure for cars" on the website of the Land Transport Authority of Singapore, at www.lta.gov.sg/content/ltaweb/en/roads-and-motoring/owning-a-vehicle/costs-of-owning-a-vehicle/tax-structure-for-cars.html (accessed January 3, 2018).
[138] See Justin Gundlach, *To Negotiate a Carbon Tax: A Rough Map of Policy Interactions, Tradeoffs, and Risks* (Sabin Center for Climate Change Law, June 2017).
[139] See, for instance, Government of British Columbia, *June Update: Budget and Fiscal Plan 2013/14–2015/16* (February 19, 2013), at 64, citing a principle of "revenue neutrality" for the introduction of a carbon tax.
[140] China, *INDC* (June 30, 2015), at 14.
[141] Such as the Partial Risk Guarantee Fund for Energy Efficiency and the Venture Capital Fund for Energy Efficiency in India, as described in India's *INDC* (June 30, 2015), at 11.

Rather than taxes, some States have sought to impose a price on GhG emissions through market-based mechanisms, also called cap-and-trade mechanisms or emission trading schemes. Many States indicated either the existence, or their desire to develop a market-based mechanism in their INDCs.[142] In particular, the European Union has implemented a market-based mechanism since 2005;[143] China, after testing seven local pilot projects, established a nationwide market-based mechanism in the power sector at the end of 2017.[144] A market-based mechanism requires economic actors to acquire emission allowances for each unit of GhG emissions caused by their activities within a certain period of implementation. Emission allowances are issued by the national authorities. While taxes impose a fixed price for emission allowances, market-based mechanisms let demand contribute to the determination of the price of these allowances.

In a market-based mechanism, emission allowances can typically be acquired in three different ways.

1. Some emission allowances are distributed for free to economic actors based on levels of emissions which are deemed unavoidable. Since individual quotas are often determined based on their emissions in previous years (as well as on a prediction of sectorial trends), this initial distribution is called "grandfathering." Provisions are typically made to ensure that grandfathering does not distort competition, including for new entrants.

2. National authorities may also auction some allowances on a primary market for emission allowance. While a new market-based mechanism relies mostly on grandfathering, auctioning typically assumes an increasing importance over time. Unlike grandfathering, auctioning allowances generates some income for national authorities, rather like a tax system.

3. Economic actors can buy and sell emission allowances on secondary markets. Initial auctioning and further trade on secondary markets ensures that economic actors gain from reducing their GhG emissions and need to pay to maintain or increase their GhG emissions, thus creating an economic incentive for climate change mitigation.

The price of emission allowances on primary or secondary markets is determined by supply and demand. Supply consists in the number of emission allowances issued by national authorities. By reducing the number of allowances, a government can increase the pressure on economic actors; by issuing more allowances, it allows the price to reduce. Unless the government intervenes, the price of emission allowances within an implementation period is determined by demand. Technological innovation or slower economic growth may thus lead to a decrease in the price of emission allowances. During the first phase of the EU Emission Trading Scheme from 2005 to 2007, the price of emission allowances varied from over 30 EUR/tCO_2eq before collapsing to just a few cents.[145]

---

[142] UNFCCC Secretariat, *supra* note 57, para. 162. See also World Bank, *State and Trends of Carbon Pricing 2016* (October 2016).

[143] EU Directive 2003/87/EC establishing a scheme for GhG emission allowance trading within the Community (October 13, 2003), doc. 32003L0087.

[144] See e.g. U.S.-China Joint Announcement on Climate Change, November 12, 2014; U.S.-China Joint Presidential Statement on Climate Change, September 25, 2015. See generally Shen Ying, "Crossing the river by groping for stones: China's pilot emissions trading schemes and the challenges for a national scheme" (2016) 18 *Asia Pacific Journal of Environmental Law* 1; Anatole Boute, "The impossible transplant of the EU emissions trading scheme: the challenge of energy market regulation" (2017) 6:1 *Transnational Environmental Law* 59; Duan Maosheng, "From carbon emissions trading pilots to national system: the road map for China" (2015) 9:3 *Carbon & Climate Law Review* 231; Zhang Hao, "Designing the regulatory framework of an emissions trading programme in China: lessons from Tianjin" (2012) 6:2 *Carbon & Climate Law Review* 329.

[145] Beat Hintermann, "Allowance price drivers in the first phase of the EU ETS" (2010) 59:1 *Journal of Environmental Economics & Management* 43.

Although a price-based incentive may incentivize climate change mitigation, it is certainly not a silver bullet.[146] Any price-based mechanism involves transaction costs borne by the government and by economic actors to measure, report and verify levels of GhG emissions.[147] These incentives for incremental changes are insufficient to trigger structural changes, for instance, in infrastructure, often determined by collective choices.[148] Overall, price-based mechanisms raise ethical questions about the rightness of selling and buying permits to harm the global environment – as if harming the environment could even be a "right" or an "entitlement."[149] Every society recognizes that some things cannot be traded – life or fundamental rights, for example.[150] And any price-based mechanism, by definition, affects the poorest the most.

C. Leadership

Governments do more than regulate; they may also play a leading economic, scientific and cultural role. This role presents multiple opportunities for governments to promote climate change mitigation.

States define collective development orientations, for instance, through land-use and development policies, or through investments in public infrastructure and energy production systems.[151] Such collective decisions create collective pathways which often lock a society into a development pathway for years or decades to come. Multiple relevant orientations can be taken, for instance, through national development strategies, city planning and land-use regulation (including forest management, among others).[152] Infrastructure investments should

---

[146] See generally Stefan E. Weishaar, *Emissions Trading Design: A Critical Overview* (Edward Elgar, 2014).

[147] Peter Heindl, "The impact of administrative transaction costs in the EU emissions trading system" (2017) 17:3 *Climate Policy* 314.

[148] David M. Driesen, "The limits of pricing carbon" (2014) 4:1–2 *Climate Law* 107.

[149] See e.g. Andrew Hedges, "Carbon units as property: guidance from analogous common law cases" (2016) 13:3 *Carbon & Climate Law Review* 190; Kelvin F.K. Low and Jolene Lin, "Carbon credits as EU like it: property, immunity, traciCO2medy? (2015) 27:3 *Journal of Environmental Law* 377; Hope Johnson et al., "Towards an international emissions trading scheme: legal specification of tradeable emissions entitlements" (2017) 34:1 *Environment & Planning Law Journal* 3; Sabina Manea, "Defining emissions entitlements in the constitution of the EU emissions trading system" (2012) 1:2 *Transnational Environmental Law* 303. Regarding a similar reasoning on the international plane, see *supra* note 45.

[150] See Robert E. Goodin, "Selling environmental indulgences" (1994) 47:4 *KYKLOS International Review for Social Sciences* 573, reproduced in Stephen M. Gardiner et al. (eds.), *Climate Ethics: Essential Readings* (Oxford University Press, 2010) 231; Gerd Winter, "The climate is no commodity: taking stock of the emissions trading system" (2010) 22:1 *Journal of Environmental Law* 1; Glen Lehman, "Environmental accounting: pollution permits or selling the environment" (1996) 7:6 *Critical Perspective on Accounting* 667. See also Douglas A. Kysar, "Global environmental constitutionalism: getting there from here" (2012) 1:1 *Transnational Environmental Law* 83; Edward A. Page, "Cashing in on climate change: political theory and global emissions trading" (2011) 14:2 *Critical Review of International Social & Political Philosophy* 259; Clive L. Spash, "The brave new world of carbon trading" (2010) 15:2 *New Political Economy* 169.

[151] See e.g. Israel Solorio and Helge Jörgens (eds.), *A Guide to EU Renewable Energy Policy: Comparing Europeanization and Domestic Policy Change in EU Member States* (Edward Elgar, 2017); Marjan Peeters and Thomas Schomerus, "Modifying our society with law: the case of EU renewable energy law" (2014) 4:1–2 *Climate Law* 131; Leonie Reins et al., "China's climate strategy and evolving energy mix: policies, strategies and challenges" (2015) 9:3 *Carbon & Climate Law Review* 256; and special issue in (2016) 6:3–4 *Climate Law* 1; Gary Bryner, with Robert J. Duffy, *Integrating Climate, Energy, and Air Pollution Policies* (MIT Press, 2012).

[152] See China, *INDC* (June 30, 2015), at 10; Brazil, *INDC* (September 28, 2015), at 3. See also Susanna Hecht and Alexander Cockburn, *The Fate of the Forest: Developers, Destroyers, and Defenders of the Amazon*, updated edn (University of Chicago Press, 2010).

promote more sustainable forms of transportation, for instance, through the development of railways and waterways in priority over roads and airports,[153] and through the development of adequate public transportation networks.[154] Local government initiatives can often go a long way towards this by promoting the development of low-carbon development areas instead of industrial areas.[155] Electricity transmission grids can be adapted to support the development of renewable energies.[156]

Leadership can be exercised through ownership policies. China, Russia, India, Saudi Arabia and many others have a strong share of ownership in some key sectors – the extraction, refinement and distribution of fossil fuels, the generation of electricity, the production of cement, public transportation, metallurgy, etc. A transnational advocacy campaign led by 350.org has pushed many investors, including public institutions, to divest from fossil fuel industries.[157] Short of divestment, State-owned enterprises could be used to demonstrate the ability of companies strongly involved in the GhG economy to reconvert and seize opportunities in the green economy – for instance, by converting an oil and gas conglomerate into a geothermal and geoengineering major. State-owned enterprises can thus be a channel for States to lead by example.[158]

Finally, States can generate change, over time, through scientific and cultural leadership. States often exercise some influence over school curriculums and the priorities of research institutions, museums, libraries and even the media. Including climate change in school syllabuses is a must;[159] museums and exhibitions on climate change can further promote public awareness.[160] Supporting research and innovation may promote technological progress and facilitate a transition to a greener economy. As China indicated in its INDC, a State could, through a mix of targeted interventions, advocate for a "green, low-carbon, healthy and civilized way of life and consumption patterns and to promote low-carbon consumption throughout society."[161] Overall, promoting education and awareness of climate change may contribute to a shift toward more "sustainable lifestyles" and more "sustainable patterns of consumption."[162]

---

[153] India, *INDC* (October 1, 2015), at 15.

[154] See *ibid.*, at 15; China, *INDC* (June 30, 2015), at 10; Brazil, *INDC* (September 28, 2015), at 4.

[155] See Dave Sawyer and Hubert Thieriot, *Policy Trends and Drivers of Low-Carbon Development in China's Industrial Zones* (IISD, March 2015).

[156] See e.g. India, *INDC* (October 1, 2015), at 9–10; China, *INDC* (June 30, 2015), at 7; Brian Scaccia, "California's Renewable Energy Transmission Initiative as a model for state renewable resource development and transmission planning" (2012) 3:1 *Climate Law* 25.

[157] See the website of the initiative, https://gofossilfree.org (accessed January 3, 2018). For a critical discussion, see Jeff Tollefson, "Fossil-fuel divestment campaign hits resistance" (2015) 521:7550 *Nature* 16.

[158] See Benoit Mayer and Mikko Rajavuori, "National fossil fuel companies and climate change mitigation under international law" (2016) 44 *Syracuse Journal of International Law and Commerce* 55; Benoit Mayer, Mikko Rajavuori and Fang Meng, "The contribution of state-owned enterprises to climate change mitigation in China" (2017) 7:2–3 *Climate Law* 97.

[159] See e.g. UNFCCC, *supra* note 10, art. 6; Kyoto Protocol, *supra* note 10, art. 10(e); Paris Agreement, *supra* note 10, art. 12.

[160] The first museum on climate change was opened on December 16, 2013 by the Hong Kong Jockey Club, hosted within the Chinese University of Hong Kong. See CUHK Jockey Club Initiative Gaia, www.gaia.cuhk.edu.hk/index.php/en/mocc (accessed January 3, 2018).

[161] China, *INDC* (June 30, 2015), at 4.

[162] Paris Agreement, *supra* note 10, recital 17. See also Paul G. Harris and Taedong Lee, "Compliance with climate change agreements: the constraints of consumption" (2017) 17:6 *International Environmental Agreements* 779.

V. CONCLUSION

States must ensure that activities conducted under their jurisdiction do not result in significant transboundary harm. To promote compliance, international negotiations have led to successive international climate agreements which adopted collective objectives and defined national commitments. Objectives and commitments have also been adopted under other regimes. States have resorted to a variety of methods to reduce GhG emissions under their jurisdiction, including not only classical command and control regulation, but also price-based incentives and measures of economic, scientific and cultural leadership. Although national commitments and measures of implementation have certainly had a non-negligible result, these efforts remain insufficient to fulfill the obligation of States not to cause significant transboundary harm. More is needed.

The two following chapters explore attempts to enhance mitigation action in two different ways. Chapter 8 retraces the creation of flexibility mechanisms, whereby a State may implement its mitigation commitment by supporting projects which are carried out on the territory of other States. Chapter 9 then explores the ethical and legal issues raised in relation to somewhat "radical" methods of enhancing sinks and reservoirs in GhGs, as well as other extreme engineering techniques gathered under the concept of geoengineering.

8

Flexibility Mechanisms

Chapter 7 described the general architecture of the provisions of international climate agreements on climate change mitigation. These agreements typically define a collective objective and national commitments that States pledge to implement within their territory. During the implementation of the agreement, however, some parties may face unexpected difficulties in achieving their commitment. Others may be able to achieve more than they had committed, especially if they receive some support to carry out additional efforts. Flexibility mechanisms allow a party to fulfill its own mitigation commitment by supporting measures implemented under the jurisdiction of another party beyond its own territory.

The main rationale for a flexibility mechanism is to make mitigation action cheaper and more effective. Like other forms of international trade, flexibility mechanisms seek to take advantage of differences in relative costs between countries. Opportunities for cost-effective mitigation action depend in no small part on national circumstances and can vary among States and over time. While developed parties agreed to take the lead in mitigating climate change,[1] they could not ignore cost-effective short-term mitigation opportunities in countries with an emerging economy. As Daniel Bodansky noted in 1992:

> If greenhouse gas emissions can be reduced more cheaply in country A than in country B, then allowing B to take advantage of this cost differential by funding an emission reduction in A is more efficient than requiring B to achieve the same reduction at home.[2]

In addition, flexibility mechanisms emerged with a political objective: reassuring States that they would be able to fulfill their commitment, especially when those commitments consist of achieving a QELRC under the Kyoto Protocol. When negotiating the Kyoto Protocol, Annex I parties could not reliably predict their ability to achieve specific mitigation outcomes. The evolution of their GhG emissions would depend in large part on economic conditions – rapid economic growth, for instance, would likely translate into an increase in their GhG emissions – as well as their ability to develop and deploy relevant

---

[1] United Nations Framework Convention on Climate Change, May 9, 1992, 1771 *UNTS* 107 (hereinafter UNFCCC), art. 3:1.
[2] Daniel Bodansky, "The United Nations Framework Convention on Climate Change: a commentary" (1993) 18:2 *Yale Journal of International Law* 451, at 520.

technologies before and during the commitment period. Achieving pre-determined QELRCs could reveal to be more expensive than predicted. Flexibility mechanisms allowed Annex I parties to reduce such risk by giving them the option of relying on projects conducted beyond their territory to achieve the mitigation outcome they pledged to make. Meanwhile, flexibility mechanisms transferred funds and technologies in support of climate change mitigation in emerging economies.

Yet, there were multiple risks associated with the creation of flexibility mechanisms. A party could report reductions in GhG emissions which would have occurred without its intervention, thus reducing the overall impact of international climate agreements. Under the Kyoto Protocol, what was reported as a reduction in GhG emissions in a non-Annex I country (without quantified mitigation commitments) could conceal the displacement of a source of GhG emissions to another site – a factory closes, but another one opens to conduct the same activity in the same conditions. Moreover, just like financial investments, a quest for the most cost-effective mitigation project carried out abroad could lead to human rights abuses or impacts on the local environment, especially in developing States with weaker laws and institutions. In turn, extensive regulation aimed at reducing these risks imposed complex procedures and a relatively close international oversight on mitigation projects carried out under flexibility mechanisms.

This chapter recounts the development of flexibility mechanisms in the UNFCCC and (most importantly) the Kyoto Protocol. It then turns to their continued implementation under the Doha Amendment and the Cancún Agreements, as well as ongoing negotiations under the Paris Agreement. Lastly, it assesses the contribution of flexibility mechanisms to climate change mitigation and sustainable development in general.

I. JOINT IMPLEMENTATION UNDER THE UN FRAMEWORK CONVENTION ON CLIMATE CHANGE

As a matter of principle, the UNFCCC provides that "[e]fforts to address climate change may be carried out cooperatively by interested Parties."[3] More specifically, it also allows developed parties to "implement [mitigation] policies and measures jointly with other Parties" and to "assist other Parties in contributing to the achievement of the objective of the Convention."[4] The modalities for the implementation of these provisions were to be decided at the first session of the COP.[5]

It was understood that a full-fledged mechanism would be established under a forthcoming Protocol, but a pilot phase for joint implementation under the Convention (JI/UNFCCC) was immediately established by COP1, in 1995, in order to encourage "learning by doing."[6] In this pilot phase, a developed party could collaborate either with another developed party or with a developing one.[7] The activities supported by one State and implemented under the jurisdiction of another State had to "bring about real, measurable and long-term" mitigation achievements "which would not have occurred in the absence of such activities,"[8] whereas the financial support provided was to be additional to the Financial Mechanism of the UNFCCC or to pre-existing

---

[3] UNFCCC, *supra* note 1, art. 3:3.
[4] *Ibid.*, art. 4:2(a).
[5] *Ibid.*, art. 4:2(d).
[6] Decision 5/CP.1, "Activities implemented jointly under the pilot phase" (April 7, 1995).
[7] Decision 5/CP.21, "Long-term climate finance" (December 10, 2015), para. 1(a).
[8] *Ibid.*, para. 1(d).

official development assistance.[9] During the pilot phase, no credit could be reported by any party from activities implemented jointly because of a lack of credible methodology.[10]

This framework provided mixed results. Between 1995 and 2002, a total of 157 projects involving 42 States were carried out under this framework, typically medium-scale renewable energy or district heating projects.[11] These activities were implemented in developed as well as developing States. According to State reports, around 500 $MtCO_2eq$ could have been avoided,[12] a figure which would have represented about 1.5 percent of global GhG emissions in 2002.[13] Yet, mitigation outcomes were not rigorously measured, reported and verified, and this figure could have been significantly overstated. There was no agreed-upon methodological approach to determining whether a mitigation outcome generated by an activity under JI/UNFCCC would genuinely not have occurred in the absence of such activity.

From 2002, due to the anticipated implementation of the Kyoto Protocol's flexibility mechanisms under the Marrakesh Accords, the pilot phase of the JI/UNFCCC entered a terminal idle phase.[14] It was formally discontinued in 2012.[15]

II. FLEXIBILITY MECHANISMS UNDER THE KYOTO PROTOCOL

The Kyoto Protocol established three flexibility mechanisms: Joint Implementation, the CDM and Emissions Trading. The COP adopted further modalities of application through multiple decisions, in particular a series of decisions adopted through the Marrakesh Accords in 2001.[16] Each of these mechanisms provides some flexibility to Annex B parties in fulfilling their QELRCs under the Kyoto Protocol.[17] The role of these mechanisms was to be complementary to domestic efforts on climate change mitigation: relying on flexibility mechanisms did not exonerate States from making efforts to reduce emissions from activities within their own territory.[18]

"Joint Implementation" under the Kyoto Protocol (JI/KP) is distinct from "Joint Implementation" under the UNFCCC (JI/UNFCCC). Whereas JI/UNFCCC allows projects to be carried out on the territory of any party, JI/KP is limited to Annex I parties – it is completed by the CDM, under which a developed party may support a mitigation project within the territory of a developing party. Moreover, what is to be jointly implemented is different: in contrast to

---

[9] *Ibid.*, para. 1(e).

[10] *Ibid.*, para. 1(f).

[11] See UNFCCC, *Activities Implemented Jointly under the Pilot Phase, Seventh Synthesis Report* (September 13, 2006), doc. FCCC/SBSTA/2006/8, paras. 6–9.

[12] Estimate based on partial reporting in UNFCCC, *Activities Implemented Jointly under the Pilot Phase, Fifth Synthesis Report* (September 12, 2001), doc. FCCC/SBST/2001/7, at 8. This figure cannot be compared with reporting on the use of flexibility mechanisms under the Kyoto Protocol (below, notes 37 and 38), whose methodology is more rigorous and conservative.

[13] About 34.3 $MtCO_2eq$ were emitted in 2002, according to WRI, CAIT Climate Data Explorer, "Total GHG emissions excluding land-use change and forestry" (2013).

[14] See decision 17/CP.7, "Modalities and procedures for a clean development mechanism, as defined in Article 12 of the Kyoto Protocol" (November 10, 2001), para. 1.

[15] See decision 22/CP.18, "Activities implemented jointly under the pilot phase" (December 7, 2012).

[16] See Chapter 3, section II.B.

[17] See Kyoto Protocol to the United Nations Framework Convention on Climate Change, December 11, 1997, 2303 *UNTS* 162 (hereinafter Kyoto Protocol), arts. 6.1 (joint implementation), 12.2 (clean development mechanism) and 17 (emissions trading) in conjunction with art. 3.1 and Annex B.

[18] See *ibid.*, arts. 6.1(d), 12.3(b) and 17. See also decision 3/CMP.7, "Emissions trading and the project-based mechanisms" (December 11, 2001), para. 1.

the vague mitigation commitments defined by the UNFCCC,[19] JI/KP relates to the implementation of the QELRCs of Annex I parties.[20] Accordingly, in order to enable Annex I parties to report mitigation outcomes and fulfill their emission limitation and reduction commitments, these outcomes had to be quantified in a more rigorous and reliable way under the Kyoto Protocol than they had been under the UNFCCC pilot project.

JI/KP provides a mechanism through which an Annex I party may support a mitigation project in another Annex I party, with the resulting Emission Reduction Units (ERUs) reported as the former party's own emission reduction.[21] For instance, country A may subsidize the upgrading of a power plant in country B; although this reduces GhG emissions in country B, it belongs to country A to report the emission reduction under JI/KP. Projects under JI/KP seek either to reduce GhG emissions by sources (such as a power plant) or to enhance sinks and reservoirs of GhGs (such as forests). Each project must be approved by the parties involved and it must result in emission reductions that are "additional to any that would otherwise occur."[22]

Whereas JI/KP relates to mitigation projects taking place within the territory of Annex I parties bound to their own QELRCs, the CDM relates to mitigation projects taking place on the territory of non-Annex I parties. The purpose of the CDM is to assist Annex I parties in achieving compliance with their QELRCs while also engaging non-Annex I parties – especially emerging economies – in mitigation action and promoting clean development strategies in the developing world.[23] An Annex I party can thus report Certified Emission Reductions (CERs) from project activities taking place on the territory of a non-Annex I party to contribute to the fulfillment of its emission limitation and reduction commitment.[24] A project activity is eligible to the CDM subject to the "voluntary participation" of each party involved, demonstration of "real, measurable, and long-term benefits related to the mitigation of climate change," and evidence that the mitigation outcome is additional to any that would have occurred in the absence of the project activity.[25]

The Marrakesh Accord of 2001 defined provisional guidelines for the implementation of the CDM and JI/KP. These guidelines were later formally adopted by the Meeting of the Parties to the Kyoto Protocol following the Protocol's entry into force in 2005.[26] Accordingly, the JI Supervisory Committee and the CDM Executive Board were tasked to monitor implementation and to further develop the complex regulatory system necessary to ensure the functioning of these mechanisms. It was subsequently agreed that developed parties would refrain from using ERUs or CERs generated from nuclear facilities to fulfill their commitment.[27] While the CDM could accommodate carbon dioxide capture and storage in geological formations, no such project activity was registered.[28]

[19] UNFCCC, *supra* note 1, art. 4.2 (a).
[20] Kyoto Protocol, *supra* note 17, art. 3.1 and Annex B.
[21] *Ibid.*, art. 6.1.
[22] *Ibid.*, art. 6.1(b).
[23] See *ibid.*, art. 12.2.
[24] See *ibid.*, art. 12.3.
[25] *Ibid.*, art. 12.5.
[26] See decision 16/CP.7, "Guidelines for the implementation of Article 6 of the Kyoto Protocol" (November 10, 2001); and decision 9/CMP.1, "Guidelines for the implementation of Article 6 of the Kyoto Protocol" (November 30, 2005).
[27] Decision 16/CP.7, *supra* note 26, recital 5; and decision 17/CP.7, *supra* note 14, recital 6.
[28] See decision 7/CMP.6, "Carbon dioxide capture and storage in geological formations as clean development mechanism project activities" (December 10–11, 2010), para. 1; decision 10/CMP.7, "Modalities and procedures for carbon dioxide capture and storage in geological formations as Clean Development Mechanism project activities" (December 9, 2011). See also Chapter 9, section III.B.3.

JI/KP and the CDM both require that the mitigation outcome of a project be additional to what would have occurred without the project. Yet, additionality to a hypothetical scenario revealed difficult to demonstrate. The Executive Board and the JI Supervisory Committee approved or reviewed multiple methodologies specific to particular types of projects,[29] for instance, by defining sectorial emissions scenarios as baselines for the determination of additionality.[30] Independent institutions were accredited to determine the eligibility of a project or project activity and to validate its compliance with the requirements for JI/KP or the CDM: the "Accredited Independent Entities" for projects under the JI/KP and the "Designated Operational Entities" for project activities under the CDM.[31]

Emissions Trading is the third flexibility mechanism under the Kyoto Protocol. This mechanism allows Annex I parties to sell and purchase emission allowances from one another for the purpose of fulfilling their QELRC.[32] Unlike the CDM and JI/KP, Emissions Trading is not based on particular projects and, therefore, it requires less technical regulation. One of the very few limitations is that parties must maintain a "commitment period reserve" of at least 90 percent of its assigned amount of GhG emissions under the Kyoto Protocol or 100 percent of its current emission rate extrapolated on the whole commitment period, whichever is lower, as a way to ensure that they comply with their commitment.[33] All emission units can be traded, including the Assigned Amount Units (AAUs) comprised within each Annex I party's emission limitation and reduction commitment, the ERUs resulting from projects under JI/KP, and the CERs from project activities under the CDM.[34]

Emissions Trading under the Kyoto Protocol needs to be distinguished from market-based mechanisms (sometimes called "emissions trading schemes") that some States or regional entities (like the European Union) have established as domestic implementation measures.[35] Emissions Trading as a flexibility mechanism under the Kyoto Protocol is trading mechanism between parties (i.e. States or regional entities); it does not need to be connected to a domestic (or regional) market-based mechanism. A party may allow legal entities to transfer or acquire emissions units, but it remains its exclusive responsibility to ensure fulfillment of its mitigation commitment.[36] Any Emission Trading transaction is recorded in an international transaction log.

The flexibility mechanisms of the Kyoto Protocol have avoided significant amounts of GhG emissions. As of mid-2017, more than 1.8 $GtCO_2eq$ of emission reduction had been reported under the CDM,[37] about 0.9 $GtCO_2eq$ of emission reductions had taken place through JI/KP,[38]

---

[29] Decision 3/CMP.1, "Modalities and procedures for a clean development mechanism as defined in Article 12 of the Kyoto Protocol" (November 30, 2005), para. 5(d); decision 9/CMP.1, *supra* note 26, Annex, para. 3(d).

[30] Decision 9/CMP.1, *supra* note 26, Annex, Appendix B, para. 1.

[31] *Ibid.*, Annex, paras. 32–34; and decision 3/CMP.1, *supra* note 29, Annex, para. 27(e).

[32] Kyoto Protocol, *supra* note 17, art. 17. See also *ibid.*, arts. 3.10. and 3.11; and decision 11/CMP.1, "Modalities, rules and guidelines for emissions trading under Article 17 of the Kyoto Protocol" (November 30, 2005).

[33] Decision 11/CMP.1, *supra* note 32, Annex, para. 6.

[34] *Ibid.*, Annex, para. 1. Another type of units, not discussed here, is the Removal Unit (RMU) under the Kyoto Protocol, *supra* note 17, arts. 3.3 and 3.4, resulting from sink enhancements through land-use change under the party's own jurisdiction. See also decision 3/CMP.1, *supra* note 29.

[35] See Chapter 7, section IV.

[36] Decision 11/CMP.1, *supra* note 32, Annex, para. 5.

[37] Information retrieved from the CDM Registry Issuance Report, http://cdm.unfccc.int/Registry/index.html (accessed January 5, 2018).

[38] Decision 4/CMP.12, "Guidance on the implementation of Article 6 of the Kyoto Protocol" (November 17, 2016).

and about 0.25 $GtCO_2eq$ had been traded through the Emissions Trading.[39] Projects under the CDM have mostly been implemented in emerging economies, in particular China, while JI/KP has largely focused on increasing energy efficiency in countries that are undergoing the process of transition to a market economy. It appears that nine out of the thirty-six Annex I parties to the Kyoto Protocol during the whole commitment period would not have been able to comply with their QELRC without using flexibility mechanisms.[40]

III. FLEXIBILITY MECHANISMS UNDER THE DOHA AMENDMENT AND THE CANCÚN AGREEMENTS

The Kyoto Protocol's initial commitment period extended from 2008 to 2012. When this came to an end, a second commitment period ranging from 2013 to 2020 was defined by the Doha Amendment, adopted *in extremis* in late 2012.[41] The three flexibility mechanisms, along with other provisions of the Kyoto Protocol, were kept unchanged under this second commitment period.[42] The parties recognized the need for "continued smooth implementation of the Kyoto Protocol, including its [flexibility] mechanisms."[43] Although the Doha Amendment had not yet entered into force as of early 2018, the parties to the Kyoto Protocol agreed to implement their commitments on a provisional basis.[44] Specific arrangements were adopted to allow Annex I parties with a QELRC under the second commitment period to make use of flexibility mechanisms, provided that they had deposited their instrument of acceptance of the Doha Amendment.[45] Three Annex I parties – Japan, New Zealand and Russia – decided not to make any new emission limitation or reduction commitment under the second commitment period.[46] The decision of the Meeting of the Parties adopting the Doha Amendment clarifies that only the Annex I parties with a new commitment under the second commitment period could become eligible to transfer or acquire any emissions allowances under the three flexibility mechanisms.[47]

In addition to the second commitment period under the Kyoto Protocol, the Copenhagen Accord and the Cancún Agreements introduced mitigation pledges under the UNFCCC

[39] R. Stavins *et al.*, "International cooperation: agreements & instruments" in O. Edenhofer *et al.* (eds.), *Climate Change 2014: Mitigation of Climate Change. Contribution of Working Group III to the Fifth Assessment Report of the Intergovernmental Panel on Climate Change* (Cambridge University Press, 2014) 1001, at 1044.

[40] These are: Austria, Denmark, Iceland, Japan, Lichtenstein, Luxembourg, Norway, Spain and Switzerland. See S. Ignor, M. Romain and B. Valentin, "Compliance of the parties to the Kyoto Protocol in the first commitment period" (2016) 16:6 *Climate Policy* 770.

[41] Doha Amendment to the Kyoto Protocol, December 8, 2012, in the annex of decision 1/CMP.8, "Amendment to the Kyoto Protocol Pursuant to its Article 3, paragraph 9 (Doha Amendment)" (December 8, 2012) (hereinafter Doha Amendment).

[42] This was already agreed in decision 1/CMP.6, "The Cancún Agreements: outcome of the work of the Ad Hoc Working Group on Further Commitments for Annex I Parties under the Kyoto Protocol at its fifteenth session" (December 10–11, 2010), para. 6(b).

[43] Doha Amendment, *supra* note 41, recital 9.

[44] Decision 1/CMP.8, *supra* note 41, paras. 5 and 6.

[45] See e.g. decision 7/CMP.9, "Modalities for expediting the establishment of eligibility for Parties included in Annex I with commitments for the second commitment period whose eligibility has not yet been established" (November 22, 2013).

[46] See Doha Amendment, *supra* note 41, modifying Annex B of the Kyoto Protocol.

[47] *Ibid.*, para. 15.

applicable to the same period.[48] These documents contain very little detail about the flexibility allowed to parties in implementing their respective national commitments on climate change mitigation. The parties which agreed to QELRCs under the second commitment period of the Kyoto Protocol announced the same quantified targets under the Cancún pledges, which were equally taken note of by the Meeting of the Parties to the Kyoto Protocol.[49] It is therefore generally assumed that the parties which have agreed to QELRCs under the second commitment period of the Kyoto Protocol can make use of the Kyoto Protocol's flexibility mechanism to fulfill not only their national commitments under the revised Kyoto Protocol, but also their national commitments under the Cancún Agreements.[50]

The UNFCCC parties which are not parties to the Kyoto Protocol (the United States and Canada) or have decided not to make any new commitment under the Second Commitment Period (Japan, New Zealand and Russia) have nevertheless pledged to implement quantified economy-wide emission reduction targets under the Cancún Agreements.[51] Likewise, while developing parties are not bound by any additional commitment under the Kyoto Protocol,[52] many have pledged to implement nationally appropriate mitigation actions under the Cancún Agreements.[53] These national commitments by Annex I parties not party to the Kyoto Protocol, Annex I parties without commitments under the second commitment period of the Kyoto Protocol, and non-Annex I parties are based on various assumptions, in particular with regard to flexibility (for Annex I parties) and international financial or technical support (for non-Annex I parties). For instance, Canada has declared that it "does not assume or provide for significant use of Kyoto Protocol mechanisms for its 2020 target"[54] under the Cancún Agreements, although it has not totally excluded the use of internationally transferred mitigation outcomes. By contrast, the United States has suggested the possibility of a mechanism that could be created under domestic law and would "meet high standards for environmental integrity and transparency."[55]

Neither the Copenhagen Accord nor the Cancún Agreements contain any substantive provision on flexibility. The Copenhagen Accord only suggests the use of market mechanisms to provide mitigation incentives to developing parties, but does not clearly link such incentives to the

---

[48] Copenhagen Accord, in the annex of decision 2/CP.15 (December 18–19, 2009), paras. 4 and 5; decision 1/CP.16, "The Cancún Agreements: outcome of the work of the Ad Hoc Working Group on Long-Term Cooperative Action under the Convention" (December 10–11, 2010), paras. 36–37 and 48.

[49] Decision 1/CMP.6, *supra* note 42, para. 3, referring to the exact same document as decision 1/CP.16, *supra* note 48, para. 36.

[50] See the assumptions made by Annex B parties and reported in UNFCCC, *Quantified Economy-Wide Emission Reduction Targets by Developed Country Parties to the Convention: Assumptions, Conditions and Comparison of the Level of Emission Reduction Efforts* (June 3, 2011), doc. FCCC/TP/2011/1. Yet, the parties which have agreed to QELRCs under the second commitment period of the Kyoto Protocol may have different assumptions regarding flexibility for the implementation of their Cancún pledges, the scope of which could possibly differ. For instance, the QELRC under the Kyoto Protocol only apply to sectors and GhGs listed in Annex A, and they do not apply to GhG emissions from aviation and marine bunker fuels. By contrast, it is left to each party to define the scope of its quantified economy-wide emission reduction target under the Cancún Agreements.

[51] See UNFCCC, *Compilation of Economy-Wide Emission Reduction Targets to Be Implemented by Parties Included in Annex I to the Convention* (June 7, 2011), doc. FCCC/SB/2011/INF.1/Rev.1.

[52] Kyoto Protocol, *supra* note 17, art. 10.1.

[53] See UNFCCC, *Compilation of Information on Nationally Appropriate Mitigation Actions to Be Implemented by Parties Not Included in Annex I to the Convention* (March 18, 2011), doc. FCCC/AWGLCA/2011/INF.1.

[54] See UNFCCC, *Quantified Economy-Wide Emission Reduction Targets by Developed Country Parties to the Convention: Assumptions, Conditions, Commonalities and Differences in Approaches and Comparison of the Level of Emission Reduction Efforts* (August 23, 2012), doc. FCCC/TP/2012/5, para. 32.

[55] See *ibid.*, para. 43.

implementation of mitigation commitments by developed parties.[56] The Cancún Agreements, on the other hand, called for further consultations in order "to clarify the assumptions and the conditions related to the attainment" of quantified economy-wide emission reduction targets, "including the use of carbon credits from the market-based mechanisms."[57] Subsequent consultations shed light on the view of many developed parties that "the use of carbon credits from market-based mechanisms is essential in order to achieve cost-efficiency of the mitigation effort to attain the targets and to enhance their stringency."[58] Yet, it was also recognized that there was "little clarity on the anticipated use of such credits or on their sources and scale of contribution to attaining the targets."[59] By the end of 2014, these consultations stalled on the acknowledgment of differences of approach regarding the use of flexibility to fulfill commitments under the Cancún Agreements.[60] In addition, the Cancún Agreements launched negotiations toward "the establishment ... of one or more market-based mechanisms to enhance the cost-effectiveness of, and to promote, mitigation actions."[61] While a "new market-based mechanism" was formally established at COP17 in 2011,[62] no modalities or procedures were ever adopted.[63] Soon, negotiations on flexibility under the Cancún Agreements subsided as the Durban Platform for Enhanced Action made significant progress toward the adoption of the Paris Agreement.

IV. FLEXIBILITY MECHANISMS UNDER THE PARIS AGREEMENT

There remains – at the time of writing this book – great uncertainty regarding the nature of flexibility under the Paris Agreement.[64] Article 6 only contains very broad outlines which will need to be implemented through decisions of the COP serving as the Meeting of the Parties to the Paris Agreement. Thus, Article 6.1 recognizes the role of "voluntary cooperation" among its parties,[65] and Article 6.2 permits the transfer of mitigation outcomes to achieve the mitigation objective contained in NDCs.[66] Such cooperation shall "promote sustainable development

---

[56] Copenhagen Accord, *supra* note 48, para. 7.

[57] Decision 1/CP.16, *supra* note 48, para. 38. See also decisions 2/CP.17, "Outcome of the work of the Ad Hoc Working Group on Long-Term Cooperative Action under the Convention" (December 11, 2011), para. 5; and 1/CP.18, "Agreed outcome pursuant to the Bali Action Plan" (December 8, 2012), para. 8.

[58] UNFCCC, *Quantified Economy-Wide Emission Reduction Targets by Developed Country Parties to the Convention: Assumptions, Conditions, Commonalities and Differences in Approaches and Comparison of the Level of Emission Reduction Efforts*, October 18, 2013, doc. FCCC/TP/2013/7, para. 24.

[59] *Ibid.*

[60] UNFCCC SBSTA, *Work Programme on Clarification of Quantified Economy-Wide Emission Reduction Targets of Developed Country Parties, Draft Conclusions Proposed by the Chair*, December 5, 2014, doc. FCCC/SBSTA/2014/L.22, para. 5.

[61] Decision 1/CP.16, *supra* note 48, para. 80.

[62] Decision 2/CP.17, *supra* note 57, para. 83.

[63] See IISD, "Summary of the Warsaw Climate Change Conference: 11–23 November 2013" (2013) 12:594 *Earth Negotiations Bulletin* 28 (right), and "Summary of the Lima Climate Change Conference: 1–14 December 2014" (2014) 12:629 *Earth Negotiations Bulletin* 10 (right).

[64] Two approaches – the allowance approach and the credit approach – are distinguished in Charlotte Streck and Moritz von Unger, "Creating, regulating and allocating rights to offset and pollute: carbon rights in practice" (2016) 10:3 *Carbon & Climate Law Review* 178.

[65] Paris Agreement, December 12, 2015, in the annex of decision 1/CP.21, "Adoption of the Paris Agreement" (December 12, 2015), art. 6.1.

[66] *Ibid.*, art. 6.3.

and ensure environmental integrity and transparency."[67] It may involve not only mitigation, but also adaptation actions.[68] Additional modalities are to be defined by the COP Serving as the Meeting of the Parties to the Paris Agreement.[69]

The parties of the Paris Agreement will have the option to situate their transfer of mitigation outcomes within the "mechanism to contribute to the mitigation of greenhouse gas emissions and support sustainable development"[70] established under Article 6.4. The "Article 6.4 Mechanism" (for lack of a better name) is essentially the reincarnation of the "New Market Mechanism" considered under the Cancún Agreements.[71] It allows support and transfer of mitigation outcome for mitigation taking place either in a developed party like JI/KP did, or in a developing one, similarly to the CDM. However, while the CDM and JI/KP were limited to well-defined mitigation projects, the Article 6.4 Mechanism could also apply to entire programs or sectorial policies.[72]

Lastly, Article 6.9 of the Paris Agreement establishes a "framework for non-market approaches to sustainable development."[73] This framework seeks to promote "integrated, holistic and balanced non-market approaches" to climate change mitigation.[74] As for other flexibility mechanisms, the framework for non-market approaches could allow for mitigation outcomes to be reported by a country other than the country on whose territory the relevant mitigation action is implemented.

V. AN ASSESSMENT OF FLEXIBILITY MECHANISMS

Flexibility mechanisms have not only allowed parties to save money, but have also facilitated more ambitious national commitments by making implementation less expensive. By a rough estimate circulated in 2012, the CDM allowed developed parties to save USD 3.6 billion while complying with their QELRCs under the first commitment period of the Kyoto Protocol.[75] Project operators and States with emerging economies have also achieved substantial profits.[76] China, which hosted a large share of the project activities under the CDM, imposed a tax of up to 65 percent of the value of CERs earned by project activities taking place within its

---

[67] *Ibid.*, art. 6.2.

[68] *Ibid.*, art. 6.1.

[69] See generally Daniel Bodansky *et al.*, "Facilitating linkage of climate policies through the Paris outcome" (2016) 16:8 *Climate Policy* 956.

[70] *Ibid.*, art. 6.4.

[71] Decision 2/CP.17, *supra* note 57, para. 83. Some observers have called the Article 6.4 Mechanism a "Sustainable Development Mechanism," although the name has not been endorsed by the parties.

[72] See informal information note by the co-facilitators, "SBSTA 46 agenda item 10 (b) – rules, modalities and procedures for the mechanism" (May 17, 2017), http://unfccc.int/files/meetings/bonn_may_2017/in-session/application/pdf/sbsta_10b_informal_note_final.pdf (accessed January 5, 2018), at 3. See also Joséluis Samaniego and Christiana Figueres, "Evolving to a sector-based Clean Development Mechanism" in Kevin A. Baumert *et al.* (eds.), *Building on the Kyoto Protocol: Options for Protecting the Climate* (WRI, 2002) 89; David Hone, "Paris Agreement: developing Article 6," *The Energy Collective* (February 22, 2016), www.theenergycollective.com/davidhone/2322758/developing-article-6 (accessed January 5, 2018).

[73] Paris Agreement, *supra* note 65, art. 6.9.

[74] *Ibid.*, art. 6.8.

[75] Randall Spalding-Fecher *et al.*, *Assessing the Impact of the Clean Development Mechanism: Report Commissioned by the High-Level Panel on the CDM Policy Dialogue* (CDM Policy Dialogue, 2012) 5, 8.

[76] Jiang Xiaoyi, *Legal Issues for Implementing the Clean Development Mechanism in China* (Springer, 2013) 78–79. See also Maosheng Duan, "Clean Development Mechanism development in China" in Yan Jinyue *et al.* (eds.), *Handbook of Clean Energy Systems* (Wiley, 2015) 3427.

territory.[77] China's CDM Fund, which collects the proceeds of this tax, reported a total income of CNY 12.15 billion (USD 1.92 billion) in 2012.[78]

Flexibility mechanisms have also raised complex regulatory challenges to ensure that formal compliance with international climate agreements result in genuine mitigation outcomes. A great difficulty is to ensure that projects conducted under a flexibility mechanism are genuinely additional to other ongoing changes in the receiving States.[79] Despite all efforts in this sense, some studies have suggested that many reported emission reductions would have occurred without the flexibility mechanisms.[80] For instance, as former socialist countries underwent an economic transformation, they made significant gains in efficiency which led to the issuance of abundant ERUs under JI/KP, even though international support was rarely necessary to achieve such outcomes.[81] Likewise, evidence suggests that emerging economies may have invested in renewable energies, even without the support of the CDM, purely out of considerations for economic development, environmental protection (i.e. reducing air pollution) and energy independence.[82] The overall contribution of the Kyoto Protocol to climate change mitigation would be lower than expected if Annex I parties were indeed able to rely on non-additional mitigation outcomes to comply with the QELRCs.

Another regulatory challenge relates to the adverse incentives that the anticipation of flexibility mechanisms may have created. Under the CDM, projects do not qualify if they do not produce mitigation outcomes additional to what would follow from the implementation of "mandatory applicable legal and regulatory requirements."[83] This provision could induce countries where projects could be carried out to delay the introduction of any mandatory requirement on climate change mitigation in order to secure foreign funding.[84]

Furthermore, the economic efficiency of flexibility mechanisms has been hindered by unpredictable variations in the market price of emission allowances.[85] The market value of ERUs and CERs has been influenced by erratic variation in the price of emission allowances within the dominant EU Emissions Trading Scheme, which influenced demand for flexibility.[86] In 2016, extremely low market values led the CDM Executive Board to encourage the voluntary

---

[77] See Measures for the Operation and Management of CDM Projects in China (2005), art. 24.

[78] Duan, *supra* note 76, at 3326.

[79] See Stavins *et al.*, *supra* note 39, at 1045–1046.

[80] See e.g. Anja Kollmuss *et al.*, *Has Joint Implementation Reduced GHG Emissions? Lessons Learned for the Design of Carbon Market Mechanisms* (SEI Working Paper No. 2015-07, 2015) 116, suggesting that "about three-quarters of JI offsets are unlikely to represent additional emissions reductions."

[81] See e.g. Gernot Klepper and Sonja Peterson, "Trading hot-air. The influence of permit allocation rules, market power and the US withdrawal from the Kyoto Protocol" (2015) 32:2 *Environmental & Resource Economics* 205; and Richard B. Stewart and Jonathan B. Wiener, *Reconstructing Climate Policy: Beyond Kyoto* (AEI Press, 2003) 117–118.

[82] See Jiang, *supra* note 76, at 44–45. See also International Rivers & Friends of the Earth, *Trading in Fake Carbon Credits: Problems with the Clean Development Mechanism (CDM)* (October 2008).

[83] See CDM Executive Board, *Tool for the Demonstration and Assessment of Additionality*, 3rd version (29th session, 2007), at 4, sub-step 1b, para. 2.

[84] See Christiana Figueres, "Sectoral CDM: opening the CDM to the yet unrealized goal of sustainable development" (2006) 2:1 *McGill International Journal of Sustainable Development Law & Policy* 5, at 12–13.

[85] See e.g. Claudia Kettner, "The EU emission trading scheme: first evidence on Phase 3" in Larry Kreiser *et al.* (eds.), *Carbon Pricing: Design, Experiences and Issues* (Edward Elgar, 2015) 63; Sean Healy, Verena Graichen and Sabine Gores, *Trends and Projections in the EU ETS in 2016: The EU Emissions Trading System in Numbers* (European Environment Agency, 2016).

[86] See the synthesis in S. Agrawala *et al.*, "Regional development and cooperation" in O. Edenhofer *et al.* (eds.), *Climate Change 2014: Mitigation of Climate Change. Contribution of Working Group III to the Fifth Assessment Report of the Intergovernmental Panel on Climate Change* (Cambridge University Press, 2014) 1083, at 1112–1114.

cancellation of CERs in order to increase prices.[87] The unpredictable evolution of the price of CERs may hinder investments relying on such forms of revenue.

While complex rules were developed to ensure that flexibility mechanisms would produce genuine environmental benefits, loopholes inevitably appeared and, before they could be corrected, were immediately exploited by unscrupulous actors. The most significant example relates to HFC-23, a by-product in the synthesis of HCFC-22, a commonly used propellant and refrigerant. HFC-23 and HCFC-22 are both extremely potent GhGs.[88] Because HFC-23 has little commercial use and no toxicity, it used to be simply released into the atmosphere during the production of HCFC-22. Such emissions of HFC-23 were a significant source of GhG emissions. To counter the release of HFC-23, the CDM Executive Board approved a methodology for the incineration of HFC-23 waste streams in 2003, thus allowing such project activities to be granted CERs under the CDM as well as ERUs under JI/KP.[89] This created a tremendous economic opportunity for HCFC-22 producers because the cost of incinerating HFC-23 was negligible compared with the substantial income generated by selling such units.[90] China, where many HCFC-22 producers operated, established a 65 percent tax on income from CERs in such activities.[91] Rapidly, nineteen projects operating under this methodology received CERs and ERUs representing 0.5 billion tCO_2e – about a fifth of all emission units generated under the CDM and JI/KP as a whole. Soon, however, evidence appeared that some operators had substantially increased the production of HCFC-22 in order to produce more HFC-23, incinerate it, and thus receive CERs or ERUs from which they could draw their main income.[92] In other words, the flexibility mechanisms had unintentionally created an industry whose main source of profit was the sale of CERs or ERUs. Tragically, this not only meant a financial loss for the CDM and JI/KP, but also incentivized the production of HCFC-22, with adverse impacts both for the ozone layer and for GhG emissions. When this methodology was finally discontinued, several HCFC-22 producers decided to vent their stocks of HFC-23 into the atmosphere, making these relatively small factories "some of the largest point source emitters of greenhouse gases in the world."[93]

The HFC-23 scandal illustrates the limits of market-based mechanisms to address specific sources of GhG emissions. A command and control approach – the gradual prohibition of the release of HFC-23 into the atmosphere – was clearly more adequate to compel emitters to take relatively inexpensive measures. This approach was finally adopted in October 2016

---

[87] World Bank Group and Ecofys, *Carbon Pricing Watch 2017* (World Bank, 2017) 9.

[88] HCFC-22, which is also an ozone-depleting substance, is regulated under the Montreal Protocol on Substances that Deplete the Ozone Layer, September 16, 1987, 1522 *UNTS* 3 (hereinafter Montreal Protocol), as modified by the Copenhagen amendment, November 25, 1992, 1785 *UNTS* 517. See generally Chapter 4, section I.

[89] CDM Executive Board, "Incineration of HFC 23 waste streams," approved baseline methodology AM0001, September 26, 2003.

[90] See Michael Wara, "Is the global carbon market working?" (2007) 445:7128 *Nature* 595.

[91] Environmental Investigation Agency, *Two Billion Tonne Climate Bomb: How to Defuse the HFC-23 Problem* (June 2013) 5.

[92] See the "Form for submission of requests for revisions of approved methodologies to the Methodologies Panel" filed by Det Norske Veritas Certification AS on March 8, 2010, regarding methodology AM0001 "Incineration of HFC23 waste streams," https://cdm.unfccc.int/methodologies/PAmethodologies/revisions/58215 (accessed January 5, 2018). See also Lambert Schneider and Anja Kollmuss, "Perverse effects of carbon markets on HFC-23 and SF6 abatement projects in Russia" (2015) 5 *Nature Climate Change* 1061.

[93] Environmental Investigation Agency, *supra* note 91, at 2.

through the adoption of the Kigali Amendment to the Montreal Protocol on Substances that Deplete the Ozone Layer, which imposes the gradual phasing out of the production and consumption of HFC-23, among other HFCs, in order to preserve the climate system.[94] In hindsight, a systematic reliance on a market-based mechanism rather than command and control regulation wasted considerable time and resources while causing additional emissions in HCFC-22.

Flexibility mechanisms may also have adverse impacts on the human rights of local communities or on the local environment wherever activities are implemented, especially, through the CDM, in developing States with less stringent standards or less systematic implementation. Inherent to a market-based approach is a quest for the cheapest mitigation projects available, notwithstanding their adverse consequences abroad. Statements according to which the CDM would assist developing States "in achieving sustainable development"[95] and that the Article 6.4 Mechanism would "support sustainable development"[96] are insufficient if they are not accompanied by more concrete guarantees,[97] which, in practice, were rarely adopted.[98]

Thus, about a third of all CERs were issued to hydropower projects.[99] Such projects have often led to the forced relocation of large numbers of people.[100] Safeguard policies that have progressively been adopted by international aid agencies, in particular to prevent wanton displacement and relocation of populations,[101] do not apply to financial support provided through flexibility mechanisms.[102] This allowed the issuance of CERs to hydropower projects which had serious and unmitigated impacts on some of the most basic human rights of the local populations.[103] In addition to a mention of human rights law in the Cancún Agreements[104] and again in the Paris Agreement,[105] more concrete provisions will need to be adopted to ensure that flexibility mechanisms respect the rights of local stakeholders and their environment.[106]

---

[94] Amendment to the Montreal Protocol on Substances that Deplete the Ozone Layer, adopted in Kigali on October 15, 2016, reproduced in (2017) 56 *ILM* 196 (hereinafter Kigali Amendment). See generally Chapter 4, section I.

[95] Kyoto Protocol, *supra* note 17, art. 12.2.

[96] Paris Agreement, *supra* note 65, art. 6.4.

[97] The modalities of application of the Article 6.4 Mechanism are to be defined in accordance with a work program defined in decision 1/CP.21, *supra* note 65, para. 37, which contains no reference to sustainable development.

[98] A parallel can be drawn with REDD+ (discussed in Chapter 12, section III), where recommendations for equitable outcomes benefiting the local communities were rarely included in actual projects. See Stephanie Venuti, "REDD+ in Papua New Guinea and the protection of the REDD+ safeguard to ensure the full and effective participation of indigenous peoples and local communities" (2015) 17 *Asia Pacific Journal of Environmental Law* 131.

[99] CDM Watch, *Hydro Power Projects in the CDM: Policy Brief* (February 2012).

[100] On hydroelectric dams, see, for instance, WCED, *Dams and Development: A Framework for Decision Making* (Earthscan, November 2000), and, for a classical case study, David Howarth, *The Shadow of the Dam* (Macmillan, 1961).

[101] See e.g. *The World Bank Environmental and Social Framework* (World Bank, 2017).

[102] Nature Code & Carbon Market Watch, *Social and Environmental Accountability of Climate Finance Instruments* (policy brief, September 2015).

[103] See e.g. Annette Gartland, "Mega-dams 'create the illusion of climate action'" *Byline* (January 19, 2016); John Vidal, "Why is Latin America so obsessed with mega dams?" *The Guardian* (May 23, 2017); Philip M Fearnside, "Tropical hydropower in the clean development mechanism: Brazil's Santo Antônio Dam as an example of the need for change" (2015) 131:4 *Climatic Change* 575.

[104] Decision 1/CP.16, *supra* note 48, recital 8 and para. 8.

[105] Paris Agreement, *supra* note 65, recital 12. See also Chapter 16, section I.

[106] See Damilola S. Olawuyi, *The Human Rights-Based Approach to Carbon Finance* (Cambridge University Press, 2016).

VI. CONCLUSION

Flexibility mechanisms have made it easier for States to agree to more ambitious commitments and have contributed to leveraging financial support on climate change mitigation. Yet, ensuring that flexibility does not come at the expense of environmental integrity requires very complex regulation; the HFC-23 scandal shows that any loophole can rapidly be exploited by unscrupulous actors.

9

Geoengineering

States have recognized the importance of preventing "dangerous anthropogenic interference with the climate system"[1] and they have, under general international law, an obligation not to cause serious damage to the global environment. However, efforts to mitigate climate change through classical means – reducing GhG emissions and enhancing natural sinks of GhGs – have not fulfilled collective objectives. More could surely be done, but at a cost that societies and their political representatives remain, by and large, unwilling to pay. It is consequently tempting to look at other ways of reconciling economic development and environmental sustainability – technological fixes that could make it easier to address the ongoing climate crisis.

Geoengineering refers to the development and potential deployment of technologies which are intended to alter the climate system or the global environment.[2] While climate change mitigation seeks to reduce the impacts of human activities on the climate system, geoengineering aims to balance it through opposite forces. Geoengineering could affect the climate system either through the management of solar radiations or through the removal of carbon dioxide. Solar radiation management would consist in altering the intake of solar radiation by the Earth system, for instance, through the deployment of large shades in space or the injection of aerosols in the higher atmosphere (stratosphere). By contrast, the removal of carbon dioxide seeks to create artificial GhG sinks and reservoirs, for instance, by injecting it deep underground in stable geological formations, thus reducing GhG concentrations in the atmosphere.

There are different ways of approaching the conceptual relation between climate change mitigation and geoengineering:

- One possible view is that the two concepts are distinct. Climate change mitigation seeks to treat the very causes of climate change by altering our conduct – reducing our GhG emissions or preserving forests. By contrast, geoengineering addresses the symptoms, either by balancing carbon dioxide concentrations in the atmosphere through carbon dioxide removal or by balancing increased greenhouse effect through solar radiation management. Accordingly, while

---

[1] United Nations Framework Convention on Climate Change, May 9, 1992, 1771 *UNTS* 107 (hereinafter UNFCCC), art. 2.

[2] S. Planton, "Glossary" in T.F. Stocker *et al.* (eds.), *Climate Change 2013: The Physical Science Basis. Contribution of Working Group I to the Fifth Assessment Report of the Intergovernmental Panel on Climate Change* (Cambridge University Press, 2013) 1447, at 1454.

preventing deforestation is a mitigation strategy, large reforestation programs should rather be considered as geoengineering.

- In another view, climate change mitigation and geoengineering are not mutually exclusive: geoengineering may be used as a technique to advance climate change mitigation. Carbon dioxide removal, whether through large reforestation programs or through capture and storage of carbon dioxide, advances the same goal as any other action on climate change mitigation: reducing GhG concentrations in the atmosphere. Alternatively, some forms of geoengineering, such as solar radiation management, could be approached as action on climate change adaptation at a global scale. Geoengineering, in this view, relates to the way an action is carried out – through large-scale techniques impacting planetary systems as a whole – rather than the objective of this action.

However, this conceptualization of the relationship between geoengineering and climate change mitigation should not distract from what is properly unique about geoengineering: the intended manipulation of planetary systems, with consequences which are immense, sudden, not fully predictable and possibly irremediable.[3] Beyond a range of technical and scientific issues, geoengineering raises crucial ethical and legal questions. A central question is whether it is ever ethically justified to purposefully alter planetary systems. The case for such interventions is easier to make for carbon dioxide removal, which simply seeks to balance anthropogenic emissions in GhGs, than for solar radiation management, which would further alter the climate system. It appears unlikely that the objective of the Paris Agreement of "achiev[ing] a balance between anthropogenic emissions by sources and removals by sinks of greenhouse gases in the second half of this century"[4] could be achieved without resorting to some large-scale program of carbon dioxide removal.[5] By contrast, if solar radiation management could revert global average temperatures to a pre-industrial level, it would create a fundamentally different climate where, for instance, days would be cooler and nights warmer than in pre-industrial times.[6] This would alter water cycles dramatically, and many other impacts of climate change such as ocean acidification and sea-level rise would not be avoided.[7] If such extreme forms of geoengineering were ever justified, it would be as a last resort if international action on climate change mitigation were to fail dramatically, leading to a situation of absolute emergency.[8]

To situate the debate on geoengineering in a historical perspective, the next section explores the precedent of weather modification techniques and their early regulation in international law. Then, the chapter turns to a description of existing geoengineering options, including the

[3] See generally Clive Hamilton, *Earthmasters: The Dawn of the Age of Climate Engineering* (Yale University Press, 2013) 20; R.K. Pachauri *et al.*, *Climate Change 2014: Synthesis Report. Contribution of Working Groups I, II and III to the Fifth Assessment Report of the Intergovernmental Panel on Climate Change* (IPCC, 2015) 125.

[4] Paris Agreement, December 12, 2015, in the annex of decision 1/CP.21, "Adoption of the Paris Agreement" (December 12, 2015), art. 4.1.

[5] The Fifth Assessment Report of the IPCC included bioenergy with carbon capture and storage in its mitigation scenario. See O. Edenhofer *et al.*, "Technical summary" in O. Edenhofer *et al.* (eds.), *Climate Change 2014: Mitigation of Climate Change. Contribution of Working Group III to the Fifth Assessment Report of the Intergovernmental Panel on Climate Change* (Cambridge University Press, 2014) 31, at 60.

[6] This is because less solar radiation would reach the Earth's surface during the day, while less radiation from the Earth's surface would be released into space during the night.

[7] See e.g. David Reichwein *et al.*, "State responsibility for environmental harm from climate engineering" (2015) 5:2–4 *Climate Law* 142.

[8] David G. Victor *et al.*, "The geoengineering option: a last resort against global warming?" *Foreign Affairs* (March–April 2009).

risks associated with their deployment. Lastly, it explores some of the central moral and legal questions that have arisen with regard to the regulation of geoengineering.

I. THE PRECEDENT OF WEATHER MODIFICATION TECHNIQUES

Attempts have long been made to develop techniques to control the weather. From the 1950s to the 1970s, research was conducted around the idea of cloud seeding: by adding small particles in the air (typically silver iodide), it was thought that condensation could occur and that precipitation would ensue. Such research was supported by strategic military budgets; some experiments were carried out by the United States during the Vietnam War as part of Operation Popeye based on a vague intuition that controlling weather could give the US military some sort of strategic advantage.[9] By and large, however, the development of relevant techniques has not met early expectations and the political currency of research on weather modification techniques has considerably declined since the 1980s. A 2003 report of the US National Research Council concluded that "every assessment of weather modification ... has found that scientific proof of the effectiveness of cloud seeding was lacking (with a few notable exceptions, such as the dispersion of cold fog)."[10] Nevertheless, there are still some repeated attempts to use weather modification techniques in developing countries such as China and Indonesia.[11]

Even though technical developments did not fully come to fruition, the debates on weather modification techniques triggered moral and legal debates which form a background for the current consideration for geoengineering technologies. In particular, it was clearly understood that weather modification in one State would likely have consequences in others. Causing rain to fall at one place would divert it from another place. As the WMO observed in a 1963 report, "the complexity of the atmospheric processes is such that a change in the weather induced artificially in one part of the world will necessarily have repercussions elsewhere," but scientific knowledge was "far from sufficient to enable us to forecast with confidence the degree, nature or duration of the secondary effects."[12] This suggested that weather modification techniques were to be used in careful and cooperative ways.

International negotiations could more easily be conducted after the United States abandoned the idea of using weather modification techniques during the Vietnam War.[13] The United States and the Union of Soviet Socialist Republics (USSR) initiated negotiations which led, in 1976, to the adoption of the Convention on the Prohibition of Military or Any Other Hostile Use of Environmental Modification Techniques.[14] Each party to this Convention, including most developed States, undertook "not to engage in military or any other hostile use of

---

[9] James R. Fleming, "The pathological history of weather and climate modification: three cycles of promise and hype" (2006) 37:1 *Historical Studies in the Physical & Biological Sciences* 3, at 13.

[10] Committee on the Status and Future Direction in US Weather Modification Research and Operations, *Critical Issues in Weather Modification Research* (National Research Council, 2003) 1.

[11] See Clifford Coonan, "How Beijing used rockets to keep opening ceremony dry" *The Independent* (August 10, 2008); and Rongxing Guo, *How the Chinese Economy Works* (Palgrave Macmillan, 2017) 332.

[12] World Weather Watch and WMO, *Second Report on the Advancement of Atmospheric Sciences and their Application in the Light of Developments in the Outer Space* (WMO, 1963) 19.

[13] Fleming, *supra* note 9, at 14.

[14] Convention on the Prohibition of Military or Any Other Hostile Use of Environmental Modification Techniques, December 10, 1976, 1108 *UNTS* 151.

environmental modification techniques having widespread, long-lasting or severe effects as the means of destruction, damage or injury to any other State Party."[15] The material scope of the Convention is broad as it applies to "any technique for changing – through the deliberate manipulation of natural processes – the dynamics, composition or structure of the Earth, including its biota, lithosphere, hydrosphere and atmosphere, or of outer space."[16] Yet, the Convention does not apply to "the use of environmental modification techniques for peaceful purposes" and it is "without prejudice to the generally recognized principles and applicable rules of international law concerning such use."[17]

Beyond their military or other hostile use, States also recognized the need to cooperate actively in understanding weather modification techniques generally. Already in 1961, the UN General Assembly had noted the need for States to "study ... measures ... to advance the state of atmospheric science and technology so as to provide greater knowledge of basic physical forces affecting climate and the possibility of large-scale weather modification."[18] Two decades later, the Governing Council of the UN Environment Programme adopted provisions for international cooperation on weather modification which highlighted the obligation of States to cooperate.[19] It affirmed the central principle that "weather modification should be dedicated to the benefit of mankind and the environment,"[20] further highlighting that:

> exchange of information, notification, consultation and other forms of co-operation regarding weather modification should be carried out on the basis of good faith, in the spirit of good neighbourliness and in such a way as to avoid any unreasonable delay either in such forms of co-operation or in carrying out weather modification activities.[21]

The requirement that weather modification techniques be used to the benefit of mankind and the environment clearly excluded unilateral action promoting the interest of one State at the expense of another, including for military use. Yet, a more positive definition of the "benefit of mankind and the environment" would have been challenging. As triggering rain in one area would likely reduce it in another, weather modification techniques would often have created zero-sum games. Questions of liability were raised, in the United States during the 1960s and more recently in China, in cases regarding droughts affecting regions adjacent to those using weather modification techniques and for damage caused by thunder and rains within the latter regions. These claims were, it seems, systematically rejected based on a series of evidentiary issues starting with the lack of evidence that weather modification techniques had any effect at all.[22] Many more

---

[15] *Ibid.*, art. I.1.

[16] *Ibid.*, art. II.

[17] *Ibid.*, art. III.

[18] UN General Assembly Resolution 1721 (XVI) C, "International co-operation in the peaceful uses of outer space" (December 20, 1961), para. 1(a).

[19] UNEP Governing Council decision 8/7 on provisions for cooperation between States in weather modification, April 29, 1980, reproduced in doc. A/35/25, 117.

[20] *Ibid.*, para. 1(a).

[21] *Ibid.*, para. 1(b).

[22] Regarding cases arising in the United States during the 1960s, see Donald Frenzen, "Weather modification: law and policy" (1971) 12:4 *Boston College Law Review* 503; and Charles F. Cooper and William C. Jolly, "Ecological effects of silver iodide and other weather modification agents: a review" (1970) 6:1 *Water Resources Research* 88.

disputes would certainly have arisen, domestically as well as internationally, if weather modification techniques had proven to be effective.

As we will see in the next two sections, these questions formed a prelude to current debates on geoengineering. Yet, the debate on geoengineering is likely to become increasingly current during the twenty-first century as the impacts of climate change are severely affecting many communities throughout the world and some forms of geoengineering appear to be within reach. Small island developing States whose territorial existence is directly threatened by sea-level rise, coastal States losing a significant part of densely inhabited territories or States suffering drought, wars or famines, or even powerful non-State actors may become tempted to resort to quick technological "fixes," notwithstanding the dramatic and largely unpredictable impact that these techniques may have on planetary systems. As questions become more concrete and national interests better understood, disagreements will also become more difficult to resolve.

II. GEOENGINEERING OPTIONS

In contrast to attempts to modify the weather at a given place and time, geoengineering seeks to alter entire planetary systems, in particular the climate system. This requires efforts at a greater scale, with unpredictable consequences on the global environment. Some techniques of solar radiation management could be implemented at limited cost but with significant consequences for planetary systems.

A. Solar Radiation Management

Solar radiation management refers to a set of techniques seeking to reduce the Earth system's intake of energy from the Sun. As defined by the IPCC, it consists in "the intentional modification of the Earth's shortwave radiative budget with the aim to reduce climate change."[23] Some incoming visible light, in particular, would be reflected into space in order to lower the Earth's average surface temperature.

For instance, solar radiation could be reduced by putting large shades somewhere between the Earth and the Sun. Yet, the few studies which explored this option confirmed that it is most likely to remain within the realm of science fiction for the years to come.[24] It appeared that, in order to have a significant effect in countering climate change, shades should be deployed over hundred millions of square kilometers in the Earth's orbit or, for greater efficiency, even further from the Earth, at a point of equal gravitational pull from the Earth and the Sun.

Other techniques would increase the albedo of the Earth, which is its ability to reflect incoming solar radiations. While a black surface warms when exposed to sunlight, a white surface reflects a larger proportion of incoming radiations and thus stays cooler. Likewise, the

---

Regarding cases arising more recently in China, see Deng Haifeng, "人工降雨法律问题研究" ["On the legal issues of rainmaking"] (2007) 5 法商研究 [*Studies in Law and Business*] 26; and for instance Intermediate People's Court of Shaoxing, China, see *Yu Feng et al.* v. *Meteorological Bureau of Shangyu* (September 18, 2009).

[23] Planton, *supra* note 2, at 1462.

[24] See generally O. Boucher *et al.*, "Clouds and aerosols" in Stocker *et al.* (eds.), *supra* note 2, at 571; Naomi E. Vaughan and Timothy M. Lenton, "A review of climate geoengineering proposals" (2011) 109:3-4 *Climatic Change* 745, at 762.

Earth as a whole could be cooled by increasing its albedo (that is to say, by "whitening" its surface).[25] Painting deserts, favoring whiter vegetation or selecting more reflective materials to build roofs or pave roads would have a rather negligible impact on the climate system as a whole. Increasing the proportion of the Earth covered by clouds at any given moment could make a difference because clouds reflect more of the incoming solar radiation than most of Earth's surface, but constantly regenerating clouds in our skies, for instance by spraying seawater into the air at a global scale, would be nearly impossible.

A more realistic way to increase the albedo of the Earth would be to inject light aerosol particles, such as sulfites, into the stratosphere (i.e. at an altitude between 10 and 50 km). Volcanic eruptions have repeatedly demonstrated the transient cooling effect of massive injections of sulfite particles in the stratosphere. The eruption of Mount Pinatubo in the Philippines in 1991 cooled the planet by about 0.5 degrees Celsius over a year.[26] The 1815 eruption of Mount Tambora in Indonesia significantly cooled the Earth's climate in what was often called "the year without a Summer," causing great upheaval throughout the world in the years that followed.[27] Other volcanic eruptions had similar consequences.[28] To achieve the same result, it would not be straightforwardly unrealistic to rely on artillery guns, balloons or high-flying airplanes to inject sufficient quantities of aerosols into the stratosphere to have an impact on the climate system as a whole.[29] As sulfite particles fall down within a few years, they would need to be replaced constantly.

Solar radiation management could limit or even counter the increase in global average temperature, but it would not counter climate change. The cooling effect of solar radiation management would certainly not cool different places on the planet in the same way as increased concentrations in GhG concentrations warms them. Rain patterns and regional patterns would continue to change, although differently. Droughts would occur, although not necessarily in the same regions. Increased carbon dioxide concentrations in the atmosphere would continue to affect marine ecosystems.[30] By analogy, volcanic eruptions were shown to have wide-ranging consequences on weather patterns and global food production.[31] In addition to this, the sulfite particles constantly injected into the stratosphere would have extremely serious adverse environmental consequences, first in the stratosphere by depleting the ozone layer, and then in the biosphere as falling sulfite particles would cause acid rains.[32] Any suspension in the process of injecting particles into the stratosphere would translate into a more sudden change in our climate to which natural and social systems would have less time to adapt, and

---

[25] See Boucher *et al.*, *supra* note 24, at 627.

[26] Stephen Self *et al.*, "The atmospheric impact of the 1991 Mount Pinatubo eruption" in Christopher G. Newhall and Raymundo S. Punongbayan (eds.), *Fire and Mud: The Eruptions and Lahars of Mount Pinatubo, Philippines* (University of Washington Press, 1996) 1089.

[27] J. Luterbacher and C. Pfister, "The year without a summer" (2015) 8:4 *Nature Geoscience* 246.

[28] K.R. Briffa *et al.*, "Influence of volcanic eruptions on Northern Hemisphere summer temperature over the past 600 years" (1998) 393:6684 *Nature* 450.

[29] Vaughan and Lenton, *supra* note 24, at 764–765.

[30] See generally Katharine L. Ricke, M. Granger Morgan and Myles R. Allen, "Regional climate response to solar-radiation management" (2010) 3:8 *Nature Geoscience* 537; Ming Tingzhen *et al.*, "Fighting global warming by climate engineering: is the Earth radiation management and the solar radiation management any option for fighting climate change?" (2014) 31 *Renewable & Sustainable Energy Reviews* 792.

[31] See e.g. Victor *et al.*, *supra* note 8.

[32] Paul J. Crutzen, "Albedo enhancement by stratospheric sulphur injections: a contribution to resolve a policy dilemma?" (2006) 77:221 *Climatic Change* 211.

thus potentially much more harmful.[33] Solar radiation management is no cure; as a symptomatic treatment, it comes with extremely undesirable side-effects, including a rapid addiction.

B. Carbon Dioxide Removal

Besides radical solar radiation management options, more reasonable geoengineering techniques could help to remove carbon dioxide from the atmosphere. Some of these techniques seek to enhance natural sinks and reservoirs of GhGs such as forests, soils and seas. Others aim to store carbon dioxide in artificial reservoirs, for instance, in deep geological foundations.[34]

1. Enhancing Natural Reservoirs of GhGs

Discussions on techniques to remove carbon dioxide have explored the potentials of all major natural reservoirs of GhGs, including forests, soils and oceans.

Firstly, forests are a great reservoir of carbon. This is because trees and plants develop through photosynthesis, a process through which carbon dioxide and water are transformed into sugar and oxygen. Forest management relates both to climate change mitigation and geoengineering: whereas avoiding deforestation is clearly within the realm of climate change mitigation, one could consider reforestation (of lands that used to be a forest) or *a fortiori* afforestation (of lands that were not recently forests) would qualify as geoengineering.[35] Yet, efforts to maintain, restore, extend or use forests are constrained by a shortage of available land and the existence of competing uses of scarce arable land, including for food production.

Secondly, soils are partly made of decomposed organic materials rich in carbon; they constitute a reservoir of carbon. This reservoir can be increased through appropriate land-use techniques, in particular through the use of bio-char. Bio-char is a charcoal produced by pyrolysis (i.e. the combustion without oxygen and without carbon dioxide emissions) of any organic waste. Because bio-char contains high concentrations of carbon as well as nutrients, it can be used both as a fertilizer and as a long-term deposit of carbons in soils. Over time, rainwater pushes carbonate into sediments in the bottom of the oceans, thus ensuring a durable sequestration of carbon.[36] While bio-char is increasingly being used throughout the world, its capacity to reduce GhG concentrations in the atmosphere is rather limited.[37]

Thirdly, oceans can be exploited as a natural reservoir of carbon dioxide. About a third of anthropogenic carbon dioxide emissions are absorbed by the ocean through dissolution as carbon acid.[38] Some techniques could enhance this natural phenomenon. Putting aside the direct injection of carbon acids into the ocean, the injection of finely ground limestone or olivine could also increase the alkalinity of the seawater and its ability to absorb carbon

[33] Andy Jones *et al.*, "The impact of abrupt suspension of solar radiation management (termination effect) in experiment G2 of the Geoengineering Model Intercomparison Project (GeoMIP)" (2013) 118:17 *Atmospheres: Journal of Geophysical Research* 9743.
[34] J.C. Abanades *et al.*, "Summary for policymakers" in B. Metz *et al.*, *IPCC Special Report on Carbon Dioxide Capture and Storage. Prepared by Working Group III of the Intergovernmental Panel on Climate Change* (Cambridge University Press, 2005) 1, at 3.
[35] See Vaughan and Lenton, *supra* note 24, at 751.
[36] See *ibid.*, at 751–752.
[37] See e.g. Dominic Woolf *et al.*, "Sustainable biochar to mitigate global climate change" (2010) 1:1 *Nature Communications* 56.
[38] See L.V. Alexander *et al.*, "Summary for policymakers" in Stocker *et al.* (eds.), *supra note 2*, 3, at 11.

acids without affecting its acidity.[39] However, rather than the chemical sink, most studies have focused on the biological sink through which oceans absorb carbons. By fostering more life in the oceans, for instance, by adding nutrients such as iron, nitrogen or phosphorous where they are lacking or by increasing ocean upwelling (the upward movement of deep water rich in nutrients), more plankton could develop and, through photosynthesis, more carbon would be absorbed. These techniques have never been implemented on a large scale. While their effectiveness is uncertain, their impact on the marine environment could be serious. Increasing the acidity of the oceans through directly adding carbon acids would endanger many species and create havoc in the entire marine ecosystem.[40] Other techniques to enhance chemical or biological carbon sinks, if they were to be deployed at a sufficient scale to have any significant impact on the atmosphere chemistry, could have diverse impacts on marine ecosystems which are largely unpredictable, but are likely to be serious and almost certainly irreversible.[41]

2. Carbon Capture and Storage

As natural carbon reservoirs are already reaching their limits, alternative approaches to carbon dioxide removal could seek to create large-scale artificial reservoirs of GhGs. This would require the wide-scale capture and the durable sequestration of large quantities of carbon dioxide.

Technology for the capture of carbon dioxide has significantly improved over time. The use of sorbent materials followed by chemical reactions make it possible to extract carbon dioxide from the air, most efficiently in air with a high concentration of carbon dioxide, such as fumes emitted from power plants. When carbon dioxide is removed from fumes produced by the combustion of fossil fuels and then stored, carbon capture and storage (CCS) prevents additional GhG emissions, but it does not reduce GhG concentrations in the atmosphere. However, GhG concentrations in the atmosphere can be reduced if carbon dioxide is captured from fumes produced by the combustion of biomass, such as wood, which is being regenerated. Thus, bioenergy with carbon capture and storage (BECCS) would make it possible to reduce the concentration of GhGs in the atmosphere through a relatively long process where trees are planted and grown, then cut and replaced, and where the carbon dioxide produced during combustion is separated and stored.

There are several ways to store carbon dioxide durably. It is technologically possible to create artificial minerals containing carbon dioxide, but the process is expensive and energy-intensive.[42] Alternatively, some studies have suggested the possibility of injecting liquefied carbon dioxide on the ocean floor, at a depth of 3,000 meters below sea level, where pressure is sufficient to keep carbon dioxide denser than water. This technique has largely been disregarded, however, because of the uncertainty regarding the durability of such storage, for instance, in the event of a significant earthquake, and its potential impact on the deep marine environment.[43]

[39] Vaughan and Lenton, *supra* note 24, at 754.
[40] M. Rhein *et al.*, "Observations: ocean" in Stocker *et al.* (eds.), *supra note 2*, 255, at 295.
[41] L. Clarke, "Assessing transformation pathways" in Edenhofer *et al.* (eds.), *supra note 5*, 413, at 485.
[42] A. Sanna *et al.*, "A review of mineral carbonation technologies to sequester CO2" (2014) 43 *Chemical Society Reviews* 8049.
[43] Rattan Lal, "Carbon sequestration" (2008) 363:1492 *Philosophical Transactions of the Royal Society B: Biological Sciences* 815.

Instead, the most reliable way to store large quantities of carbon dioxide appeared to be by injecting it into deep geological formations such as depleted oil and gas reservoirs. Under adequate conditions, compressed carbon dioxide trapped in such formations reacts with surrounding elements and, over years, becomes part of stable minerals.[44] In 2005, a special report of the IPCC concluded that, through such a storing technique, "it is considered *likely* that 99 per cent or more of the injected CO_2 will be retained for 1000 years."[45] More recently, a working group of the IPCC observed that "all of the components of integrated CCS systems exist and are in use today by the hydrocarbon exploration, production, and transport, as well as the petrochemical refining sectors."[46]

BECCS with injection into deep geological formations appears to be, by far, the most attractive large-scale geoengineering option. Yet, it is not a silver bullet and it must not replace more efficient steps to reduce GhG emissions in the first place. To make any significant difference in GhG concentrations in the atmosphere, BECCS will need to be deployed on a large scale over several decades.[47] The production of bioenergy requires the use of land which can be in competition with food production, although it could create a much faster carbon sink than forests.[48] Implementing BECCS will require international cooperation. International and domestic regulations will also need to be developed to ensure a close monitoring of the safety and durability of storage sites.

III. REGULATING GEOENGINEERING

Geoengineering, as a deliberate attempt to alter the global environment as a whole, raises unique ethical, legal and geopolitical questions. These questions have become more pressing because of the failure of mitigation action and the development of geoengineering techniques such as BECCS. Thus, discussions on geoengineering have gradually shifted from technical debates about what *could* be done to ethical debates about what *should* be done, and to regulatory debates about *how* it should be done and about *who* should decide these questions.

Each of these debates unfolds differently with regard to given forms of geoengineering. While moral philosophers tend to focus on solar radiation management techniques as a "clear case" of geoengineering,[49] concrete regulatory developments relate more frequently to CCS. There are clear ethical reasons not to deploy solar radiation management techniques (except, perhaps, if absolutely necessary), but well-regulated BECCS appears as an attractive, arguably even a necessary complement to climate change mitigation policies.

---

[44] See e.g. Lei Li *et al.*, "A review of research progress on CO2 capture, storage, and utilization in Chinese Academy of Sciences" (2013) 108 *Fuel* 112; Matthew E. Boot-Handford, "Carbon capture and storage update" (2014) 7 *Energy & Environmental Sciences* 130.

[45] S. Benson *et al.*, "Underground geological storage" in Metz *et al.*, *supra note 34*, 195, at 197.

[46] T. Bruckner *et al.*, "Energy systems" in Edenhofer *et al.* (eds.), *supra note 5*, 511, at 532.

[47] P. Ciais *et al.*, "Carbon and other biogeochemical cycles" in Stocker *et al.* (eds.), *supra note 2*, 465, at 546.

[48] See Clarke, *supra note 41*, at 485.

[49] Stephen Gardiner, "Is 'arming the future' with geoengineering really the lesser evil?" in Stephen Gardiner *et al.* (eds.), *Climate Ethics: Essential Readings* (Oxford University Press, 2010) 284, at 285. See also Christopher Preston (ed.), *Engineering the Climate: The Ethics of Solar Radiation Management* (Lexington, 2012).

A. The Opportunity to Resort to Some Forms of Geoengineering

There is something utterly terrifying in the idea that we, human societies, could further alter our global environment, given how little we know about the extremely complex responses of interacting planetary systems. Solar radiation management would reduce global warming without preventing further changes in the chemistry of our atmosphere and oceans. The injection of sulfites into the stratosphere would produce serious environmental harm through acid rain. Those in favor of further research on solar radiation management present such techniques as potentially a lesser evil in the event that we enter a runaway climate change scenario.[50] There is however a risk that developing a technique may feed social expectation that it should be used, or otherwise divert attention from much-needed efforts to promote climate change mitigation.[51] Technology development would also create an interested community of investors promoting new vested interests, thus increasing the temptation for our societies to rely on palliative treatments rather than taking urgent measures to reduce GhG emissions.[52] Although these considerations should not necessarily bar any research on solar radiation management, they should command great caution from the scientific community when engaging in research and communicating on such extreme geoengineering options.

By contrast, there are fewer general objections against carbon dioxide removal. Efforts directed at land-use policies and forest management were among the first policies implemented to mitigate climate change. Instead, ethical issues are more likely to be raised with regard to the modalities of such actions. If CCS in deep geological formation is to be implemented on a large scale, great caution will be needed to ensure that regulatory frameworks do not "inadvertently perpetuate, rather than reduce, fossil fuel reliance."[53] BECCS should complement rather than replace efforts to mitigate climate change.[54] There is no doubt that it is far more economical and far less impactful on our environment not to emit carbon dioxide in the first place rather than trying to remove and store it afterwards.[55] At present, there seems to be little alternative to BECCS in order to achieve the objective defined by the Paris Agreement of "achiev[ing] a balance between anthropogenic emissions by sources and removals by sinks of greenhouse gases in the second half of this century."[56] However, regulation is needed to ensure the durability and safety of carbon transportation and sequestration, as well as to arbitrate between alternative uses of land for bioenergy or food production.[57]

---

[50] See e.g. Crutzen, *supra* note 32; David Keith, *A Case for Climate Engineering* (MIT Press, 2013).

[51] C. Kolstad *et al.*, "Social, economic, and ethical concepts and methods" in Edenhofer *et al.* (eds.), *supra note 5*, 207 at 219. See also Gardiner, *supra* note 49; Nils Markusson, "Review of 'A Case for Climate Engineering' by David Keith" (2014) 4:3–4 *Climate Law* 365.

[52] Dale Jamieson, "Ethics and intentional climate change" (1996) 33:3 *Climatic Change* 323, at 333. See also Hamilton, *supra* note 3, at 162. See generally William C.G. Burns and Andrew L. Strauss (eds.), *Climate Change Geoengineering: Philosophical Perspectives, Legal Issues, and Governance Frameworks* (Cambridge University Press, 2015).

[53] Jennie C. Stephens, "Carbon capture and storage: a controversial climate mitigation approach" (2015) 50:1 *International Spectator: Italian Journal of International Affairs* 74.

[54] Kevin Anderson and Glen Peters, "The trouble with negative emissions" (2016) 354:6309 *Science* 182.

[55] Hamilton, *supra* note 3, at 20–50.

[56] Paris Agreement, *supra* note 4. The Fifth Assessment Report of the IPCC included BECCS in its mitigation scenario. See Edenhofer *et al.*, *supra* note 5, at 60.

[57] Robert Amon, "Bioenergy carbon capture and storage in global climate policy: examining the issues" (2016) 10:4 *Carbon & Climate Law Review* 187.

B. The Regulation of Geoengineering

Geoengineering activities are regulated by norms of general international law as well as some rules established by treaties. The relevant norms of general international law include the no-harm principle and the precautionary principle, the obligation to conduct an environmental impact assessment and, in the event of a breach of a primary obligation, secondary obligations under the law of State responsibility.[58] Treaty-based regimes clarify some aspects of these norms in relation to geoengineering from the perspective of the protection of the marine environment, that of biological diversity, or within the UNFCCC regime. Further regulatory frameworks are developed in domestic contexts.

1. Protection of the Marine Environment

Several techniques of carbon dioxide removal would have a great impact on the marine environment, whether through attempts to enhance natural carbon sinks in the ocean or through the injection of liquefied carbon dioxide on or below the ocean floor. In this regard, several treaties provide rules and forums for further developments relevant to such geoengineering activities.

Article 192 of the UN Convention on the Law of the Sea defines a general obligation for States "to protect and preserve the marine environment."[59] More specific treaties include the 1972 London Convention on the Prevention of Marine Pollution by Dumping of Wastes and Other Matters and its 1996 Protocol,[60] complemented regionally by the Convention for the Protection of the Marine Environment of the North-East Atlantic (OSPAR Convention).[61] The parties to these treaties adopted rules to regulate seaborne geoengineering activities.

Regarding ocean fertilization – techniques which aim to increase biological sinks of carbon dioxide in the oceans – the Contracting Parties to the London Protocol and to the London Convention agreed in 2008 that "given the present state of knowledge," such activities should be limited to "legitimate scientific research."[62] Two years later, they adopted an "Assessment Framework for Scientific Research Involving Ocean Fertilization" in order to determine, on a case-by-case basis, what constitutes legitimate scientific research.[63] An amendment to the London Protocol adopted

---

[58] See Kerryn Brent, Jeffrey McGee and Amy Maguire, "Does the 'no-harm' rule have a role in preventing transboundary harm and harm to the global atmospheric commons from geoengineering?" (2015) 5:1 *Climate Law* 35; David Reichwein *et al.*, "State responsibility for environmental harm from climate engineering" (2015) 5:2–4 *Climate Law* 142; Karen N. Scott, "International law in the Anthropocene: responding to the geoengineering challenge" (2013) 34:2 *Michigan Journal of International Law* 309; Ralph Bodle, "Geoengineering and international law: the search for common legal ground" (2010) 46:2 *Tulsa Law Review* 305; Neil Craik, "International EIA law and geoengineering: do emerging technologies require special rules?" (2015) 5:2–4 *Climate Law* 111. See also CBD decision XI/20, "Climate-related geoengineering" (Hyderabad, October 8–19, 2012), para. 11; UN General Assembly Resolution 66/288, "The future we want" (July 27, 2012), para. 167.

[59] United Nations Convention on the Law of the Sea, December 10, 1982, 1833 *UNTS* 3, art. 192 and generally Chapter 12.

[60] London Convention on the Prevention of Marine Pollution by Dumping of Wastes and Other Matter, December 29, 1972, 1046 *UNTS* 120 (hereinafter London Convention); Protocol to the Convention on the Prevention of Marine Pollution by Dumping of Wastes and Other Matter, November 7, 1996, 1046 *UNTS* 138 (hereinafter London Protocol).

[61] Convention for the Protection of the Marine Environment of the North-East Atlantic, September 22, 1992, 2354 *UNTS* 67.

[62] London Convention/London Protocol resolution LC-LP.1(2008) on the regulation of ocean fertilization (October 31, 2008), para. 8.

[63] London Convention/London Protocol resolution LC-LP.2(2010) on the assessment framework for scientific research involving ocean fertilization (October 14, 2010).

in 2013 transposed these provisions into treaty law.[64] This moratorium on ocean fertilization was supported by a resolution adopted by the UN General Assembly in 2012, based on the outcome of the United Nations Conference on Sustainable Development, reflecting States' "concern about the potential environmental impacts of ocean fertilization" and calling for "utmost caution."[65]

In addition, a 2006 amendment to the London Protocol makes it clear that the sequestration of carbon dioxide in geological formations situated *under* the ocean floor (i.e. underground) would not amount to illegal dumping prohibited by the Protocol.[66] This would allow the storage of carbon dioxide in deep geological formations situated under the seabed, thus providing an extra layer of safety in case of any leakage. By implication, this amendment confirms that any injection of carbon dioxide *above* the ocean floor (i.e. in the water column) would amount to dumping and would constitute a breach of the Protocol.[67] At the regional level, in 2007 the Contracting Parties to the OSPAR Convention committed not to store carbon dioxide in ocean waters and adopted guidelines for risk assessment and management of carbon storage in sub-seabed geological formations.[68]

2. The Convention on Biological Diversity

Important steps toward the regulation of geoengineering have also taken place under the Convention on Biological Diversity. A 2008 decision of the COP/CBD noted the adoption of the decision of the parties to the London Protocol laying a moratorium on ocean fertilization and, in similar language, it called upon States:

> to ensure that ocean fertilization activities do not take place until there is an adequate scientific basis on which to justify such activities, including assessing associated risks, and a global, transparent and effective control and regulatory mechanism is in place for these activities; with the exception of small scale scientific research studies within coastal waters.[69]

The COP/CBD went further in 2010 when it adopted a decision expanding this moratorium beyond the marine environment to any "climate-related geo-engineering activities ... that may affect biodiversity ... until there is an adequate scientific basis on which to justify such

---

[64] London Protocol resolution LP.4(8) on the amendment to the London Protocol to regulate the placement of matter for ocean fertilization and other marine geoengineering activities (October 18, 2013). See generally Harald Ginzky and Robyn Frost, "Marine geo-engineering: legally binding regulation under the London Protocol" (2014) 9:2 *Carbon & Climate Law Review* 82.

[65] UN General Assembly Resolution 66/288, *supra* note 58, para. 167. See also *infra* note 69.

[66] London Protocol resolution LP.1(1) on the amendment to include CO_2 sequestration in sub-seabed geological formations in Annex 1 to the London Protocol (November 2, 2006). See also London Protocol resolution LP.3(4) on the amendment to Article 6 of the London Protocol (October 30, 2009), regarding the export of carbon dioxide for sequestration in sub-seabed geological formation; and *Specific Guidelines for the Assessment of Carbon Dioxide for Disposal into Sub-seabed Geological Formations* (November 2, 2012), London Convention doc. LC 34/15, Annex 8.

[67] See London Protocol, *supra* note 60, art. 4.1.

[68] See OSPAR Convention decision 2007/1 to Prohibit the Storage of Carbon Dioxide Streams in the Water Column or on the Sea-Bed (June 25–29, 2007); OSPAR Convention decision 2007/2 on the Storage of Carbon Dioxide Streams in Geological Formations (June 25–29, 2007); and *OSPAR Guidelines for Risk Assessment and Management of CO2 in Sub-seabed Geological Formations*, document 2007–12.

[69] CBD decision IX/16, "Biodiversity and climate change" (Bonn, May 19–30, 2008), para. C.4. See also resolution LC-LP.1(2008), *supra* note 62.

activities."[70] Although the parties could not agree on a definition of "geo-engineering activities," a footnote indicated that this moratorium applied to solar radiation management as well as CCS activities, except CCS "from fossil fuels when it captures carbon dioxide before it is released into the atmosphere."[71] This moratorium should hold "until there is an adequate scientific basis on which to justify such activities and appropriate consideration of the associated risks for the environment and biodiversity and associated social, economic and cultural impacts."[72] "Small scale scientific research studies" are not affected.[73] This moratorium does not affect geoengineering activities that are unlikely to have any impact on biological diversity.

Following this decision, the Executive Secretary of the CBD prepared two reports: one on the impacts of climate-related geoengineering on biological diversity; and the other on the regulatory framework for climate-related geoengineering.[74] On these bases, the COP/CBD reaffirmed the moratorium of 2010 in a 2012 decision, noting in particular "the lack of science-based, global, transparent and effective control and regulatory mechanisms for climate-related geoengineering."[75] Similar decisions were adopted in 2014 and 2016.[76]

3. The UNFCCC Regime

Relatively little regulation on geoengineering was adopted under the UNFCCC regime due to a lack of consensus on the role that such techniques should play in responses to climate change. Nevertheless, the UNFCCC defined national commitments on climate change mitigation as involving not only efforts to limit and reduce GhG emissions, but also "protecting and enhancing ... greenhouse gas sinks and reservoirs."[77] In 1992, these were understood as references to forests and other natural sinks and reservoirs rather than any artificial CCS process. Based on the same understanding, the Kyoto Protocol allowed Annex I parties to use "net changes in greenhouse gas emissions by sources and removals by sinks resulting from direct human-induced land-use change and forestry activities" to comply with their mitigation commitments.[78] Such activities could qualify for project activities under the CDM.[79] Beyond this, the Bali Action Plan initiated a program on "Reducing Emissions from Deforestation and Forest Degradation in Developing Countries" and on "the role of conservation, sustainable management of forests and enhancement of forest carbon stocks in developing countries" (REDD+).[80]

---

[70] CBD decision X/33, "Biodiversity and climate change " (Nagoya, October 18–29, 2010), para. 8(w).

[71] *Ibid.*, note 3.

[72] *Ibid.*, para. 8(w).

[73] *Ibid.*

[74] See respectively CBD Subsidiary Body on Scientific, Technical and Technological Advice, *Impacts of Climate-Related Geoengineering on Biological Diversity* (April 5, 2012), doc. UNEP/CBD/SBSTTA/16/INF/28; CBD Subsidiary Body on Scientific, Technical and Technological Advice, *Regulatory Framework for Climate-Related Geoengineering Relevant to the Convention Biological Diversity* (April 2, 2012).

[75] CBD decision XI/20, *supra* note 58, para. 6.

[76] CBD decision XII/20, "Biodiversity and climate change and disaster risk reduction" (Pyeongchang, October 6–17, 2014), para. 1; CBD decision XIII/14, "Climate-related geoengineering" (Cancún, December 4–17, 2016), para. 1.

[77] UNFCCC, *supra* note 1, art. 4.2(a).

[78] Kyoto Protocol to the United Nations Framework Convention on Climate Change, December 11, 1997, 2303 *UNTS* 162 (hereinafter Kyoto Protocol), art. 3.3.

[79] See decision 5/CMP.1, "Modalities and procedures for afforestation and reforestation project activities under the clean development mechanism in the first commitment period of the Kyoto Protocol" (November 30, 2005).

[80] Decision 1/CP.13, "Bali Action Plan" (December 14–15, 2013), para. 1(b)(iii). See generally Chapter 12, section III.

By contrast, avoided emissions from CCS and negative emissions from BECCS were not immediately admitted. In 2006, the revised guidelines for national GhG inventories of the IPCC included a new chapter on "carbon dioxide transport, injection and geological storage."[81] After years of debate,[82] it was finally agreed in 2010 that, as a matter of principle, the CDM could lend support to CCS in geological formations.[83] Modalities and procedures were adopted in 2011, including detailed requirements regarding the selection and characterization of the geological storage site, risk and safety assessment, monitoring, financial provisions and liability, and environmental and socio-economic impact assessments.[84] To date, however, no methodology has been approved by the CDM Executive Board.[85]

The objective of the Paris Agreement to achieve "a balance between anthropogenic emissions by sources and removals by sinks of greenhouse gases in the second half of this century"[86] reflects, for the first time, an emerging consensus on the need for "negative emissions" which are unlikely to be limited to natural sinks and reservoirs, but would most likely include BECCS.[87] Several States mentioned CCS as part of the mitigation action described in their INDCs.[88] A major challenge for the UNFCCC regime in the years to come will be to channel such growing interest for CCS toward the development of safe, reliable and affordable technologies for BECCS activities, while ensuring that such developments do not divert attention from the urgent need to reduce GhG emissions.

4. Domestic Regulation

The development and implementation of technical regulation are instrumental to ensuring that CCS techniques are safe and effective. The transportation and storage of carbon dioxide poses risks to human and animal lives because leaks can displace breathable air and cause asphyxiation.[89] Slower leaks from storage locations would reduce the benefit of the whole process. In the case of storage

---

[81] Sam Holloway et al., "Carbon dioxide transport, injection and geological storage" in Simon Eggleston et al., 2006 IPCC Guidelines for National Greenhouse Gas Inventories (IGES, 2006) Chapter 5.

[82] See Global Change Strategies International, Use of the Clean Development Mechanism for CO2 Capture and Storage (IEA GhG R&D Programme, July 15, 2004); Cédric Philibert, Jane Ellis and Jacek Podkanski, Carbon Capture and Storage in the CDM (OECD and IEA, December 2007); Tim Dixon et al., "CCS projects as Kyoto Protocol CDM activities" (2013) 37 Energy Procedia 7596.

[83] Decision 7/CMP.6, "Carbon dioxide capture and storage in geological formations as clean development mechanism project activities" (December 10–11, 2010), para. 1. For an overview of the arguments made by different parties and observers, see UNFCCC SBSTA, Synthesis of Views on Technological, Methodological, Legal, Policy and Financial Issues Relevant to the Consideration of Carbon Dioxide Capture and Storage in Geological Formations as Project Activities under the Clean Development Mechanism (September 25, 2008), doc. FCCC/SBSTA/2008/INF.3.

[84] Decision 10/CMP.7, "Modalities and procedures for carbon dioxide capture and storage in geological formations as Clean Development Mechanism project activities" (December 9, 2011). See generally Meinhard Doelle and Emily Lukaweski, "Carbon capture and storage in the CDM: finding its place among climate mitigation options?" (2012) 3:1 Climate Law 49.

[85] See UNFCCC CDM, CDM Methodology Booklet, 8th edn (November 2016) 5.

[86] Paris Agreement, supra note 4, art. 4.1. That the IPCC included BECCS in its mitigation scenarios is also symptomatic of a growing consensus. See also Edenhofer et al., supra note 5, at 60.

[87] Joeri Rogelj et al., "Paris Agreement climate proposals need a boost to keep warming well below 2°C" (2016) 534:7609 Nature 631.

[88] See, for instance, the INDCs communicated by Bahrein, Canada, China, Malawi, Norway, Saudi Arabia, South Africa and the United Arab Emirates.

[89] The risk to populations exposed to sudden release of carbon dioxide was tragically illustrated by a natural disaster which occurred in Cameroon in 1986. See George W. Kling et al., "The 1986 Lake Nyos gas disaster in Cameroon, West Africa" (1987) 236:4798 Science 169.

in deep geological formations, potential sites need to be carefully selected, safety precautions need to be taken during the process and site monitoring should be carried out for decades. In legal terms, this raises difficult questions regarding the liability of operators over long periods of time.

EU Directive 2009/31/EC on the geological storage of carbon dioxide is one of the most advanced regulations on CCS. It provides a detailed set of rules on the selection of storage sites, on the issuance of storage permits and on operation, closure and post-closure obligations.[90] Following the completion of operations, the Directive provides for a transfer of responsibility for post-operation monitoring and maintenance to the competent national authorities, with a financial contribution of the operator to cover what national authorities estimate as "the anticipated cost of monitoring for a period of 30 years."[91]

C. Decision–Making

The development of geoengineering and national regulation raises new questions relating to international cooperation. Deciding whether we should resort to particular geoengineering technologies and under which conditions requires more than technical expertise; such decisions unavoidably involve pondering competing interests, diverging values and conflicting legal principles. Virtually any form of geoengineering would benefit some interests, pursue some values and implement some legal principles while acting against others. For instance, solar radiation management technologies would reduce global temperature but would also affect global environmental systems in many ways which cannot entirely be predicted. Likewise, a large-scale deployment of BECCS would reduce GhG concentrations in the atmosphere, but the production of bioenergy would compete with other uses of scarce lands, with a global impact on the price of food, while carbon dioxide transportation and storage could pose risks to public health and to the environment if it is not properly monitored.

More generally, debates on geoengineering suggest a world where intended transformations can be effected on the climate system, raising formidable challenges for international cooperation. Each State is likely to have a different perspective on what the Earth's climate should be like: some countries might have great interest in reducing global average temperature back to pre-historical levels, while others might satisfy themselves with maintaining slightly higher temperatures, or even push for lower temperatures. Some States might be ready to pay a higher cost, for instance, in terms of environmental pollution or biological diversity, in order to avoid any further warming, while others may not. Any geoengineering option would benefit some countries but harm others. As David Victor and colleagues observed:

> One nation's emergency can be another's opportunity, and it is unlikely that all countries will have similar assessments of how to balance the ills of unchecked climate change with the risk that geoengineering could do more harm than good.[92]

---

[90] EU Directive 2009/31/EC on the geological storage of carbon dioxide (April 23, 2009), doc. 02009L0031. See generally Michael G. Faure and Roy A. Partain, *Carbon Capture and Storage: Efficient Legal Policies for Risk Governance and Compensation* (MIT Press, 2017); and Michael B. Gerrard and Tracy Hester (eds.), *Climate Engineering and the Law: Regulation and Liability for Solar Radiation Management and Carbon Dioxide Removal* (Cambridge University Press, 2018).

[91] Directive 2009/31/EC, *supra* note 90, art. 20.

[92] Victor *et al.*, *supra* note 8.

From a technological and economic perspective, it could rapidly become possible for one or a few States, or some powerful non-State actors, to implement solar radiation management techniques such as the injection of sulfites into the stratosphere, with catastrophic environmental consequences globally.[93] The States most vulnerable to climate change impacts, those whose existence is threatened or whose development is significantly disturbed by sea-level rise, could lose patience with the slow progress of international cooperation on climate change mitigation. The use or the threat of use of geoengineering could exacerbate latent conflicts that closely relate to competition for access to environmental resources. International cooperation and trust therein will be essential to avoid the availability of radical geoengineering techniques that severely affect relations between States. Rather than a technological challenge, geoengineering is primarily a challenge to international cooperation.

IV. CONCLUSION

Coming straight from science fiction, geoengineering has now entered the realm of geopolitics. The options it offers are often frankly undesirable; none of them is a panacea to climate change. Solar radiation management technologies such as the injection of sulfites into the stratosphere could rapidly be deployed, but they would represent an incomplete, temporary and extremely dangerous "fix." Carbon dioxide removal, on the other hand, is slower to operationalize, but it has clearer benefits. BECCS could be a useful technique, in the decades to come, to reduce carbon dioxide concentrations in the atmosphere, alongside drastic cuts in GhG emissions.

---

[93] Vaughan and Lenton, *supra* note 24, at 782.

10

International Action on Climate Change Adaptation

Despite all present efforts to reduce GhG emissions and enhance sinks and reservoirs of GhGs, climate change is already happening. Global average temperature is estimated to have increased by around 1.1 degrees Celsius above pre-industrial levels.[1] Present GhG concentrations in the atmosphere and continued emissions will cause further warming, which is likely, following current trends in GhG emissions, to exceed 2 degrees Celsius above pre-industrial levels during the twenty-first century.[2] This is and will increasingly be affecting the Earth system in various ways, from the sheer effect of higher temperatures on the water cycle (e.g. droughts and changes in precipitation patterns) to a rise in sea level, an acidification of seawater and an increase in the frequency and severity of some extreme weather events (e.g. tornadoes, storms and wild fires). These physical impacts of climate change will have far-reaching consequences for human societies, impacting food production, public health, economic development, peace, security and human mobility, among other things.[3] Sometimes, measures can be taken to reduce or manage these adverse consequences.

The adverse impacts of climate change can be approached from two alternative legal perspectives: as a matter of protection or as a matter of remediation.[4] From a protection perspective, the adverse impacts of climate change are viewed as challenges that States need to address within their territory, for instance, through the protection of human rights and the environment, or more generally through the promotion of sustainable development. From a remedial perspective, by contrast, these impacts are perceived as the consequences of a wrongful act – the failure of States to prevent excessive GhG emissions – entailing the responsibility of the responsible States and the obligation to pay adequate reparation to the States affected.[5] The protection perspective has largely dominated the debates on adaptation to climate change within the UNFCCC regime and beyond, and it is explored in this chapter.

---

[1] World Meteorological Organization, *Statement on the State of the Global Climate in 2016* (WMO, 2017) 2.

[2] L.V. Alexander *et al.*, "Summary for policymakers" in T.F. Stocker *et al.* (eds.), *Climate Change 2013: The Physical Science Basis. Contribution of Working Group I to the Fifth Assessment Report of the Intergovernmental Panel on Climate Change* (Cambridge University Press, 2013) 3, at 20.

[3] See generally Chapter 1, section I.C.

[4] In addition, the impacts of climate change have also been emphasized in support of arguments for more stringent action on climate change mitigation, with the risk that simplistic causal attribution of events to climate change could backfire against the same advocates. See, for instance, Michelle Nijhuis, "What's missing from 'An Inconvenient Sequel,' Al Gore's new climate-change documentary" *New Yorker* (July 29, 2017).

[5] See generally Chapter 5, section IV.

Chapter 11 turns to the concept of "loss and damage," approached as a new attempt to develop a remedial perspective on the adverse impacts of climate change.

As defined by the IPCC, adaptation to climate change refers to "the process of adjustment to actual or expected climate and its effects."[6] More specifically, Neil Adger and colleagues explain that climate change adaptation is fundamentally "about adjusting to risks, either in reaction to or in anticipation of changes arising from changing weather and climate."[7] Adaptation efforts become more urgent as mitigation action remains insufficient and our societies continue to interfere with the climate system.[8] While the UNFCCC regime was initially confined to mitigation efforts,[9] adaptation has progressively been recognized as an international priority, in particular since the adoption of the 2007 Bali Action Plan.[10] Today, mitigation and adaptation are generally considered as the two main pillars of international responses to climate change.

National governments have sought to promote adaptation in a great variety of ways. Adaptation action may concern, for instance, food and water, public health, cultural heritage and biodiversity; measures can relate to infrastructure development, land-use planning, construction law or social protection and insurance, among other things.[11] Adaptation action is highly context-specific as it seeks to address the particular impacts of climate change which impact the territory of a country according to its particular circumstances. In addition, adaptation involves value-based judgments about what is worth protecting and what is worth sacrificing. As such, the determination of adaptation strategies is ultimately political and deeply rooted in national development strategies, and it is often challenging to disentangle adaptation efforts from such development strategies.[12] It is unsurprising, therefore, that the international law on climate change imposes few unconditional rights or obligations on States in relation to adaptation within their own territory.[13] Rather, the international law on climate change

---

[6] R.K. Pachauri *et al.*, *Climate Change 2014: Synthesis Report. Contribution of Working Groups I, II and III to the Fifth Assessment Report of the Intergovernmental Panel on Climate Change* (IPCC, 2015) 118.

[7] Neil Adger *et al.*, "Cultural dimensions of climate change impacts and adaptation" (2013) 3:2 *Nature Climate Change* 112, at 112.

[8] Lisa Schipper, "Conceptual history of adaptation in the UNFCCC process" (2006) 15:1 *Review of European, Comparative & International Environmental Law* 82, at 84.

[9] See United Nations Framework Convention on Climate Change, May 9, 1992, 1771 *UNTS* 107 (hereinafter UNFCCC), art. 2.

[10] Decision 1/CP.13, "Bali Action Plan" (December 14–15, 2007), para. 1(c).

[11] This chapter does not contain a section on domestic measures of implementation. Instead, concrete illustrations can be found, for instance, in Jonathan Verschuuren (ed.), *Research Handbook on Climate Change Adaptation Law* (Edward Elgar, 2013); He Xiangbai, "Setting the legal enabling environment for adaptation mainstreaming into environmental management in China: applying key environmental law principles" (2014) 17 *Asia Pacific Journal of Environmental Law* 23; Sushil Vachani and Jawed Usmani (eds.), *Adaptation to Climate Change in Asia* (Edward Elgar, 2014); Koh Kheng-Lian *et al.* (eds.), *Adaptation to Climate Change* (World Scientific, 2015); Meinhard Doelle, Steven Evans and Tony George Puthucherril, "The role of the UNFCCC regime in ensuring effective adaptation in developing countries: lessons from Bangladesh" (2014) 4:3–4 *Climate Law* 327; Daniel A. Farber, "The challenge of climate change adaptation: learning from national planning efforts in Britain, China, and the USA" (2011) 23:3 *Journal of Environmental Law* 359; Anja Bauer, Judith Feichtinger and Reinhard Steurer, "The governance of climate change adaptation in 10 OECD countries: challenges and approaches" (2012) 14:3 *Journal of Environmental Policy & Planning* 279; and Justin Gundlach and P. Dane Warren, *Local Law Provisions for Climate Change Adaptation* (Sabin Center for Climate Change Law, May 2016).

[12] J.D. Ford *et al.*, "Adaptation tracking for a post-2015 climate agreement" (2015) 5:11 *Nature Climate Change* 967, at 967.

[13] See, for instance, Paris Agreement, December 12, 2015, in the annex of decision 1/CP.21, "Adoption of the Paris Agreement" (December 12, 2015), art. 7 which, over two pages, does not define any unconditional right or obligation

adaptation is largely limited to an attempt at promoting awareness and attention to adaptation challenges within national governments and at enhancing the provision of dedicated support to developing States.

Following further exploration of the concept of adaptation, this chapter documents the relevant developments which have taken place within the UNFCCC regime and in other regimes. Finally, it identifies five principles which have gradually emerged and ought to guide international action on climate change adaptation.

I. THE CONCEPT OF ADAPTATION

States have certain obligations under international law which apply in relation to the impacts of climate change within their jurisdiction. Under international human rights law, they must take all reasonable measures within their capacity to protect and fulfill the rights of anyone under their jurisdiction.[14] More generally, they should create the conditions for a human development,[15] promote an effective management of disaster risks[16] and protect the environment.[17] States must fulfill these obligations individually but also, if need be, through cooperation.[18]

Often, climate change hinders the achievement of these objectives. A recent assessment report by the IPCC recognized clear evidence that "many of the observed changes" recently observed in planetary systems "are unprecedented over decades to millennia."[19] The consequences are largely concentrated on some of the poorest nations – the very same nations whose institutions are the least equipped to face additional challenges.[20] Droughts, floods, heatwaves, cyclones and famines occur in the countries least able to face such challenges. The impacts of climate change raise new protection challenges, to which international cooperation on climate change adaptation seeks to provide at least some partial responses.

In its fifth assessment report, the IPCC has developed a conceptual framework that is useful in terms of understanding the complementary entry points in addressing the impacts of climate change. This framework approaches the impacts of climate change as determined by

---

other than procedural obligations. See also Lavanya Rajamani, "The 2015 Paris Agreement: interplay between hard, soft and non-obligations" (2016) 28:2 *Journal of Environmental Law* 337, at 344. For a more general reflection, see Jan McDonald, "The role of law in adapting to climate change" (2011) 2:2 *Wiley Interdisciplinary Reviews: Climate Change* 283; J.B. Ruhl, "Climate change adaptation and the structural transformation of environmental law" (2010) 40:2 *Environmental Law* 363.

[14] See e.g. Charter of the United Nations, June 26, 1945, 1 *UNTS* XVI, art. 1.3; UN General Assembly Resolution 217 A, "Universal Declaration of Human Rights" (December 10, 1948); International Covenant on Economic, Social and Cultural Rights, December 16, 1966, 993 *UNTS* 3; and International Covenant on Civil and Political Rights, December 16, 1966, 999 *UNTS* 171.

[15] See e.g. UN General Assembly Resolution 41/128, "Declaration on the Right to Development" (December 4, 1986); UN General Assembly Resolution 70/1, "Transforming our world: the 2030 Agenda for Sustainable Development" (September 25, 2015).

[16] See in particular Sendai Framework for Disaster Risk Reduction 2015–2030, in UN General Assembly Resolution 69/283 (June 3, 2015), Annex II.

[17] See e.g. UNCED, Rio Declaration on Environment and Development (June 3–14, 1992), available in (1992) 31 *ILM* 874 (hereinafter Rio Declaration).

[18] See Chapter 5, section II.D.

[19] Alexander *et al.*, *supra* note 2, at 4.

[20] Christopher B. Field *et al.*, "Summary for policymakers" in Christopher B. Field *et al.* (eds.), *Climate Change 2014: Impacts, Adaptation, and Vulnerability. Part A: Global and Sectoral Aspects. Working Group II Contribution to the Fifth Assessment Report of the Intergovernmental Panel on Climate Change* (Cambridge University Press, 2014) 1, at 6.

three elements: hazard, exposure and vulnerability. Hazard refers to the potential occurrence of a physical event at a given place. Exposure relates to the presence of people or things in places and settings where the event may occur. Lastly, vulnerability is the tendency of exposed people or things to be adversely and severely affected by the event. While climate change mitigation aims to limit the increase in hazards, adaptation action seeks to address exposure and vulnerability to such hazards in order to prevent or reduce harm.[21]

Efforts to reduce exposure may include resettlement and migration, but also land-use planning, infrastructure investments or various financial incentives to avoid human settlement in high-risk areas. Vulnerability, on the other hand, can be decreased through context-specific risk management strategies such as storing freshwater to reduce vulnerability to drought or building dikes to protect population from floods. Vulnerability can also be decreased through building resilience – the ability of a system to recover from an external stress – for instance, through social protection or insurance mechanisms allowing affected individuals or companies to rapidly recover their loss following a disaster.

A distinction can also be made between incremental and structural measures of adaptation. Incremental adaptation keeps the essential features of the system unchanged, but it may be insufficient in response to the most consequential impacts of climate change.[22] For instance, the construction of a dike to protect a community from risks of floods, which can be considered as an incremental measure for the adaptation of a community to sea-level rise, may become insufficient if the risk of flooding is particularly high. Structural adaptation, such as resettlement, generally involves greater immediate economic and human costs, but also possibly more long-term benefits.

"Maladaptation" refers to ill-conceived measures of adaptation, leading to undesirable outcomes. Maladaptation occurs, for instance, because of "[p]oor planning, overemphasizing short-term outcomes, or failing to sufficiently anticipate consequences."[23] To avoid maladaptation, experts sought to define ways to "enable climate-resilient pathways"[24] through a series of decisions and actions which would promote adaptation to climate change in an efficient way and thus minimize the harm resulting from climate change.

Yet, adaptation decisions cannot be made by experts alone. Central to adaptation to climate change are value judgments about what constitutes adverse impacts, which impacts should be addressed in priority, and how many resources should be dedicated to this.[25] For instance, there could be reasonable disagreements, based on alternative values, on whether the protection of an endangered species of human predators, such as sharks or tigers, would be an adverse impact of climate change on biodiversity or a positive impact in terms of human safety. Any adaptation action comes at a cost and often involves some other adverse consequences. The resettlement of a community may, for instance, place it away from physical risk, but it could also uproot it from a place which has cultural value and affect traditional livelihoods. In this

---

[21] *Ibid.*, at 5.
[22] Christopher B. Field *et al.*, "Technical summary" in Field *et al.* (eds.), *supra note 20,* 35, at 40.
[23] Field *et al.*, "Summary for policymakers," *supra* note 20, at 28.
[24] *Ibid.*, at 29.
[25] See generally Siri H, Eriksen, Andrea J. Nightingale and Hallie Eakin, "Reframing adaptation: the political nature of climate change adaptation" (2015) 35 *Global Environmental Change* 523; Olanrewaju Fagbohun, "Cultural legitimacy of mitigation and adaptation to climate change: an analytical framework" (2011) 6:3 *Carbon & Climate Law Review* 308; and Thoko Kaime, *International Climate Change Law and Policy: Cultural Legitimacy in Adaptation and Mitigation* (Routledge, 2014).

example and in others, there is ample room for reasonable disagreements, based on alternative values, about the action that should be carried out in adapting to climate change.

Adaptation takes place at many different levels of decision-making. Whether they know it or not, individuals and families, companies, local and national governments, regional entities and international organizations all make decisions relevant to climate change adaptation. Individuals, families and companies are often most able to determine what they value the most and how best to respond when their environment changes. However, regulation may be necessary, for instance, to make up for individuals' lack of knowledge, rationality or resources, and to ensure that everyone can, as far as possible, be out of harm's way. Although States have the primary responsibility for the protection of their population and the environment within their jurisdiction, international support may be instrumental in successful adaptation strategies.

The role of international law with regard to adaptation to climate change is limited. Unlike climate change mitigation, there is no overarching collective action issue that requires international cooperation in adapting to climate change. In general, whether a State can protect the population and the environment within its jurisdiction does not depend on the conduct of any other State.[26] The general and abstract obligation of a State with regard to human rights or environmental protection applies whether or not the State is affected by particular impacts of climate change. Therefore, despite the adoption of multiple COP decisions and other international documents, the international law on climate change defines few substantive rights and obligations concerning adaptation. Rather, as will be discussed in the following sections, the relevant provisions are often limited to recommendations couched in a "should" language.

II. DEVELOPMENTS WITHIN THE UNFCCC REGIME

Throughout a quarter of a century of international negotiations on climate change, adaptation was gradually elevated from little more than an afterthought to a global priority formally on an equal footing with mitigation. In addition to encouraging international support, successive agreements and decisions have determined some priority areas and created opportunities for exchange of practices.

A. The UNFCCC

Although the UNFCCC focuses largely on climate change mitigation, it contains a few provisions on climate change adaptation. Within the "ultimate objective" defined by Article 2 of the UNFCCC, adaptation could be construed as a complementary strategy to prevent "anthropogenic interference with the climate system" from becoming "dangerous."[27] Efforts should be sufficient, according to Article 2, in order "to allow ecosystems to adapt naturally to climate change, to ensure that food production is not threatened and to enable economic development to proceed in a sustainable manner."[28]

The parties to the UNFCCC are committed to take measures to promote adaptation to climate change, although these commitments are phrased in very general terms. Unlike climate change

---

[26] There are some exceptions, for instance concerning situations which could lead to international migration or the protection of species whose habitat spans over the territory of more than one country.

[27] UNFCCC, *supra* note 9, art. 2.

[28] *Ibid.*, art. 2.

mitigation, national commitments on climate change adaptation are generally applicable in the same way to all parties: it belongs to each State, rich or poor, notwithstanding its rate of GhG emissions, to protect as far as it can the people and ecosystems within its jurisdiction. Accordingly, all parties are to "formulate, implement, publish and regularly update national and, where appropriate, regional programmes containing ... measures to facilitate adequate adaptation to climate change."[29] All parties must also "[c]ooperate in preparing for adaptation to the impacts of climate change," in particular by "develop[ing] and elaborat[ing] appropriate and integrated plans for coastal zone management, water resources and agriculture, and for the protection and rehabilitation of areas, particularly in Africa, affected by drought and desertification, as well as floods."[30] Lastly, all parties further committed to "[t]ake climate change considerations into account, to the extent feasible, in their relevant social, economic and environmental policies and actions."[31]

Although adaptation efforts must be made by all States, some may lack the necessary resources to implement the appropriate measures. In the final days of the negotiations, the Alliance of Small Island States (AOSIS) secured the inclusion of a provision on financial support to adaptation in developing parties. According to Article 4.4, Annex II parties "shall ... assist the developing country Parties that are particularly vulnerable to the adverse effects of climate change in meeting costs of adaptation to those adverse effects."[32] Some contemporary commentators saw in this provision "an implicit acceptance by developed country Parties of responsibility for causing climate change."[33] Yet, as Daniel Bodansky then observed, "costs" did not mean "all the costs."[34] The commitment of developed States was to cover *some* of the costs of adaptation in developing States – a fraction which, in subsequent practice, turned out to be particularly small.[35]

The parties to the UNFCCC further committed to "give full consideration to what actions are necessary under the convention, including actions related to funding, insurance and the transfer of technology, to meet the specific needs and concerns of developing country Parties arising from the adverse effects of climate change,"[36] and in particular "to take full account of the specific needs and special situations of the least developed countries."[37] Discussions on the implementation of these provisions led to the creation of a work program for the least developed States and the development of National Adaptation Programmes of Action, which are to be supported by a least developed country fund.[38]

---

[29] *Ibid.*, art. 4.1(b).

[30] *Ibid.*, art. 4.1(e). See UNCED, *Agenda 21* (June 3–14, 1992), paras. 12.40; United Nations Convention to Combat Desertification in Those Countries Experiencing Serious Drought and/or Desertification, Particularly in Africa, October 14, 1994, 1954 *UNTS* 3 (hereinafter Convention against Desertification).

[31] UNFCCC, *supra* note 9, art. 4.1(f).

[32] *Ibid.*, art. 4.4.

[33] Philippe Sands, "The United Nations Framework Convention on Climate Change" (1992) 1 *Review of European Community & International Environmental Law* 270, at 275.

[34] See Daniel Bodansky, "The United Nations Framework Convention on Climate Change: a commentary" (1993) 18:2 *Yale Journal of International Law* 451, at 528.

[35] See Chapter 12.

[36] UNFCCC, *supra* note 9, art. 4.8.

[37] *Ibid.*, art. 4.9.

[38] See decision 5/CP.7, "Implementation of Article 4, paragraphs 8 and 9, of the Convention" (November 10, 2001), para. 11(c); and generally decision 28/CP.7, "Guidelines for the preparation of national adaptation programmes of action" (November 10, 2001).

Generally, however, the objectives and national commitments that the UNFCCC defines on climate change adaptation are less ambitious and less specific than those on climate change mitigation.[39] The emphasis on climate change mitigation rather than adaptation was natural at a time when the effects of climate change, although already more than speculative, remained of an uncertain nature. In this context, "adaptation was viewed as a 'defeatist' option, and support for it was considered an acknowledgement that climate change impacts would require adjustments beyond normal behaviour, as well as an admission that mitigation would be insufficient or ineffective."[40] Developed States saw any mention of adaptation as a possible entry point for questions of responsibility which they were – and, by and large, remain – unwilling to discuss.[41]

B. The Kyoto Protocol

The Kyoto Protocol does little to balance the emphasis of the UNFCCC on climate change mitigation: it is primarily an agreement on mitigation. It was adopted on the basis of Article 4.2(d) of the UNFCCC, allowing the COP to review the adequacy of the commitments made by developed States with regard, exclusively, to climate change mitigation.[42] Negotiations were conducted based on the preliminary agreement that no new commitments would be introduced for developing States,[43] although nothing prevented a reaffirmation of what had already been agreed upon. Thus, under Article 10 of the Kyoto Agreement, the parties "reaffirm ... existing commitments" made under the UNFCCC and pledge to "continue ... to advance the implementation of these commitments."[44] Article 10 further recalls the obligation for all States to "[f]ormulate, implement, publish and regularly update ... programmes containing .... measures to facilitate adequate adaptation to climate change."[45]

Yet, at the insistence of developing States, the CDM was also created to support climate change adaptation.[46] The Kyoto Protocol provided that a share of the proceeds from certified project activities would be used "to assist developing country Parties that are particularly vulnerable to the adverse effects of climate change to meet the costs of adaptation."[47] From the perspective of developing States, this provision was an opportunity not only to enhance financial support to adaptation, but also to ensure that at least a share of such support would be provided through multilateral funds, thus avoiding the imposition of political conditions to such financial support.[48] The modalities of implementation of this provision were decided by the COP as part of the Marrakesh Accords of 2001. Accordingly, an Adaptation Fund was

---

[39] See Chapter 7, section II.
[40] Schipper, *supra* note 8, at 84.
[41] *Ibid.*, at 83–84.
[42] See UNFCCC, *supra* note 9, art. 4.2(d).
[43] Decision 1/CP.1, "The Berlin Mandate: review of the adequacy of Article 4, paragraph 2 (a) and (b), of the Convention, including proposals related to a protocol and decisions on follow-up" (April 7, 1995), para. 2(b).
[44] Kyoto Protocol to the United Nations Framework Convention on Climate Change, December 11, 1997, 2303 *UNTS* 162 (hereinafter Kyoto Protocol), art. 10.
[45] *Ibid.*, art. 10(b). See also UNFCCC, *supra* note 9, art. 4.1(b).
[46] See Chapter 8, section II.
[47] Kyoto Protocol, *supra* note 44, art. 12.8.
[48] See Sebastian Oberthür and Hermann E. Ott, *The Kyoto Protocol: International Climate Policy for the 21st Century* (Springer, 1999) 170.

created to receive funding representing 2 percent of the proceeds of the CDM as well as any additional funding from voluntary donors.[49] The goal of the adaptation fund is to "finance concrete adaptation projects and programmes in developing country Parties that are Parties to the Protocol."[50]

C. The Bali Action Plan

The 2007 Bali Action Plan is a milestone in the international negotiations on climate change, not only for mitigation, but also for international action on adaptation to climate change. A few months before the adoption of the Bali Action Plan, the publication of the fourth Assessment Report of the IPCC conveyed alarming scientific evidence of unfolding impacts of climate change. Negotiations were described as long and "fierce."[51] Developing States wanted support for domestic action on climate change; developed States wanted at least some developing States to pledge some concrete action on climate change mitigation.

The Bali Action Plan launched a "comprehensive process"[52] toward the adoption of an instrument to succeed the Kyoto Protocol. Although greater emphasis was put on climate change mitigation, this roadmap contains the strongest provisions on adaptation so far.[53] The parties agreed to issue a call to "enhanced action on adaptation," which should include five components:

(i) International cooperation to support urgent implementation of adaptation actions, including through vulnerability assessments, prioritization of actions, financial needs assessments, capacity-building and response strategies, integration of adaptation actions into sectoral and national planning, specific projects and programmes, means to incentivize the implementation of adaptation actions, and other ways to enable climate-resilient development and reduce vulnerability of all Parties ...;

(ii) Risk management and risk reduction strategies ...;

(iii) Disaster reduction strategies and means to address loss and damage associated with climate change impacts in developing countries that are particularly vulnerable to the adverse effects of climate change;

(iv) Economic diversification to build resilience;

(v) Ways to strengthen the catalytic role of the Convention in encouraging multilateral bodies, the public and private sectors and civil society, building on synergies among activities and processes, as a means to support adaptation in a coherent and integrated manner.[54]

[49] See decision 10/CP.7, "Funding under the Kyoto Protocol" (November 10, 2001), para. 2; decision 17/CP.7, "Modalities and procedures for a clean development mechanism, as defined in Article 12 of the Kyoto Protocol" (November 10, 2001), para. 15. See also decision 28/CMP.1, "Initial guidance to an entity entrusted with the operation of the financial mechanism of the Convention, for the operation of the Adaptation Fund" (December 9–10, 2005), recital 7; decision 1/CMP.8, "Amendment to the Kyoto Protocol pursuant to its Article 3, paragraph 9 (the Doha Amendment)" (December 8, 2012), paras. 20 and 22.

[50] Decision 10/CP.7, *supra* note 49, para. 1. See also decision 28/CMP.1, *supra* note 49, para. 1.

[51] Hermann E. Ott, Volfgang Sterk and Rie Watanabe, "The Bali roadmap: new horizons for global climate policy" (2008) 8:1 *Climate Policy* 91, at 92.

[52] Decision 1/CP.13, *supra* note 1, para. 1.

[53] Compare with decision 1/CP.10, "Buenos Aires programme of work on adaptation and response measures" (December 17–18, 2004).

[54] Decision 1/CP.13, *supra* note 10, para. 1(c).

As a "roadmap" for negotiations, the Bali Action Plan unfolded through further negotiations conducted by the Ad Hoc Working Group on Long-Term Cooperative Action under the Convention (AWG-LCA), leading in particular to the adoption of the Cancún Agreements.

D. The Copenhagen Accord and the Cancún Agreements

The Copenhagen Accord reaffirmed the importance of international action on climate change adaptation. The 141 parties that agreed to this document "recognize[d] the critical impacts of climate change ... on countries particularly vulnerable to its adverse effects and stress the need to establish a comprehensive adaptation programme including international support."[55] Noting that "[a]daptation ... is a challenge faced by all countries," they affirmed that "[e]nhanced action and international cooperation on adaptation is urgently required."[56] The signatories of the Copenhagen Accord further "agree[d] that developed countries shall provide adequate, predictable and sustainable financial resources, technology and capacity-building to support the implementation of adaptation action in developing countries."[57] They concluded that scaled-up financial support should be provided to support climate change adaptation, along with mitigation, in developing States.[58]

The 2010 Cancún Agreements represent the formal outcome of the process initiated by the Bali Action Plan. In the "shared vision for long-term cooperative action" that this document defined, the provisions on "enhanced action on adaptation" appear – for the first time – before those on "enhanced action on mitigation,"[59] in what appears to be a recognition of the increasing importance of adaptation under the UNFCCC regime. Some provisions are similar to the content of the Copenhagen Accord, in particular the recognition that "adaptation is a challenge faced by all Parties."[60] In addition, general language was agreed upon to sketch a common vision of action on climate change adaptation, in particular the aspiration that adaptation:

> should follow a country-driven, gender-sensitive, participatory and fully transparent approach, taking into consideration vulnerable groups, communities and ecosystems, and should be based on and guided by the best available science and, as appropriate, traditional and indigenous knowledge, with a view to integrating adaptation into relevant social, economic and environmental policies and actions.[61]

More concretely, under the so-called "Cancún Adaptation Framework,"[62] each party was encouraged to follow a particular course of action, including "planning, prioritizing and implementing adaptation actions";[63] assessing impacts, vulnerabilities and adaptation

---

[55] Copenhagen Accord, in the annex of decision 2/CP.15 (December 18–19, 2009), para. 1.
[56] *Ibid.*, para. 3.
[57] *Ibid.*
[58] *Ibid.*, para. 8. Support to adaptation is discussed in Chapter 12.
[59] See decision 1/CP.16, "The Cancún Agreements: outcome of the work of the Ad Hoc Working Group on Long-Term Cooperative Action under the Convention" (December 10–11, 2010), para. 11, parts II and III.
[60] *Ibid.*, para. 11.
[61] *Ibid.*, para. 12.
[62] *Ibid.*, para. 13.
[63] *Ibid.*, para. 14(a).

needs;[64] "building resilience ... including through economic diversification";[65] "enhancing climate change related disaster risk reduction strategies";[66] and taking measures to understand and respond to the incidence of climate change on human mobility.[67] An Adaptation Committee was created in order "to promote the implementation of enhanced action on adaptation in a coherent manner under the Convention,"[68] for instance, through support and guidance, information-sharing and advocacy. Least developed and other developing States are encouraged to "formulate and implement national adaptation plans ... as a means of identifying medium- and long-term adaptation needs and developing and implementing strategies and programmes to address those needs."[69] Complementary provisions on financial support, technology and capacity building were also adopted.[70]

E. The Paris Agreement

The Paris Agreement endorsed the approach of the Bali Action Plan and the Cancún Agreements by situating adaptation and mitigation formally on an equal footing. Article 2.1 defined the objective of the Paris Agreement as including not just climate change mitigation, but also "increasing the ability to adapt to the adverse impacts of climate change and foster climate resilience and low greenhouse gas emissions development, in a manner that does not threaten food production."[71] While Article 4 seeks to advance international action on climate change mitigation, Article 7 is dedicated to adaptation action.

In particular, Article 7 "establish[es] the global goal on adaptation of enhancing adaptive capacity, strengthening resilience and reducing vulnerability to climate change."[72] This article insists on the need to protect "people, livelihoods and ecosystems"[73] from the impacts of climate change and to recognize "the adaptation efforts of developing country Parties."[74] It further emphasizes the relations between mitigation outcomes, adaptation needs and adaptation costs,[75] as well as "the importance of support for and international cooperation on adaptation efforts."[76] Lastly, as a matter of principle, the parties "acknowledge that adaptation action should follow a country-driven, gender-responsive, participatory and fully transparent approach, taking into consideration vulnerable groups, communities and ecosystems."[77]

In addition, the parties to the Paris Agreement pledged, although in a "should" language, to share information and good practices, strengthen institutional arrangements, strengthen scientific knowledge, assist developing country parties and, more generally, "improve the

---

[64] *Ibid.*, para. 14(b).
[65] *Ibid.*, para. 14(d).
[66] *Ibid.*, para. 14(e).
[67] *Ibid.*, para. 14(f).
[68] *Ibid.*, para. 20.
[69] *Ibid.*, para. 15.
[70] *Ibid.*, paras. 95–137.
[71] Paris Agreement, *supra* note 13, art. 2.1(b).
[72] *Ibid.*, art. 7.1.
[73] *Ibid.*, art. 7.2.
[74] *Ibid.*, art. 7.3.
[75] *Ibid.*, art. 7.4.
[76] *Ibid.*, art. 7.6.
[77] *Ibid.*, art. 7.5.

effectiveness and durability of adaptation actions."[78] Each party "shall" further, "as appropriate," implement adaptation action, formulate adaptation plans, prioritize adaptation needs, monitor and learn from experience, and "build ... the resilience of socioeconomic and ecological systems."[79] Parties are encouraged to "submit and update periodically an adaptation communication."[80] Under Article 3 of the Agreement, all parties "are to undertake and communicate" action on climate change adaptation, along with mitigation and other aspects of climate action, "as nationally determined contributions to the global response to climate change."[81]

Although the Paris Agreement is a treaty, it hardly creates any clear legal obligation for States in relation to climate change adaptation.[82] This is because most of the provisions of the Paris Agreement on adaptation "are formulated as recommendations or expectations rather than legal obligations."[83] Some commentators regretted the inability of the provisions of the Paris Agreement on adaptation to establish "a concrete link" between adaptation need and "a country's responsibility to pay."[84] Others hoped that a global goal on adaptation could be further defined, in a measurable way, with tracking criteria which could then be implemented systematically through national action with international support.[85]

III. RELEVANT DEVELOPMENTS IN OTHER INTERNATIONAL REGIMES

Efforts to adapt to climate change are not confined to any narrow field of governance. Rather, the impacts of climate change need to be accounted for in many relevant international regimes beyond the UNCCC. International human rights law and international environmental law are particularly relevant given the impacts of climate change, respectively, on populations and ecosystems.

A. Human Rights and Development

The obligation of a State to advance the rights and the welfare of individuals within its jurisdiction has been recognized by numerous instruments, including the two International Covenants of 1966.[86] While the responsibilities of States are primarily territorial, international human rights treaties have recognized the obligation of each State to engage in international assistance and cooperation "to the maximum of its available resources."[87] International institutions have made repeated attempts to foster and facilitate such assistance and cooperation.[88] Thus,

---

[78] *Ibid.*, art. 7.7.

[79] *Ibid.*, art. 7.9.

[80] *Ibid.*, art. 7.10.

[81] *Ibid.*, art. 3.

[82] See also Rajamani, *supra* note 13.

[83] Daniel Bodansky, "The legal character of the Paris Agreement" (2016) 25:2 *Review of European Comparative & International Environmental Law* 142, at 146.

[84] Anju Sharma, "Precaution and post-caution in the Paris Agreement: adaptation, loss and damage and finance" (2017) 17:1 *Climate Policy* 33, at 43.

[85] See Alexandre K. Magnan and Teresa Ribera, "Global adaptation after Paris" (2016) 352:6291 *Science* 1280; Alexandra Lesnikowski *et al.*, "What does the Paris Agreement mean for adaptation?" (2017) 17:7 *Climate Policy* 825.

[86] See references *supra* note 14. See also, generally, Amitai Etzioni, "Sovereignty as responsibility" (2006) 50:1 *Orbis* 71.

[87] International Covenant on Economic, Social and Cultural Rights, *supra* note 14, art. 2.1.

[88] See in particular UN General Assembly Resolution 70/1, *supra* note 15.

multilateral development banks have lent financial support for development, the UN Office for the Coordination of Humanitarian Affairs (OCHA) has encouraged humanitarian assistance, and agencies such as the UN High Commissioner for Refugees (UNHCR) or the Food and Agriculture Organization have addressed specific protection needs.

Yet, many of the impacts of climate change challenge the ability of States to provide effective protection within their jurisdiction and hinder the work of international institutions. For instance, droughts make it more difficult for States to ensure that everyone within their territory can eat. Often, the impacts of climate change highlight the shortcomings of international cooperation and assistance to provide subsidiary protection when a State's own resources appear insufficient to provide effective protection to the rights and welfare of its population. Famines, disasters, conflicts, mass migration and many other social impacts of climate change are nothing new, but climate change reveals the inability of existing institutions to do enough to reduce and address such risks.

Climate change impacts led to renewed efforts to improving the protection of human rights and to fostering development. In particular, efforts were made to reduce and manage the risk of disasters through consultations held in Yokohama (1994),[89] Kobe (2005)[90] and Sendai (2015),[91] whereby some good practices were exchanged and some non-binding guidelines were adopted. As is often the case, these initiatives do not distinguish between climate change impacts and other similar protection needs. For instance, the Sendai Framework for Disaster Risk Reduction (2015) notes that "many [disasters] are exacerbated by climate change,"[92] but it does not seek to differentiate between "climate disasters" and others. All disasters should be addressed, whether related to climate change or not, even though climate change makes it more urgent to address the shortcomings of existing governance mechanisms.

Even more distinctively, the impact of climate change on migration has led to renewed debates on the shortcomings of international migration governance. The 1951 Convention Relating to the Status of Refugees, as amended by its 1967 Protocol, only protects those unable to return to their home country "owing to [a] well-founded fear of being persecuted for reason of race, religion, nationality, membership of a particular social group or political opinion."[93] This definition does not formally include the persons unable to return to their home country because, for instance, of generalized violence, extreme poverty or any sort of environmental impact, in relation to whom equally pressing humanitarian consideration would oppose deportation.[94] Nor do international laws and institutions provide much protection to internally displaced persons.[95] Advocacy for the protection of so-called "climate refugees" or

---

[89] World Conference on Natural Disaster Reduction, *Yokohama Strategy and Plan of Action for a Safer World* (May 23–27, 1994).

[90] World Conference on Disaster Reduction, "Hyogo Framework for Action 2005–2015: building the resilience of nations and communities to disasters" (January 18–22, 2005), doc. A/CONF.206/6.

[91] UNGA resolution 69/283, "Sendai Framework for Disaster Risk Reduction 2015–2030" (June 3, 2015).

[92] *Ibid.*, Annex II, para. 4.

[93] See Convention Relating to the Status of Refugees, July 28, 1951, 189 *UNTS* 150, art. 1(A)(2); Protocol Relating to the Status of Refugees, January 31, 1967, 606 *UNTS* 267.

[94] See James Hathaway, "A reconsideration of the underlying premise of refugee law" (1990) 31:1 *Harvard International Law Journal* 129, at 132–133.

[95] See *Guiding Principles on Internal Displacement* (February 11, 1998), doc. E/CN.4/1998/53/Add.2, a progressive synthesis of the rights of internally displaced persons under international law. See also Roberta Cohen and Francis M. Deng, *Masses in Flight: The Global Crisis of Internal Displacement* (Brookings Institution, 1998); Erin Mooney,

"climate migrants" – often vaguely conceived as individuals forced to leave their country or region of origin due to climate change impacts – has highlighted one aspect of a shortcoming of existing protection institutions, but these shortcomings are not limited to the impacts of climate change.[96] All forced migrants ought to be protected, notwithstanding whether their migration is caused by political persecution, climate change or whatever else. In practice, it is often almost impossible to disentangle the causes of migration, climate change being generally an indirect factor in a cluster of causes.[97] Thus, intergovernmental consultations spurred by a conference on "climate change and displacement"[98] led in 2015 to the adoption of a general protection agenda on disaster-induced cross-border displacement, whose relevance extends far beyond the impacts of climate change.[99]

Thus, debates on the impacts of climate change may be an opportunity to raise awareness on unfulfilled protection needs in areas such as disaster risk reduction and the protection of forced migrants. With regard to these issue-areas, climate change may be an eye-opening crisis, but reforms are unlikely to be specific to climate change. It would be extremely difficult, if not downright impossible, to distinguish between disasters or migrants which are induced by climate change and others because of the complex, indirect and often probabilistic causal relation between climate change and disasters or migration. Furthermore, such a distinction would be arbitrary inasmuch as it is accepted that protection should be provided on the basis of need, not on the basis of the cause for these needs.

There are, however, instances where more specific responses may be needed for discrete impacts of climate change. Such is the case when the impacts of climate change are of a distinct nature, as in the case of sea-level rise and, possibly, other slow-onset environmental changes caused by climate change. Thus, a large part of the architectural heritage of humankind is situated in coastal regions which are threatened by sea-level rise. The World Heritage Committee, created by the 1972 World Heritage Convention,[100] recognized "genuine concerns ... relating to threats to natural World Heritage properties that are or may be the result of climate change"[101] and "encourage[d] all States Parties ... to take early action in response to these potential impacts."[102] In 2007, the General Assembly of States Parties to the World Heritage Convention adopted a "Strategy to Assist States Parties to Implement Appropriate Management Responses," among other relevant documents.[103] Likewise, the Meeting of the Parties to the 2001 Convention on the Protection of the Underwater Cultural Heritage[104]

"The concept of internal displacement and the case for internally displaced persons as a category of concern" (2005) 24:3 *Refugee Survey Quarterly* 9.

[96] See Alexander Betts, *Survival Migration: Failed Governance and the Crisis of Displacement* (2013).

[97] Benoit Mayer, *The Concept of "Climate Migration": Advocacy and its Prospects* (Edward Elgar, 2016).

[98] The Nansen Conference, *Climate Change and Displacement in the 21st Century* (Oslo, June 5–7, 2011).

[99] Nansen Initiative, *Agenda for the Protection of Cross-Border Displaced Persons in the Context of Disasters and Climate Change* (December 2015).

[100] Convention for the Protection of the World Cultural and Natural Heritage, November 16, 1972, 1037 *UNTS* 151.

[101] UNESCO World Heritage Committee, 29th Session, decision 29 COM 7B.a (Durban, July 10–17, 2005), in doc. WHC-05/29.COM/22, 36, para. 4.

[102] *Ibid.*, para. 6.

[103] UNESCO World Heritage Convention, *A Strategy to Assist States Parties to Implement Appropriate Management Responses* (2007), reproduced in UNESCO World Heritage Centre, *Climate Change and World Heritage* (World Heritage Reports, May 22, 2007). See also UNESCO World Heritage Convention, *Policy Document on the Impacts of Climate Change on World Heritage Properties* (2007), at 4.

[104] Convention on the Protection of the Underwater Cultural Heritage, November 2, 2001, 41 *ILM* 37.

highlighted "the urgent need to preserve sites against the impact of climate and sea-level changes" and "encourage[d] States Parties to take active measures"[105] to protect relevant sites.

B. Environmental Protection

Initiatives have taken place under multilateral environmental agreements to address impacts of climate change on the natural environment, in particular on biodiversity and on natural ecosystems. Some such initiatives were taken under the Convention on Biological Diversity, a treaty adopted along with the UNFCCC at the 1992 Earth Summit.[106] Starting with coral reefs,[107] the COP to the Convention on Biological Diversity has repeatedly emphasized the impacts of climate change on particular aspects of biodiversity before emphasizing, more generally, that "the loss of biodiversity and its potential damage is one impact of ... climate change."[108] The COP promoted the concept of an "ecosystem-based" adaptation to climate change, contending that "ecosystems can be managed to limit climate change impacts on biodiversity and to help people adapt to the adverse effects of climate change."[109] Likewise, since 2005, the COP to the 1979 Convention on Migratory Species[110] have emphasized the impact of climate change on migratory species and has called on all States "to implement, as appropriate, adaptation measures that would help reduce the foreseeable adverse effects of climate change" on protected species.[111] In 2014, the same COP adopted a "programme of work on climate change and migratory species" containing diverse measures aimed, among other things, "to facilitate species adaptation in response to climate change."[112]

The Convention to Combat Desertification is another multilateral environmental agreement of direct relevance to climate change adaptation.[113] The Preamble to the Convention notes "that desertification is caused by complex interactions among physical, biological, political, social, cultural and economic factors."[114] It highlights "the contribution that combating desertification can make to achieving the objectives of the [UNFCCC]."[115] Article 8 promotes the coordination of activities carried out under this Convention and under the UNFCCC.[116] The States most

---

[105] Meeting of the States Parties to the Convention on the Protection of the Underwater Cultural Heritage, resolution 5 / MSP 5 (April 28–29, 2015), reproduced in doc. UCH/15/5.MSP/11, para. 10.

[106] United Nations Convention on Biological Diversity, June 5, 1992, 1760 *UNTS* 79 (hereinafter CBD).

[107] CBD decision V/3, "Progress report on the implementation of the programme of work on marine and coastal biological diversity (implementation of decision IV/5)" (Nairobi, May 15–26, 2000), para. 5.

[108] CBD decision X/33, "Biodiversity and climate change" (Nagoya, October 18–29, 2010), para. 2.

[109] *Ibid.*, para. 8(j). See also e.g. CBD decisions XI/15, "Review of the programme of work on island biodiversity" (Hyderabad, October 8–19, 2012), paras. 2(b) and 4(b); and XII/20 "Biodiversity and climate change and disaster risk reduction" (Pyeongchang, October 6–17, 2014), paras. 7(a) and (c).

[110] Convention on the Conservation of Migratory Species of Wild Animals, June 23, 1979, 1651 *UNTS* 333 (hereinafter Convention on Migratory Species).

[111] Convention on Migratory Species resolution 8.13, "Climate change and migratory species" (Nairobi, November 20–25, 2005), para. 3. See also Convention on Migratory Species resolutions 9.7, "Climate change impacts on migratory species" (Rome, December 1–5, 2008); and 10.19, "Migratory species conservation in the light of climate change" (Bergen, November 20–25, 2011).

[112] Conference of the Parties to the Convention on Migratory Species, 11th meeting, *Programme of Work on Climate Change and Migratory Species* (Quito, November 4–9, 2014).

[113] Convention against Desertification, *supra* note 30, art. 2.

[114] *Ibid.*, recital 7.

[115] *Ibid.*, recital 24.

[116] *Ibid.*, art. 8.1.

affected by desertification have often joined their efforts to adapt to climate change with their efforts to combat desertification and mitigate the effects of drought.[117]

Likewise, the 1971 "Ramsar" Convention on Wetlands of International Importance Especially as Waterfowl Habitat[118] has recognized that such ecosystems have increasingly been affected by the impacts of climate change. The Conference of the Contracting Parties highlighted in 2002 "that climate change may substantially affect the ecological character of wetlands and their sustainable use."[119] Accordingly, it encouraged States to consider relevant measures of conservation and restoration to reduce the impact of climate change on wetlands.[120]

These different developments aim to flag the impacts of climate change and ensure appropriate level of awareness, but, in and of themselves, they do not provide any appropriate response to these impacts. Exchanges of good practices are typically encouraged in these, while international assistance and support could also play a role. Often, however, more concrete actions are needed to address the impacts of climate change on the environment.

IV. EMERGING PRINCIPLES

Several principles have emerged through the developments taking place mainly within the UNFCCC regime and beyond with respect to adaptation to climate change. Some of these principles are solidly established in existing international law. Others remain in the nebulous zone where a common practice, repeatedly affirmed by international institutions and by States, are progressively being accepted as a norm of customary international law. These emerging principles are more than mere aspirations, but often less than clear-cut binding legal principles. Thus, although they cannot always be relied upon directly as an independent source of rights or obligations, they may nevertheless play a role in interpreting other legal instruments.

Five emerging legal principles on climate change adaptation are identified in the following: the principle that adaptation action should contribute to sustainable development; that adaptation action should be determined through a participatory process; that adaptation action should be effective; that international cooperation should promote adaptation action; and that developed States, in particular, should provide support to developing States.

A. Contribution to Sustainable Development

The UNFCCC recognizes that "[t]he Parties have a right to, and should, promote sustainable development."[121] Responses to climate change, including adaptation measures, should not compromise this right; where possible, they should advance the three dimensions of sustainable

---

[117] See e.g. UNFCCC, "Synergy among multilateral environmental agreements in the context of national adaptation programmes of action" (April 6, 2005), doc. FCCC/TP/2005/3.

[118] Convention on Wetlands of International Importance Especially as Waterfowl Habitat, February 2, 1971, 996 *UNTS* 246 (hereinafter Ramsar Convention, from the name of the town where the Convention was adopted).

[119] RAMSAR Convention resolution VIII.3, "Climate change and wetlands: impacts, adaptation, and mitigation" (Valencia, November 18–26, 2002), recital 1.

[120] See *ibid.*; RAMSAR Convention resolutions X.24, "Climate change and wetlands" (Changwon, October 28–November 4, 2008); XI.14, "Climate change and wetlands : implications for the Ramsar Convention on Wetlands" (Bucharest, July 6–13, 2002); and XII.11, "Peatlands, climate change and wise use: implications for the Ramsar Convention" (Punta del Este, June 1–9, 2015).

[121] UNFCCC, *supra* note 9, art. 3.4.

development – social protection, economic development and environmental protection.[122] Thus, the COP affirmed multiple times that "responses to climate change should be coordinated with social and economic development in an integrated manner."[123] Adaptation measures should in particular be cost-effective,[124] although some parties insisted that cost-effectiveness should not come at the expense of other aspects of sustainable development.[125] The Paris Agreement confirmed that the global goal on adaptation should be implemented "with a view to contributing to sustainable development."[126]

The relationship between climate change adaptation and sustainable development was equally recognized beyond the UNFCCC regime. The parties to the Convention on Biological Diversity agreed that climate change adaptation should be achieved through a "sustainable use of biodiversity."[127] Likewise, the 2030 Agenda for Sustainable Development adopted by the UN General Assembly in 2015 acknowledged of climate change adaptation as a necessary component of sustainable development.[128] The 2030 Agenda for Sustainable Development further highlighted the connections between climate change adaptation and two other Sustainable Development Goals: the goal of ensuring food security, which could reduce the impacts of climate change on food production;[129] and the goal of "mak[ing] cities and human settlement safe, resilient and sustainable," in which successful climate change adaptation participates.[130]

The principle that adaptation action should contribute to sustainable development implies in particular that measures to promote climate change adaptation must not infringe human rights. Civil society advocacy and several resolutions of the UN Human Rights Council have highlighted the relations between responses to climate change – in particular, adaptation action – and the protection of human rights.[131] In the Cancún Agreements, the COP recognized "that the adverse effects of climate change have a range of direct and indirect implications for the effective enjoyment of human rights."[132] Five years later, the Paris Agreement "acknowledg[ed] that ... Parties should, when taking action to address climate change, respect, promote and consider

---

[122] See Chapter 5, section II.A.

[123] See e.g. decision 2/CP.11, "Five-year programme of work of the Subsidiary Body for Scientific and Technological Advice on impacts, vulnerability and adaptation to climate change" (December 9–10, 2005), recital 5; decision 1/CP.16, *supra* note 59, para. 12.

[124] Annex of decision 28/CP.7, *supra* note 38, para. 7. See also e.g. decision 5/CP.9, "Further guidance to an entity entrusted with the operation of the financial mechanism of the Convention, for the operation of the Special Climate Change Fund" (December 12, 2003), para. 1(b); "Terms of reference for the third review of the Adaptation Fund," in the annex of decision 1/CMP.12, "Third review of the Adaptation Fund" (November 17, 2016), para. 1.

[125] "Annotated Guidelines for the Preparation of National Adaptation Programmes of Action," in UNFCCC SBI, *Input of the Least Developed Countries Expert Group on the Improvement of the Guidelines for the Preparation of National Adaptation Programmes of Action*, doc. FCCC/SBI/2002/INF.14 (August 26, 2002), annotations under para. 7(h).

[126] Paris Agreement, *supra* note 13, art. 7.1.

[127] CBD decision XII/20, *supra* note 109, recital 2.

[128] UN General Assembly Resolution 70/1, *supra* note 15, para. 31.

[129] *Ibid.*, 2.4.

[130] *Ibid.*, goal 11.b.

[131] See Human Rights Council resolutions 35/20, "Human rights and climate change" (June 22, 2017); 32/33, "Human rights and climate change" (July 1, 2016); 29/15, "Human rights and climate change" (July 2, 2015); 18/22, "Climate change and human rights" (September 30, 2011); 7/23, "Human rights and climate change" (March 28, 2008); 10/4, "Human rights and climate change" (March 25, 2009); 7/23, "Human rights and climate change" (March 28, 2008).

[132] Decision 1/CP.16, *supra* note 59, recital 8.

their respective obligations on human rights."[133] Provisions on adaptation to climate change systematically emphasized the need to protect vulnerable populations, including through a "gender-sensitive"[134] or even "gender-responsive"[135] approach to adaptation, and more generally by "taking into consideration vulnerable groups, communities and ecosystems."[136] The unintended consequences of implementation measures on the human rights and welfare of populations have also come under international scrutiny.[137]

B. The Bottom-up Approach

Another emerging principle is that adaptation efforts should be determined by the relevant stakeholders (the bottom-up approach) rather than imposed upon them (the top-down approach). According to the "principle of sovereignty of States in international cooperation to address climate change"[138] affirmed in the Preamble to the UN Framework Convention on Climate Change, national governments should determine for themselves what they consider to be the best strategies to promote climate change adaptation within their jurisdiction. The importance of a "country-driven" approach to climate change adaptation was recognized many times, including in the Marrakesh Accords of 2001[139] and in the 2015 Paris Agreement.[140] Adaptation needs depend largely on the circumstances of a given community, in particular its exposure to particular hazards, its vulnerability and its resilience, but it also involves value judgments which are best left to the concerned populations through political processes taking place within each State.[141] Thus, adaptation action is also more likely to achieve co-benefits with national development priorities when it is country-driven.

This bottom-up approach of adaptation has several implications. For instance, "owing to the context-specific nature of adaptation," the COP decided that developing a common set of global indicators on climate change adaptation would not be appropriate,[142] although the approach of the Paris Agreement – a global goal on adaptation implemented through national commitments – appears to have reopened this debate.[143] Domestically, States are also called upon to implement a participatory and transparent approach to climate change adaptation.[144]

---

[133] Paris Agreement, *supra* note 13, recital 12. See also Benoit Mayer, "Human rights in the Paris Agreement" (2016) 6:1–2 *Climate Law* 109.

[134] Decision 1/CP.16, *supra* note 59, para. 12.

[135] Paris Agreement, *supra* note 13, art. 7.5. See also decisions 21/CP.22, "Gender and climate change" (November 17, 2016); and 18/CP.20, "Lima work programme on gender" (December 12, 2014).

[136] Paris Agreement, *supra* note 13, art. 7.5. See also, e.g. decision 1/CP.16, *supra* note 59, para. 12; decision 5/CP.17, "National adaptation plans" (December 11, 2011), para. 3; decision 3/CP.20, "National adaptation plans " (December 12, 2014), para. 3.

[137] See e.g. *UNFCCC, supra* note 9, art. 4.8; decision 3/CP.3, "Implementation of Article 4, paragraphs 8 and 9, of the Convention" (December 11, 1997); decision 11/CP.21, "Forum and work programme on the impact of the implementation of response measures" (December 13, 2015). See also Chapter 16, section II.B.

[138] UNFCCC, *supra* note 9, recital 10.

[139] See e.g. Guidelines for the preparation of national adaptation programmes of action in the annex of decision 28/CP.7, *supra* note 38, para. 6; decision 1/CP.16, *supra* note 59, para. 12.

[140] Paris Agreement, *supra* note 13, art. 7.5.

[141] See above, section I.

[142] "Recommendations for the Conference of the Parties," in the annex of decision 4/CP.20, "Report of the Adaptation Committee" (December 12, 2014), para. 3(a).

[143] See Magnan and Ribera, *supra* note 85.

[144] See e.g. Paris Agreement, *supra* note 13, art. 7.5; decision 1/CP.16, *supra* note 59, para. 12.

The involvement of the private sector, NGOs and civil society organizations should inform decision-making while also contributing to its legitimacy and facilitating the effective implementation of adaption plans.[145] On the other hand, the COP recognized the need to limit the scope of its own decisions, for instance, by affirming that "the national adaptation plan process should not be prescriptive."[146] Likewise, the Least Developed Countries Expert Group insisted that "[a] country should be free to choose from [adaptation] criteria as best suits their case."[147]

However, there remains a tension between the recognition of the need for a bottom-up approach of adaptation and efforts to assist developing States. This tension is epitomized by the desire once expressed by the COP to "provid[e] technical support and guidance to the Parties [while] respecting the country-driven approach."[148] Technical support and guidance are not necessarily binding, but when there is a significant power differential, the governments of developing States may be pressured to follow the suggestions made by foreign or international institutions. The desire of many governments to receive international support is likely to incentivize them to comply with international guidance, whether or not, based on their own assessment, such guidance is appropriate in their particular circumstances.

C. Effectiveness

A third emerging principle is that of effectiveness – the ability of adaptation measures to meet the objectives defined through a bottom-up approach. This requirement of effectiveness involves at least a rational planning process. Thus, all parties to the UNFCCC committed to "[f]ormulate, implement, publish and regularly update ... measures to facilitate adequate adaptation to climate change."[149] The COP "insist[ed] that action relating to adaptation follow an assessment and evaluation process ... so as to prevent maladaptation and to ensure that adaptation actions are environmentally sound and will produce real benefits in support of sustainable development."[150] In addition, the COP adopted meticulously detailed guidelines for the planning process, in particular, in the least developed States.[151] The COP insisted that "the process to formulate and implement national adaptation plans is fundamental for building adaptive capacity and reducing vulnerability to the impacts of climate change."[152] Synergies and integration with other policy fields, in particular national development priorities, was repeatedly encouraged.[153]

---

[145] See e.g. "Annotated Guidelines for the Preparation of National Adaptation Programmes of Action," *supra* note 125, annotations under para. 7(a).

[146] Decision 5/CP.17, *supra* note 136, para. 4. See also decision 12/CP.18, "National adaptation plans" (December 7, 2012), recital 10.

[147] "Annotated Guidelines for the Preparation of National Adaptation Programmes of Action," *supra* note 125, annotations above para. 15.

[148] Decision 1/CP.16, *supra* note 59, para. 20(a).

[149] UNFCCC, *supra* note 9, art. 4.1(b). See also Kyoto Protocol, *supra* note 44, art. 10(b); Paris Agreement, *supra* note 13, art. 7.9.

[150] Decision 1/CP.10, *supra* note 53, para. 4. See also e.g. decision 1/CP.16, *supra* note 59, recital 5.

[151] See the annex of decision 28/CP.7, *supra* note 38.

[152] Decision 3/CP.20, *supra* note 136, para. 2.

[153] See e.g. UNFCCC, *supra* note 9, art. 4.1(f); decision 1/CP.16, *supra* note 59, para. 34; "Summary and recommendations by the Standing Committee on Finance on the 2016 biennial assessment and overview of climate finance flows" in the annex of decision 8/CP.22 "Report of the Standing Committee on Finance" (November 18, 2016), para. 32.

Another aspect of this principle of effectiveness is that adaptation planning be based on relevant knowledge. According to the COP, States should make decisions based on sound scientific evidence, but also on "local" or "traditional" as well as "indigenous knowledge."[154] A five-year program of work was conducted under the aegis of the UNFCCC "to assist all Parties, in particular developing countries ... to make informed decisions on practical adaptation actions and measures to respond to climate change on a sound, scientific, technical and socioeconomic basis."[155] In the same line, the Paris Agreement requires that adaptation action be "guided by the best available science and, as appropriate, traditional knowledge, knowledge of indigenous peoples and local knowledge systems."[156] How scientific and traditional worldviews are to be reconciled is unclear and should perhaps best be left to national authorities to decide.

D. International Cooperation

The requirement that States strive to work together when devising and implementing measures to adapt to climate change is another emerging principle in the international law on climate change adaptation. The 2030 Agenda for Sustainable Development, for instance, called for "the widest possible international cooperation aimed at ... addressing adaptation to the adverse impacts of climate change."[157]

The justification for such cooperation is somewhat inconsistent. While the Copenhagen Accord relates the need to cooperate with the fact that adaptation "is a challenge faced by all countries,"[158] the former does not follow from the latter. Many contemporary governance issues are faced by all or most countries, and yet international cooperation is not seen as a necessity. For instance, many national governments are trying to provide universal health insurance, but they do not see any need to cooperate in doing so. Unlike climate change mitigation, adaptation to climate change does not raise a collective action problem: the success of adaptation in one country generally does not depend on the measures taken in another.[159] Rather, the general recognition that States should cooperate in adapting to climate change appears to relate to a sense that all States – especially developed ones – share some responsibility, either for their failure to avoid excessive GhG emissions under their jurisdiction or at least when a State is unable to provide effective protection to populations within its jurisdiction. This, however, would do more to justify international *support* (the next emerging principle) than international *cooperation*.

International cooperation on climate change adaptation involves, for the most part, the sharing of information and good practices. Under the UNFCCC, all States are required to communicate the measures they are taking to adapt to climate change[160] and to "cooperate in preparing for adaptation to the impacts of climate change."[161] The Marrakesh Accords further

---

[154] Decision 2/CP.11, *supra* note 123, recital 6; 1/CP.16, *supra* note 59, para. 12; and decision 5/CP.17, *supra* note 136, para. 3.
[155] "Five-year programme of work of the Subsidiary Body for Scientific and Technological Advice on impacts, vulnerability and adaptation to climate change," in the annex of decision 2/CP.11, *supra* note 123, para. 1.
[156] Paris Agreement, *supra* note 13, art. 7.5.
[157] UN General Assembly Resolution 70/1, *supra* note 15, para. 31.
[158] Copenhagen Accord, *supra* note 55, para. 3.
[159] There are some exceptions where cooperation is essential, generally in a regional context, for instance, to protect transboundary ecosystems, migratory species or transboundary freshwater resources.
[160] UNFCCC, *supra* note 9, arts. 4.1(b) and 12.1(b).
[161] *Ibid.*, art. 4.1(e).

encouraged parties "to exchange information on their experience regarding the adverse effects of climate change and on measures to meet their needs arising from these adverse effects."[162] The Cancún Agreements recognized the need for "enhancing action on adaptation, including through international cooperation."[163] The tasks of the Adaptation Committee included "strengthening, consolidating and enhancing the sharing of relevant information, knowledge, experience and good practices"[164] and "providing information and recommendations, drawing on adaptation good practices, for consideration by the Conference of the Parties."[165] The Paris Agreement called on its parties to "strengthen their cooperation on enhancing action on adaptation" through similar means.[166]

Cooperation in climate change adaptation has gradually extended beyond States to involve international institutions as well as the private sector. Thus, the Bali Action Plan encouraged parties to "strengthen the catalytic role of the Convention in encouraging multilateral bodies, the public and private sectors and civil society, building on synergies among activities and processes, as a means to support adaptation in a coherent and integrated manner."[167] Likewise, the Cancún Agreements invited "relevant multilateral, international, regional and national organizations, the public and private sectors, civil society and other relevant stakeholders to undertake and support enhanced action on adaptation at all levels."[168] The following year, the COP encouraged parties to "promot[e] synergy and strengthen ... engagement with national, regional and international organizations, centres and networks, in order to enhance the implementation of adaptation actions."[169] The Paris Agreement specifically highlights the role of UN specialized organizations and agencies.[170] Consistently, a 2016 in-session workshop on long-term climate finance concluded that "[t]he role of the private sector in adaptation finance needs to be further enhanced,"[171] and the Green Climate Fund (GCF) was encouraged to take additional steps to engage the private sector in support of adaptation.[172]

E. International Support

Another closely related emerging principle regards the requirement that all relevant actors (but developed States more specifically) provide support to adaptation in developing States. This principle has two alternative fundaments in general international law, related respectively to the protection and remedial perspectives discussed in the introduction to this chapter. If international action on climate change adaptation is viewed from a protection perspective, international support for adaptation appears essentially as a way for wealthy States to assist those facing serious difficulties because, among other things, of the impacts of climate change. However, if the international action on climate change adaptation is viewed in relation to

---

[162] Decision 5/CP.7, *supra* note 38, para. 5.
[163] Decision 1/CP.16, *supra* note 59, para. 13.
[164] *Ibid.*, para. 20(b).
[165] *Ibid.*, para. 20(d).
[166] Paris Agreement, *supra* note 13, art. 7.7.
[167] Decision 1/CP.13, *supra* note 10, para. 1(c)(v).
[168] Decision 1/CP.16, *supra* note 59, para. 34.
[169] Decision 2/CP.17, para. 93(c).
[170] Paris Agreement, *supra* note 13, art. 7.8.
[171] Decision 7/CP.22, "Long-term climate finance" (November 18, 2016), para. 7(c).
[172] See e.g. decision 10/CP.22, "Report of the Green Climate Fund to the Conference of the Parties and Guidance to the Green Climate Fund" (November 18, 2016), para. 11.

the responsibility of developed States for their failure to prevent excessive GhG emissions, support for adaptation may be conceived of as an attempt to mitigate the injury that a State has caused to another or, possibly, an atypical form of reparation.[173] Arguments advanced from a remedial perspective, once central to the negotiating position of developing States,[174] have constantly been rejected by developed States.[175] Although the workstream on loss and damage attempted to bring these arguments back to international negotiations under the UNFCCC, existing support for climate change adaptation is more convincingly interpreted from the protection perspective.

A duty of assistance is well established in general international law, although its scope and modalities remain vague. Already in the eighteenth century, Emer de Vattel wrote that "[i]f a nation is afflicted with famine, all those who have provisions to spare ought to relieve her distress, without, however, exposing themselves to want."[176] This was not only a moral duty, for Vattel added: "[t]o give assistance in such extreme necessity is so essentially conformable to humanity, that the duty is seldom neglected by any nation that has received the slightest polish of civilization."[177] The commitment of the United Nations to "employ international machinery for the promotion of the economic and social advancement of all peoples"[178] implies an acknowledgment of that same duty. The International Covenant on Economic, Social and Cultural Rights calls on each State to take steps, including "through international assistance and co-operation, especially economic and technical, to the maximum of its available resources, with a view to achieving progressively the full realization of the rights"[179] recognized in the Covenant. Multiple resolutions of the UN General Assembly emphasize the obligation of developed States to assist developing States,[180] including through the provision of financial assistance.[181] The Committee on Economic, Social and Cultural Rights (CESCR) affirmed that "international cooperation for development and thus for the realization of economic, social and cultural rights is an obligation of all States."[182]

Consistently, as discussed above, the developed parties to the UNFCCC committed to "assist the developing country Parties that are particularly vulnerable to the adverse effects of climate change in meeting costs of adaptation to those adverse effects."[183] In addition, all parties committed to give full consideration to the needs and concerns of developing countries, in particular small island countries, countries with low-lying coastal areas, with arid areas, with areas prone to natural disasters or with fragile ecosystems, and the least developed countries.[184] A particular emphasis has repeatedly been placed on need for assistance as a criteria

---

[173] See Chapter 5, sections I and IV.

[174] Caracas Declaration of the Ministers of Foreign Affairs of the Group of 77 on the Occasion of the Twenty-Fifth Anniversary of the Group (June 21–23, 1989), reproduced in doc. A/44/361.

[175] Written statement of the United States on Principle 7 of the Rio Declaration, in *Report of the United Nations Conference on Environment and Development*, UN document A/CONF.151/26 (vol. IV) (September 28, 1992).

[176] Emer de Vattel, *The Law of Nations*, Joseph Chitty trans. (Sweet, Stevens and Maxwell, 1758), at II.1, para. 5.

[177] *Ibid.*

[178] UN Charter, *supra* note 14, recital 8.

[179] International Covenant on Economic, Social and Cultural Rights, *supra* note 14, art. 2.1.

[180] See e.g. UN General Assembly Resolutions 3201 (S-VI), "Declaration on the Establishment of a New International Economic Order" (May 1, 1974); 41/128, *supra* note 15.

[181] See UN General Assembly Resolutions 2626 (XXV), "International Development Strategy for the Second United Nations Development Decade" (October 24, 1970); and 66/288, "The future we want" (July 27, 2012).

[182] CESCR General Comment No. 3, "The nature of State parties' obligations" (December 14, 1990), para. 14.

[183] UNFCCC, *supra* note 9, art. 4.4.

[184] *Ibid.*, arts. 4.8 and 4.9.

for the distribution of limited international support among many developing States. Under the Cancún adaptation framework, for instance, "funding for adaptation will be prioritized for the most vulnerable developing countries, such as the least developed countries, small island developing States and Africa."[185] Likewise, the Board of the GCF was encouraged to prioritize "the urgent and immediate needs of developing countries that are particularly vulnerable to the adverse effects of climate change, including LDCs, SIDS and African States."[186]

However, despite these repeated statements, actual support for climate change adaptation – which ought to be central to international cooperation on adaptation – has remained limited. The share of proceeds of the CDM used "to assist developing country Parties that are particularly vulnerable to the adverse effects of climate change to meet the costs of adaptation"[187] remains to date the most specific commitment to support climate change adaptation to which developed States agreed. Instead of any substantial obligations, States have only been called to "provide detailed information ... on their assistance to developing country Parties ... in meeting costs of adaptation."[188] Under the Paris Agreement, developed States committed to "provide financial resources to assist developing country Parties with respect to both mitigation and adaptation,"[189] while developing States were "encouraged to provide or continue to provide such support voluntarily."[190] Transfer of technology and capacity building have also been encouraged.[191]

V. CONCLUSION

International action on climate change adaptation remains one of the most politicized aspects of the international law on climate change. For the most part, the initial claims of developing States for reparation for the harm caused by the excessive GhG emissions in developed States have remained unaddressed. Even claims for assistance in protecting populations in those States most affected by and least able to face the adverse impacts of climate change have remained largely unaddressed. Instead of much-needed financial and technological support, international action on climate change adaptation has largely been confined to the determination of the principles which should guide domestic action and the exchange of good practices.

---

[185] Decision 1/CP.16, *supra* note 59, para. 95.

[186] Annex of decision 3/CP.17, "Launching the Green Climate Fund" (December 11, 2011), para. 52.

[187] Kyoto Protocol, *supra* note 44, art. 12.8.

[188] Decision 1/CP.9, "National communications from parties included in Annex I to the Convention" (December 12, 2003), para. 5.

[189] Paris Agreement, *supra* note 13, art. 9.1.

[190] *Ibid.*, art. 9.2.

[191] See e.g. decision 4/CP.4, "Development and transfer of technologies" (November 14, 1998), para. 6; Paris Agreement, *supra* note 13, arts. 10 and 11. See Chapter 12.

11

Loss and Damage

Adaptation to climate change can reduce the harms suffered as a consequence of climate change, but it cannot avoid any harms. Adaptation action itself often involves some economic and human costs; harms also occur when adaptation efforts are impossible or uneconomical, insufficient, poorly funded, incompletely implemented or ill-planned (maladaptation). This raises ethical questions. Often, those who contribute the least to climate change and benefit the least from industrial development are among those most severely affected by its impacts and less able to adapt. At the international level, the least developed States contribute little to global GhG emissions, but they are often the most severely affected by the impacts of climate change and can rarely invest in effective adaptation strategies. Likewise, within States, the poor often contribute the least to climate change, given their limited purchasing power, but they are also often the most exposed to environmental hazards due to a lack of financial resources to invest in their own safety.[1]

Arguments on responsibility and claims for reparations have long been voiced by representatives of developing States. At a 1989 conference of the G77, developing States declared that "[s]ince developed countries account for the bulk of the production and consumption of environmentally damaging substances, they should bear the main responsibility in the search for long-term remedies for global environmental protection."[2] Yet, such claims for "climate justice" have long been ignored by the most powerful States. Although the UNFCCC provided some support to developing States, this was primarily for mitigation action implemented in emerging economies (e.g. through the CDM) rather than to promote adaptation in the least developed States. As an observer noted, the least developed States simply "had less to offer the developed world in exchange for financial transfers."[3] Since the adoption of the UNFCCC, the demand of developing States for compensation has only marginally been addressed by international cooperation on climate change adaptation.[4] It was not until the 2007 Bali Action Plan that the COP could agree, as part of "enhanced action on adaptation," to start considering

---

[1] See, for instance, Chester Hartman and Gregory D. Squires (eds.), *There Is No Such Thing as a Natural Disaster: Race, Class, and Hurricane Katrina* (Routledge, 2006).

[2] Caracas Declaration of the Ministers of Foreign Affairs of the Group of 77 on the Occasion of the Twenty-Fifth Anniversary of the Group (June 21-23, 1989), reproduced in doc. A/44/361, para. II-34.

[3] Daniel Bodansky, "The United Nations Framework Convention on Climate Change: a commentary" (1993) 18:2 *Yale Journal of International Law* 451, at 528.

[4] See Chapter 10 on adaptation and Chapter 12 on international support to climate action.

"means to address loss and damage associated with climate change impacts in developing countries that are particularly vulnerable to the adverse effects of climate change."[5]

A decade later, there remain radically different views on what should be done to "address" loss and damage. On the one hand, if the means to address loss and damage aimed only to reduce or avoid loss and damage, they would simply replicate what is already under consideration as action on climate change adaptation. On the other hand, claims for reparation for the harms that cannot be reduced or avoided through adaptation continue to face stiff opposition from developed States. The workstream on loss and damage has become the institutional battlefield for disputes on responsibilities and reparations. While remedial obligations have not been recognized within the UNFCCC regime, such obligations are established in general international law, namely on the responsibility of States for internationally wrongful acts in relation to their failure to prevent excessive GhG emissions from activities under their jurisdiction.

This chapter first provides an overview of the concept of loss and damage. Then it explores the bases for remedial obligations under general international law before recounting attempts to implement these obligations through the UNFCCC workstream on loss and damage. Finally, it assesses the prospects of ongoing negotiations on loss and damage.

I. THE CONCEPT OF LOSS AND DAMAGE

There is no consensual definition of loss and damage associated with climate change impacts.[6] The concept emerged through international negotiations before attempts were made to make sense of it from social science and legal perspectives. In plain English and sometimes in legal English, the expression "loss and damage" refers to something akin to damages, for which compensation might be paid.[7] Unlike an "injury," however, the existence of loss and damage does not presuppose the attribution to a wrongful act – loss and damage could be recognized without prejudice of the existence of a wrongful act. In international practice, the expression has often been used in international agreements involving the payment of a lump sum without a formal admission of wrongdoing.[8] As such, the terminology could facilitate international negotiations where a party is willing to pay some form of damages, but not to formally recognize any wrongdoing, for instance, because this could hurt nationalistic feelings or, more pragmatically, for fear that an admission of wrongdoing would open the gate to far larger claims for reparation.

As such, "loss and damage" appeared as a constructive ambiguity in international negotiations on climate change, temporarily defusing the tension between the claims for reparation from some developing States and their rejection by developed States. For the representatives of the most vulnerable States, "loss and damage" is a placeholder for claims on compensation. By contrast, the representatives of industrial nations preferred to view the recognition of loss

---

[5] Decision 1/CP.13, para. 1(c)(iii).

[6] For instance, the IPCC did not define this notion in the glossaries included in its Fifth Assessment Report.

[7] See, for instance, *Lambert* v. *Bessey* (T. Ray. 421), cited in James Barr Ames, "Law and morals" (1908) 22 *Harvard Law Review* 97, at 98.

[8] See, for instance, the Additional Protocol between the USSR and Finland concerning compensation for loss and damage and for the works to be carried out by Finland in connection with the implementation of the Agreement of 29 April 1959 concerning the regulation of Lake Inari by means of the Kaitakoski hydro-electric power station and dam, April 29, 1959, 212 *UNTS* 4981, art. 1; Memorandum of Understanding between the United Nations and the Government of Pakistan contributing resources to the United Nations Special Police Unit in Kosovo, April 4, 2000, 43437 *UNTS* I-43437, paras. 17–22.

and damage as a recognition of the need to reinforce international cooperation on climate change mitigation and action on adaptation to climate change. Legal scholars, hard-pressed to make sense of a concept forged through protracted political negotiations, have generally been inclined to approach it as a euphemism for reparation.[9]

This astute constructive ambiguity led the term to achieve great political currency in about a decade. Policy-oriented institutions such as the United Nations University have initiated large research programs seeking to develop general conceptual or practical tools to avert or address loss and damage associated with climate change impacts.[10] Attempts have also been made to further define the concept. Saleemul Huq and colleagues, for instance, suggested that loss refers to "the negative impacts of climate change that are permanent," whereas damage relates to those harms "that can be reversed."[11]

Debates on loss and damage often emphasize the loss and damage – in particular, non-economic loss such as casualties, migration or loss of cultural practices and traditions – which cannot readily be avoided through climate change adaptation.[12] For instance, a few days after Typhoon Haiyan hit the Philippines, the country's lead negotiator to COP18 made an emotional plea calling on the parties to the UNFCCC to take measures in order to "stop this madness."[13] Climate-related "disasters," "refugees," "wars" and "victims" have repeatedly been invoked as loss and damage that had to be addressed through international cooperation.[14]

Yet, it is particularly difficult to identify specific instances of loss and damage which can be directly attributed to climate change impacts based on solid scientific evidence. Attributing specific harm to climate change faces two successive difficulties. Firstly, a physical event (e.g. a drought) needs to be attributed to climate change. Secondly, consequences of this physical event on individuals, societies and ecosystems need to be attributed to this physical event. The causal relation between climate change, a physical event and a specific harm is rarely straightforward.

---

[9] See e.g. Rosemary Lyster, "A fossil fuel-funded climate disaster response fund under the Warsaw International Mechanism for Loss and Damage Associated with Climate Change Impacts" (2015) 4:1 *Transnational Environmental Law* 125; Meinhard Doelle, "The birth of the Warsaw Loss & Damage Mechanism: planting a seed to grow ambition?" (2014) 8:1 *Carbon & Climate Law Review* 35; Benoit Mayer, "Whose 'loss and damage'? Promoting the agency of beneficiary states" (2014) 4:3–4 *Climate Law* 267; Emma Lees, "Responsibility and liability for climate loss and damage after Paris" (2017) 17:1 *Climate Policy* 59; Maxine Burkett, "Reading between the red lines: loss and damage and the Paris outcome" (2016) 6:1–2 *Climate Law* 118; Sumudu Atapattu, "Climate change, differentiated responsibilities and State responsibility: devising novel legal strategies for damage caused by climate change" in Benjamin J. Richardson *et al.* (eds.), *Climate Law and Developing Countries: Legal and Policy Challenges for the World Economy* (Edward Elgar, 2009) 37; Benoit Mayer, "Migration in the UNFCCC Workstream on Loss and Damage: an assessment of alternative framings and conceivable responses" (2017) 6:1 *Transnational Environmental Law* 107. See also Ivo Wallimann-Helmer, "Justice for climate loss and damage" (2015) 133:3 *Climatic Change* 469; Roda Verheyen, *Climate Change Damage and International Law: Prevention Duties and State Responsibility* (Brill, 2005).

[10] See e.g. Kees van der Geest and Markus Schindler, *Handbook for Assessing Loss and Damage in Vulnerable Communities* (UNU EHS, 2017); Erin Roberts *et al.*, *Loss and Damage: When Adaptation is Not Enough* (UNEP, 2014); Koko Warner *et al.*, *Evidence from the Frontlines of Climate Change: Loss and Damage to Communities Despite Coping and Adaptation* (UNU-EHS Report, 2012).

[11] Saleemul Huq, Erin Roberts and Adrian Fenton, "Loss and damage" (2013) 3:11 *Nature Climate Change* 947, at 948.

[12] Decision 1/CP.16, "The Cancún Agreements: outcome of the work of the Ad Hoc Working Group on Long-Term Cooperative Action under the Convention" (December 10–11, 2010), footnote 3 under para. 25.

[13] "'It's time to stop this madness': Philippines plea at UN climate talks" *Climate Home* (November 11, 2013).

[14] See, for instance, US President Barack Obama, "Remarks at United Nations Climate Change Summit" (New York, September 22, 2009).

Science rarely provides support for a direct attribution of a physical event to climate change. Since climate relates to the probability of weather patterns,[15] climate change consists in variations in such probability. Assessing that severe tornadoes, storms or droughts, for instance, are becoming more likely does not mean that any such event would not have occurred without excessive anthropogenic GhG emissions or that it is "caused by" climate change in the same way as a murderer causes the death of his victim by pulling a trigger. More often, scientific attribution of an event to climate change is a probabilistic nature. For instance, a study suggested that climate change had increased the risk of floods that occurred in England and Wales in the autumn of 2000 by 20 percent.[16] With regard to extreme events (which often cause the most serious harm), even such probabilistic attribution is hindered by limited knowledge of the likelihood of a given type of physical event at a given place before climate change and its evolution with climate change. Since extreme weather events are by definition of rare occurrence, scientists recognize that it is "challenging to detect systematic changes in their occurrence given the relative shortness of observational records."[17]

The second challenge of attribution relates to the link between a given physical event and loss and damage affecting individuals, societies and ecosystems. Physical events may trigger long chains of events. Like the concentric circles that an impact produces on the surface of water, the consequences of climate change on human societies and on ecosystems may extend *ad infinitum* and *at absurdum* in time and space. In a somewhat philosophical sense, it is hardly an exaggeration to suggest that the impacts of climate change affect virtually every individual, society and ecosystem in myriads of ways, both good and bad. But these causal waves do not flow unimpeded on a calm surface water; rather, they interact with multiple other waves, in the course of human lives, the history of human societies and the evolution of ecosystems. It is received wisdom that a "natural" disaster is never entirely natural – although disasters may be triggered by a natural phenomenon, the failure of protection mechanisms involves social and political factors determining the vulnerability of individuals, societies and, sometimes, even ecosystems.[18] In other words, the harm that a physical event causes on a society or an ecosystem depends in large part on the characteristics of the latter, in particular their exposure, vulnerability and resilience, and on measures taken to change these characteristics, such as adaptation action.

II. REPARATIONS UNDER GENERAL INTERNATIONAL LAW

While the concept of loss and damage originates in international negotiations on climate change, it closely relates to arguments on responsibility which can also be approached through the lens of general international law. Chapter 5 emphasized the relevance of customary

---

[15] Climate is usually defined as "the average weather, or more rigorously, as the statistical description in terms of the mean and variability of relevant quantities over a period of time ranging from months to thousands or millions of years." See S. Planton, "Glossary" in T.F. Stocker *et al.* (eds.), *Climate Change 2013: The Physical Science Basis. Contribution of Working Group I to the Fifth Assessment Report of the Intergovernmental Panel on Climate Change* (Cambridge University Press, 2013) 1447, at 1450.

[16] See e.g. Pardeep Pall *et al.*, "Anthropogenic greenhouse gas contribution to flood risk in England and Wales in autumn 2000" (2011) 470:7334 *Nature* 382; and for a critical review, Mike Hulme, "Attributing weather extremes to 'climate change': a review" (2014) 38:4 *Progress in Physical Geography* 499.

[17] Peter A. Stott *et al.*, "Attribution of extreme weather and climate-related events" (2016) 7:1 *WIREs Climate Change* 23, at 23.

[18] See Hartman and Squires (eds.), *supra* note 1.

international law on the responsibility of States for internationally wrongful acts. In particular, it was suggested that the failure of a State to prevent excessive GhG emissions from activities under its jurisdiction constitutes a breach of the no-harm principle.

As was noted in Chapter 5, the responsibility of States for an internationally wrongful act entails remedial obligations. The International Law Commission's Articles on the Responsibility of States for Internationally Wrongful Acts provide that a State responsible for an internationally wrongful act "is under an obligation to make full reparation for the injury caused by the internationally wrongful act."[19] Yet, the practice of States does not support the requirement of a full reparation. In relation to extensive and serious damage occurring globally and spreading over centuries, adequate reparation should be determined not only on the basis of the injury, but also based on consideration of the respective payment capacity of responsible States and the need to sanction the continuation of a wrongful act.[20] This interpretation of the law of State responsibility would require a less specific assessment of the injury caused by the failure of States to prevent excessive GhG emissions, although some broad understanding of this injury would remain necessary.

Characterizing the injury caused by the failure of States to prevent excessive GhG emissions under the law of State responsibility faces similar challenges to those discussed above regarding the determination of loss and damage. The failure of a State to prevent excessive GhG emissions from activities under its jurisdiction does not directly cause any loss and damage. Rather, this omission contributes more or less significantly (depending largely on the extent of the State's jurisdiction) to an anthropogenic interference with the climate system, whose impacts unfold over time, scattered throughout the world, often exacerbating pre-existing challenges.

One could argue that such difficulties in characterizing the injury would rule out the existence of remedial obligations. Comparable cases have never been brought before international courts. However, it may appear absurd and contrary to any principle of justice that the international law on State responsibility, imposing on States' remedial obligations even for relatively anecdotal injuries, would require absolutely no reparation when the wrongful act spreads misery throughout the world, makes parts or the whole of the territory of other States uninhabitable, and could threaten our existence as civilizations and a species. Lack of clarity on the modalities of the law should not prevent a court from applying the law.

If a court were to be seized of a dispute regarding the responsibility of a State for excessive GhG emissions and recognized its jurisdiction,[21] it may be reluctant to impose extensive remedial obligations. The priority should arguably be on the cessation of the continuing internationally wrongful act rather than on the reparation of the harm already done. Considerations of jurisprudential policy could also be at stake, as little good would be achieved by a court decision imposing heavy reparations on some States which may not even be implemented by the latter. On the other hand, a court in such a case could not reject any claims for reparation without openly betraying the project of international law of promoting justice in the relations between nations. In all likelihood, a court would at least provide a strong political message through the imposition of symbolic measures of satisfaction. The International Law

---

[19] ILC, *Draft Articles on Responsibility of States for Internationally Wrongful Acts with Commentaries*, in (2001) *Yearbook of the International Law Commission*, vol. II, part two (hereinafter *Articles on State Responsibility*), art. 31.1.

[20] See Chapter 5, section IV.B.

[21] See Chapter 14, section I.

Commission recognized the role of measures of satisfaction for injury caused by an internationally wrongful act insofar as it cannot be made good by restitution or compensation,[22] "including an acknowledgement of the breach, an expression of regret, a formal apology or another appropriate modality."[23] In a case related to climate change, such measures could include an obligation to acknowledge responsibility and adopt an apologetic attitude, but also a policy of memory which could involve education, awareness-raising, memorials, commemoration and even, for instance, museums.[24]

However, there are legal bases for a court to identify an obligation to pay compensation as reparation for the injury caused by the failure of a State to prevent excessive GhG emissions from activities under its jurisdiction. The law of State responsibility does not require a direct causal relation between an international wrongful act and the injury. The oft-cited judgment of the Permanent Court of International Justice in the case of the *Factory at Chorzów* stated that "reparation must, so far as possible, wipe out all the consequences of the illegal act,"[25] without limitation to the *direct* consequences of the act. Other international courts and tribunals have assessed remedial obligations in relation to injuries "attributable [to the wrongful act] as a proximate cause,"[26] and the International Law Commission concluded that "the requirement of a causal link is not necessarily the same in relation to every breach of an international obligation."[27] A sufficient causal link to establish an obligation of reparation could be identified when it can be foreseen that the conduct of a State – its failure to prevent excessive GhG emissions from activities under its jurisdiction – will further exacerbate serious impacts of climate change in other States, even if particular loss and damage will unfold only indirectly.

III. THE UNFCCC WORKSTREAM ON LOSS AND DAMAGE

Claims for the responsibility of industrial nations with regard to climate change have been voiced since the very start of international negotiations on climate change. As mentioned before, through the 1989 Caracas Declaration, developing States argued that developed States "should bear the main responsibility" because of their larger contribution to global environmental degradation.[28] Former Malaysian Prime Minister Mahathir Mohamad took an active role in the blame game that ensued the following years, contrasting the "pittance" that Western States offered as development assistance or promised as support to adaptation to the much greater "loss of earnings by the poor countries."[29] The AOSIS proposed in 1991 that an

---

[22] *Articles on State Responsibility, supra* note 19, art. 37.1.

[23] *Ibid.,* art. 37.2.

[24] See Benoit Mayer, "Climate change reparations and the law and practice of state responsibility" (2016) 7:1 *Asian Journal of International Law* 185. Such innovative measures of reparation have repeatedly been imposed by the Inter-American Court of Human Rights (IACrHR), for instance, in IACrHR, *Velásquez Rodríguez* v. *Honduras,* judgment of July 29, 1988, Ser. C, No. 4 (1988).

[25] PCIJ, *Factory at Chorzów (Germany* v. *Poland),* Judgment on the merits of the claim for indemnity of September 13, 1928, in Series A, No. 17, at 47.

[26] United States-Germany Mixed Claims Commission, Administrative Decision No. II, November 1, 1923, VII *UNRIAA,* 23, at 30.

[27] *Articles on State Responsibility, supra* note 19, commentary under Article 31, para. 10.

[28] See Caracas Declaration, *supra* note 2, para. II-34.

[29] Statement by Malaysia Prime Minister Mahathir Mohamad, in *Report of the United Nations Conference on Environment and Development,* UN document A/CONF.151/26/Rev.1(vol. III) (August 14, 1992) 230, at 233.

insurance mechanism be funded by industrialized States to "compensate the most vulnerable small island and low-lying coastal developing countries for loss and damage resulting from sea level rise."[30]

In lieu of such a mechanism, the UNFCCC only recognized a vague obligation for developed States to "assist the developing country Parties that are particularly vulnerable to the adverse effects of climate change in meeting costs of adaptation to those adverse effects."[31] The UNFCCC regime focused on the future by prioritizing climate change mitigation and adaptation to climate change. The dominant view was that arguments on past responsibilities were an untimely distraction from more pressing priorities. Yet, as a matter of principle rather than as a pecuniary question, different views on questions of responsibility hindered international negotiations by fueling incompatible expectations.

Relevant changes took place in around 2007. That year was marked by the publication of the fourth Assessment Report of the IPCC, which reflected growing evidence that climate change was rapidly unfolding.[32] The Bali Action Plan was also adopted that year and launched a new track of negotiations on the basis of a "shared vision for long-term cooperative action,"[33] to be conducted under the aegis of the Ad Hoc Working Group on Long-Term Cooperative Action under the Convention (AWG-LCA). At this stage of the negotiations, concessions were made to developing States in an effort to have them engage in climate change mitigation. This new stream of negotiations was to involve enhanced action on adaptation, including "consideration of ... means to address loss and damage associated with climate change impacts in developing countries that are particularly vulnerable to the adverse effects of climate change."[34] The concept of loss and damage was thus introduced for the first time in a decision adopted by the parties to the UNFCCC. It was clearly situated within the scope of enhanced action on climate change adaptation, although it would later be considered as a field of cooperation distinct from adaptation.

In the following years, however, debates on loss and damage were sidelined. Developed States were seeking to "avoid discussions related to proposals around compensation for loss and damage"[35] by proposing an alternative focus on risk management, in particular through risk-sharing mechanisms and disaster risk-reduction strategies.[36] Three years later, COP16 recognized, through the 2010 Cancún Agreements, "the need to strengthen international cooperation and expertise in order to understand and reduce loss and damage associated with

---

[30] Submission by Vanuatu, "Draft Annex Relating to Insurance," in INCFCC, *Negotiation of a Framework Convention on Climate Change: Elements Relating to Mechanisms* (December 17, 1991), doc. A/AC.237/WG.II/CRP.8, at 2. For a comparable proposal to the AWG-LCA, see AOSIS submission, "Multi-window Mechanism to Address Loss and Damage from Climate Change Impacts," in UNFCCC AWG-LCA, *Ideas and Proposals on the Elements Contained in Paragraph 1 of the Bali Action Plan: Submissions from Parties* (December 10, 2008), doc. FCCC/AWGLCA/2008/Misc.5/Add.2 (Part I), at 24. For a historical account of UNFCCC negotiations on loss and damage, see Maxine Burkett, "Loss and damage" (2014) 4:1-2 *Climate Law* 119.

[31] United Nations Framework Convention on Climate Change, May 9, 1992, 1771 *UNTS* 107 (hereinafter UNFCCC), art. 4.4. See also arts. 4.8 and 4.9.

[32] See in particular M.L. Parry *et al.* (eds.), *Contribution of Working Group II to the Fourth Assessment Report of the Intergovernmental Panel on Climate Change* (Cambridge University Press, 2007).

[33] Decision 1/CP.13, *supra* note 5, para. 1(a). See generally Chapter 3, section III.

[34] Decision 1/CP.13, *supra* note 5, para. 1(c)(iii).

[35] K. Warner and S. Zakieldeen, *Loss and Damage Due to Climate Change: An Overview of the UNFCCC Negotiations* (European Capacity Building Initiative, 2012) 4.

[36] Loss and damage was mentioned along with "disaster reduction strategies" in the Bali Action Plan. See decision 1/CP.13, *supra* note 5, para. 1(c)(iii).

the adverse effects of climate change, including impacts related to extreme weather events and slow onset events."[37] A footnote to the decision clarified that the impacts of climate change at issue included "sea level rise, increasing temperatures, ocean acidification, glacial retreat and related impacts, salinization, land and forest degradation, loss of biodiversity and desertification."[38] In order to strengthen international cooperation and expertise, the Cancún Agreements established a "work programme" to be implemented by the Subsidiary Body for Implementation (SBI), reporting to the COP, with the objective "to consider, including through workshops and expert meetings, as appropriate, approaches to address loss and damage associated with climate change impacts in developing countries that are particularly vulnerable do the adverse effects of climate change."[39]

The following year, a dedicated COP decision guided the SBI toward the exploration of three different thematic areas:

1. assessing the risk of loss and damage and current knowledge of this risk;
2. defining approaches to address loss and damage on the basis of experience;
3. exploring the options for international cooperation within the UNFCCC regime.[40]

Specific actions were recommended under each of these thematic areas.

In 2012, COP18 adopted another decision where it reaffirmed a desire "to enhance action on addressing loss and damage."[41] This decision defined eight priorities, such as "assessing the risk of loss and damage," "identifying options," "systematic observation" and "implementing comprehensive climate risk management approaches."[42] It also emphasized the need for "further work to advance the understanding of and expertise on loss and damage."[43] While encouraging research and data collection as well as coordination and collaboration between existing institutions and mechanisms, this decision generally put less emphasis than the Cancún Agreements on cooperation.[44] Most importantly, the COP decided to establish, at the following session, "institutional arrangements, such as an international mechanism, including functions and modalities ... to address loss and damage."[45]

Accordingly, COP19 established the Warsaw International Mechanism for Loss and Damage associated with Climate Change Impact (WIM).[46] The objective of the WIM is to "fulfil the role under the Convention of promoting the implementation of approaches to address loss and damage ... in a comprehensive, integrated and coherent manner," including through "enhancing knowledge and understanding," "strengthening dialogue, coordination, coherence and synergies among relevant stakeholders," and "enhancing action and support, including finance,

[37] Decision 1/CP.16, *supra* note 12, para. 25.

[38] *Ibid.*, footnote 3 under para. 25.

[39] *Ibid.*, para. 26.

[40] Decision 7/CP.17, "Work programme on loss and damage" (December 9, 2011), paras. 6–15.

[41] Decision 3/CP.18, "Approaches to address loss and damage associated with climate change impacts in developing countries that are particularly vulnerable to the adverse effects of climate change to enhance adaptive capacity" (December 8, 2012), para. 6.

[42] *Ibid.*, para. 6(a), (b), (c) and (d).

[43] *Ibid.*, para. 7.

[44] See *ibid.*, recital 4.

[45] *Ibid.*, para. 9.

[46] Decision 2/CP.19, "Warsaw International Mechanism for Loss and Damage associated with Climate Change Impacts" (November 23, 2013).

technology and capacity-building, to address loss and damage."[47] Specific arrangements were adopted the following year at COP20, including the composition of the Executive Committee of the WIM, basic rules on procedure and a two-year workplan.[48] The workplan put emphasis on spurring research and raising awareness on factors of vulnerability, risk management approaches, slow-onset impacts, non-economic loss, resilience and migration, as well as financial instruments and tools.[49]

The following year, the workstream on loss and damage was recognized by the Paris Agreement. Article 8 of the Paris Agreement emphasizes "the importance of averting, minimizing and addressing loss and damage"[50] and endorsed the creation of the WIM.[51] The parties to the Paris Agreement committed to "enhance understanding, action and support, including through the WIM, as appropriate, on a cooperative and facilitative basis with respect to loss and damage associated with the adverse effects of climate change."[52] At the insistence of developed States, the COP decision on the adoption of the Paris Agreement included a caveat to clarify "that Article 8 of the Agreement does not involve or provide a basis for any liability or compensation."[53] This clause clarifies the meaning of Article 8 of the Paris Agreement as neutral on questions of compensation; it does not imply any renunciation to claims for reparation based on general international law and does not preclude the possibility of a future agreement on reparation.

The COP decision on the adoption of the Paris Agreement requested the WIM to establish "a clearing house for risk transfer that serves as a repository for information on insurance and risk transfer"[54] as well as "a task force ... to develop recommendations for integrated approaches to avert, minimize and address displacement related to the adverse impacts of climate change."[55] The following year, COP22 approved a second workplan for the WIM, this time for a five-year period.[56] The COP further directed the Executive Committee of the WIM to initiate a "strategic workstream to guide the implementation of the Warsaw International Mechanism's function of enhancing action and support, including finance, technology and capacity-building, to address loss and damage."[57] It also requested the preparation of "a technical paper ... elaborating the sources of financial support, as provided through the Financial Mechanism, for addressing loss and damage as described in relevant decisions, as well as

---

[47] *Ibid.*, para. 5.

[48] Decision 2/CP.20, "Warsaw International Mechanism for Loss and Damage associated with Climate Change Impacts" (December 13, 2014), para. 5.

[49] UNFCCC SBSTA and SBI, *Report of the Executive Committee of the Warsaw International Mechanism for Loss and Damage Associated with Climate Change Impacts* (October 24, 2014), doc. FCCC/SB/2014/4, Annex II; and decision 2/CP.20, *supra* note 48, para. 1.

[50] Paris Agreement, December 12, 2015, in the annex of decision 1/CP.21, "Adoption of the Paris Agreement" (December 12, 2015), art. 8.1.

[51] *Ibid.*, art. 8.2.

[52] *Ibid.*, art. 8.3.

[53] Decision 1/CP.21, *supra* note 50, para. 51.

[54] *Ibid.*, para. 48. See also UNFCCC WIM, *Concept Paper: Clearing House on Risk Transfer* (September 2016).

[55] Decision 1/CP.21, *supra* note 50, para. 49. See also UNFCCC WIM, *Summary of Proceedings of the First Meeting of the Task Force on Displacement* (June 7, 2017). See, generally, Benoit Mayer, *The Concept of Climate Migration: Advocacies and its Prospects* (Edward Elgar, 2016).

[56] Decision 3/CP.22, "Warsaw International Mechanism for Loss and Damage associated with Climate Change" (November 17, 2016), para. 3.

[57] *Ibid.*, para. 4.

modalities for accessing such support."[58] These provisions could announce a progressive trans-formation of the WIM toward a more operational institution – one where guidance would be completed by effective support.

Throughout a decade of negotiations, the workstream on loss and damage has fostered research and contributed to advocacy and agenda-setting, but it is yet to achieve any of the tangible results that its proponents desire. The natural evolution of a mechanism to address loss and damage caused by some States and suffered by others lies in something akin to reparation.[59] Yet, anything akin to reparation remains a non-starter for most developed States. Relevant negotiations have drifted from a component of the long-term cooperative action under the Bali Action Plan to a "work programme" carried out by the SBI under the Cancún Agreements, then to a succession of workplans adopted by the WIM Executive Committee, which has then established several subsidiary organs. Over the decade following the Bali Action Plan, the workstream on loss and damage epitomizes the tendency of climate change negotiations to produce complexity when incompatible negotiating positions preclude sub-stantial agreements.

IV. POSSIBLE WAYS FORWARD

For the time being, many questions remain open regarding the nature and purpose of the UNFCCC workstream on loss and damage. A first question concerns the relation between this workstream and action on climate change adaptation. Under the Bali Action, "approaches to address loss and damage" were part of enhanced action on climate change adaptation.[60] Under the Cancún Agreements, provisions on loss and damage were inserted into a section on adaptation.[61] Yet, since then, approaches to address loss and damage associated with climate change impacts have been addressed in separate COP decisions,[62] separate sections of COP decisions[63] and a separate article of the Paris Agreement,[64] although some references to this workstream were occasionally inserted into discussions on adaptation.[65] Yet, there remains

---

[58] Decision 4/CP.22, "Review of the Warsaw International Mechanism for Loss and Damage associated with Climate Change Impacts" (November 17, 2016), para. 2(f).

[59] See for instance the informal note of the Co-Chairs, "Reflections on progress made at the fourth part of the second session of the Ad Hoc Working Group on the Durban Platform for Enhanced Action" (April 17, 2014), at 12, reflecting the demand of some parties for the inclusion, in the Paris Agreement, of "[a] specific commitment to provide support for financing and operationalization of the WIM for Loss and Damage." See also the submission of Nauru on behalf of AOSIS on its view on Loss and Damage in the 2015 Agreement (November 4, 2014), at 1, noting that: "Immediate financial, technical and capacity building support that is adequate, provided on a timely basis and truly accessible will be required to address loss and damage in SIDS. Financial flows from developed countries for addressing loss and damage in vulnerable developing countries should be new and additional to financing for mitigation and adaptation."

[60] Decision 1/CP.13, *supra* note 5, para. 1(c)(iii).

[61] Decision 1/CP.16, *supra* note 12, paras. 25–29.

[62] See decision 7/CP.17, *supra* note 40; decision 3/CP.18, *supra* note 41; decision 2/CP.19, *supra* note 46; decision 2/CP.20, *supra* note 48; decision 2/CP.21, "Warsaw International Mechanism for Loss and Damage associated with Climate Change Impact" (December 10, 2015); decision 3/CP.22, *supra* note 56; 4/CP.22 *supra* note 58.

[63] See decision 1/CP.21, *supra* note 50, paras. 48–52.

[64] See Paris Agreement, *supra* note 50, art. 8.

[65] See e.g. decision 1/CP.18, "Agreed outcome pursuant to the Bali Action Plan" (December 8, 2012). See also C. Ruell, "Australia wins fossil on lost and damaged position on loss and damage" *Climate Action Network International* (December 4, 2014).

resistance to the idea that loss and damage is now distinct from adaptation. As of early 2018, the website of the UNFCCC kept loss and damage within its section on adaptation,[66] apparently due to a lack of a clear consensus for doing otherwise. Beyond the UNFCCC regime, other institutions and regimes are yet to endorse the terminology of "loss and damage" in the same way as they did endorse "adaptation."[67] Developed States remain generally reluctant to recognize loss and damage as a distinct type of response to climate change, seemingly due to fears of the claims for compensation or, otherwise, for the support that this would fuel.

A second, related question concerns, precisely, the potential outcomes of international negotiations on loss and damage. It appears unlikely in the foreseeable future that international negotiations on climate change could lead to an agreement on compensation, although some limited financial support could well be provided which would not differ substantially from support for adaptation. Nevertheless, the existence of a forum where arguments on responsibility could be channeled provides the most vulnerable States with additional political leverage in ongoing negotiations. The workstream will surely continue to document the risks associated with climate change impacts and circulate information on the response measures adopted by different institutions. The workstream could further encourage parties to take specific response measures which, in its view, appear more effective or desirable. It may encourage developed State parties to provide technical and financial support to the nations most affected by the adverse impacts of climate change. It is also possible that the WIM could turn into a more operational institution, able to implement projects directly to the States which need it.

However, the operationalization of the WIM would exacerbate the risk of redundancies with international action on adaptation. There is no clear distinction between efforts to adapt to climate change on the one hand and efforts to avoid and reduce loss and damage associated with climate change impacts on the other hand. These concepts were adopted through protracted international negotiations, not because of their analytical value, but because of their ambiguity and their political convenience. Some scholars tried to distinguish adaptation as a process of *avoiding* harms and approaches to address loss and damage as a process of *managing* unavoidable impacts.[68] Yet, as an adaptation measure may never be able to avoid absolutely every harm,[69] this distinction is one of degree rather than nature.

Shifting from claims for compensation to the provision of guidance and support, the workstream on loss and damage tends to replicate the evolution of adaptation under the UNFCCC. Unlike compensation, support and guidance allow "donors" to play a role in the determination of measures implemented in the most vulnerable countries. This creates a risk that these countries will implement measures which are not in their own best interests, but in the best interests of the donors. Such trend is already at work, it seems, in the recent focus of the workstream on loss and damage on migration and insurance.[70] While the management of international migration is a clear priority of developed States, strong evidence exists that those most vulnerable to adverse impacts of climate change often lack the resources to migrate – and

---

[66] See "Groups and Committees: Adaptation" on the UNFCCC website, http://unfccc.int/adaptation/groups_committees/items/6992.php (accessed January 18, 2018).

[67] On adaptation, see Chapter 10, section III.

[68] See Kees van der Geest and Koko Warner, "Loss and damage from climate change: emerging perspectives" (2015) 8:2 *International Journal of Global Warming* 133, at 135.

[69] This is the case at least because adaptation measures have costs and unintended effects.

[70] See references *supra* notes 54 and 55 and accompanying text.

especially to migrate abroad.[71] Focusing responses to loss and damage on the protection of cross-border migrants rather than on more vulnerable populations which are not even able to migrate does not appropriately address the most urgent needs for assistance. Likewise, the promotion of insurance mechanisms as a way to address loss and damage associated with climate change impacts could favor the interests of large reinsurance companies, but insurance would not appropriately cover risks which are either too unpredictable (due to a lack of scientific understanding of the future impacts of climate change) or too predictable (as in the case of slow-onset environmental changes such as sea-level rise). Insurance remains, globally and domestically, a preserve of the wealthy,[72] and any effort to expand insurance coverage is unlikely to protect the most vulnerable in any society in the foreseeable future. Existing alternatives, such as State-led social protection, have not sufficiently been promoted in the workstream on loss and damage.

The UNFCCC workstream on loss and damage could bring about a reconsideration of claims for reparation, although it is most likely not to. Such claims, however, will not be relinquished until they are satisfied. As the impacts of climate change become more palpable year after year, creating havoc around the world, such claims will be raised again and again. Litigation has been and will continue to be explored.[73] Expert institutions such as the International Law Commission will continue their work. Civil society organizations will advocate for justice. Some acknowledgment of international responsibilities and some form of reparation for the injury caused because of excessive GhG emissions will, one day, be granted.

---

[71] See e.g. Foresight Agency: The Government Office for Science (United Kingdom), *Migration and Global Environmental Change: Final Project Report* (2011).

[72] See, for instance, Joanne Linnerooth-Bayer, Reinhard Mechler and Stefan Hochrainer-Stigler, "Insurance against losses from natural disasters in developing countries: evidence, gaps and the way forward" (2014) 7 *International Journal of Disaster Risk Reduction* 154; Swenja Surminski and Dalioma Oramas-Dorta, "Flood insurance schemes and climate adaptation in developing countries" (2014) 7 *International Journal of Disaster Risk Reduction* 154; Jyotsna Jalan and Martin Ravallion, "Are the poor less well insured? Evidence on vulnerability to income risk in rural China" (1999) 58:1 *Journal of Development Economics* 61.

[73] See Chapter 14.

12

International Support

Lack of capacity may be a major impediment to effective climate action, in particular in States which face the greatest development and environmental challenges. Other actors can help these countries to implement action to promote climate change adaptation and advance climate change mitigation. The flexibility mechanisms discussed in Chapter 8, in particular the CDM, have contributed to providing support to developing States, but many other initiatives have also taken place. Long confined to financial assistance from developed to developing States, international support has extended to other sources of support. "South-South cooperation" refers to support provided by developing States to other developing States – typically emerging economies like China helping some of the least developed countries. Non-State actors – investors, NGOs and subnational governments – have also played a growing role. Likewise, beyond financial assistance, attempts have been made to provide long-term support through transfer of technology as well as capacity building.

Developed States have pledged to levy USD 1 trillion to support climate action in developing States during the 2020s.[1] Nevertheless, international support for climate action has remained little explored from an international law perspective.[2] This general omission may relate to a lack of clear rights or obligations of States in this regard. For a quarter of a century, developing States have voiced claims for a new, additional, adequate and predictable support for climate action on the basis of common but differentiated responsibilities. While developed States have conceded that they ought to provide some support, they displayed no enthusiasm for any specific commitment. To date, international support for climate action remains, by and large, provided solely on discretionary bases, often contingent upon the economic or political interests of "donors." As developing States are making greater efforts to mitigate climate change and vast amounts of capital are being invested in these countries, legal questions related to international support will become increasingly pressing.

This chapter provides a general overview of the international law on international support for climate action. The first section singles out three distinct rationales for such support. The second section explores regulatory and institutional developments regarding climate finance, technology transfer and capacity building. The third section describes support targeted more

---

[1] See Copenhagen Accord, in the annex of decision 2/CP.15 (December 18–19, 2009), para. 8, setting a collective goal for developed countries to jointly mobilize USD 100 billion per year by 2020.
[2] See, however, Alexander Zahar, *Climate Change Finance and International Law* (Routledge, 2017).

specifically at forest policies, provided under the UNFCCC REDD+ program. Lastly, the fourth section identifies the backbone of an emerging normative discourse on international support for climate action.

I. THE RATIONALES FOR INTERNATIONAL SUPPORT FOR CLIMATE ACTION

International support for climate action can be justified in three distinct ways. Firstly, it can be approached as assistance granted to those most affected by climate change, as commanded by international human rights law and the international law on development. Secondly, international support for climate action can alternatively be viewed as an attempt to mitigate damages or even, perhaps, to provide an atypical form of reparation for a wrongful act (e.g. the failure of developed States to prevent excessive GhG emissions). Thirdly, support to mitigation action can be understood as simply a cheaper way to tackle climate change through projects in developing States. Each of these rationales suggests different forms of support. Each of them has had some influence on the negotiations on international support for climate action, although it is the third one which has been most influential in the actual practice of international support.

A. General Assistance Duties

Efforts have long been made to promote international cooperation and assistance among nations, whether simply to guarantee a durable peace or, more ambitiously, to enhance human welfare. Thus, "to achieve international cooperation in solving international problems of an economic, social, cultural, or humanitarian character" was recognized as one of the objectives of the United Nations.[3] The effective guarantee of universal and inalienable human rights calls for cooperation beyond national borders, for instance, to provide assistance to a State which, for whatever reason, is unable to provide effective protection to its own population. The Universal Declaration of Human Rights called on States "to achieve, *in cooperation with the United Nations*, the promotion of universal respect for and observance of human rights and fundamental freedoms."[4]

As decolonization opened international institutions to new voices and new ideas, underdevelopment emerged as a challenge to the effective universal enjoyment of human rights. The International Covenant on Economic, Social and Cultural Rights affirmed the obligation of each State to "take steps ... through international assistance and co-operation, especially economic and technical, to the maximum of its available resources."[5] Yet, the obligation of all States to cooperate in the realization of economic, social and cultural rights was couched in vague language, making it particularly challenging to assess compliance in any concrete case. A progressive scholarly interpretation of international human rights law suggests that States have the obligation "to create an international enabling environment," "to provide international

---

[3] Charter of the United Nations, June 26, 1945, 1 *UNTS* XVI, art. 1.3. See also *ibid.*, art. 55.

[4] UN General Assembly Resolution 217 A, "Universal Declaration of Human Rights" (December 10, 1948), recital 6 (emphasis added).

[5] International Covenant on Economic, Social and Cultural Rights, December 16, 1966, 993 *UNTS* 3, art. 2.1. See also CESCR General Comment No. 3, "The nature of State parties' obligations" (December 14, 1990), para. 14.

assistance" and, upon request for assistance, to "consider the quest in good faith."[6] Even this does not guarantee that any State in need of international assistance will receive it.

Through decades of haggling in the UN General Assembly, developing States tried to specify the obligation of developed States to address global inequalities, provide for an international economic order allowing for the development of poorer countries, and generally acknowledge "the reality of interdependence of all the members of the world community."[7] As an effort to define tangible obligations since the early 1970s, developed States were called to allocate 0.7 percent of their gross national product as official development assistance,[8] a commitment with which most have never complied. The Millennium Development Goals and the Sustainable Development Goals sought to create a momentum in support for international development assistance for States to assume their "shared responsibility ... for managing worldwide economic and social development."[9]

In this context, climate change has appeared as a factor exacerbating pre-existing challenges to the protection capacity of developing countries, especially the least developed and the most vulnerable. Accordingly, the Sustainable Development Goals called on States to "take urgent action to combat climate change and its impacts."[10] Likewise, the Human Rights Council repeatedly emphasized the implications of climate change on the effective enjoyment of human rights, especially for "those segments of the population that are already in vulnerable situations owing to factors such as geography, poverty, gender, age, indigenous or minority status, national or social origin, birth or other status and disability."[11] On these bases, the Human Rights Council emphasized "the urgent importance of continuing to address, as they relate to States' human rights obligations, the adverse consequences of climate change impacts for all."[12] States are called upon "to continue and enhance international cooperation and assistance for adaptation measures to help developing countries, especially those that are particularly vulnerable to the adverse effects of climate change as well as persons in vulnerable situations."[13]

---

[6] Maastricht Principles on Extraterritorial Obligations of States in the area of Economic, Social and Cultural Rights, adopted by a group of experts in international law and human rights (September 28, 2011), principles 29, 33 and 35.

[7] UN General Assembly Resolution 3202(S-VI), "Declaration on the Establishment of a New International Economic Order" (May 1, 1974), para. 3. See also UN General Assembly Resolution 41/128, "Declaration on the Right to Development" (December 4, 1986); World Conference on Human Rights, *Vienna Declaration and Programme of Action* (June 25, 1993), doc. A/CONF.157/23.

[8] See e.g. UN General Assembly Resolution 2626 (XXV), "International Development Strategy for the Second United Nations Development Decade" (October 24, 1970), para. 43; International Conference on Financing for Development, *Monterrey Consensus of the International Conference on Financing for Development* (March 18–22, 2002), para. 42; UN General Assembly Resolution 60/1, "2005 World Summit outcome" (October 24, 2005), para. 23(b); UN General Assembly Resolution 66/288, "The future we want" (July 27, 2012), para. 258.

[9] UN General Assembly Resolution 55/2, "United Nations Millennium Declaration" (September 8, 2000), para. 6. See also UN General Assembly Resolution 70/1, "Transforming our world: the 2030 Agenda for Sustainable Development" (September 25, 2015).

[10] UN General Assembly Resolution 70/1, *supra* note 9, goal 13.

[11] See Human Rights Council resolution 35/20, "Human rights and climate change" (June 22, 2017), recital 15. See also Human Rights Council resolutions 32/33, "Human rights and climate change" (July 1, 2016); 29/15, "Human rights and climate change" (July 2, 2015), 18/22, "Climate change and human rights" (September 30, 2011); 7/23, "Human rights and climate change" (March 28, 2008); 10/4, "Human rights and climate change" (March 25, 2009); 7/23, "Human rights and climate change" (March 28, 2008).

[12] Human Rights Council resolution 35/20, *supra* note 11, para. 2.

[13] *Ibid.*, para. 6.

Thus, the general obligation of States to provide assistance and cooperation to the effective protection of human rights calls for international support for climate action. Climate change mitigation certainly plays a role in addressing a challenge to the effective enjoyment of human rights and a hindrance to development. Most directly, however, general assistance duties under international human rights law call for the provision of international support to adaptation action in the countries that are the most vulnerable to the adverse impacts of climate change. States have consistently recognized, for instance in the Paris Agreement, that international support should be given as a priority to developing countries "that are particularly vulnerable to the adverse effects of climate change and have significant capacity constraints, such as the least developed countries and small island developing States."[14]

B. Specific Responsibilities

The anthropogenic causes of climate change lend support to more specific claims, not just for assistance to the effective protection of human rights or to development, but also on the basis of moral or legal responsibilities arising from wrongdoing. Developing States approached international climate change negotiations with the understanding that it was down to developed countries to make up for the damage that they had caused to the global environment.[15] Similar arguments can be based on general international law, suggesting that developed States have a legal obligation to pay adequate reparation for the injury caused by their failure to prevent excessive GhG emissions from activities under their jurisdiction.[16] Support differs from standard forms of reparation, in particular compensation, because it allows the "donor" to oversee the use of the funds by the "recipient." If not as reparation, international support for climate action may be approached as a way for developed States to mitigate the injury caused by their wrongful conduct.

Claims for compensation voiced by developing States largely contributed to triggering negotiations on international support to adaptation and to approaches to address loss and damage.[17] Although developed States have never fully accepted a narrative on wrong-based responsibilities, the UNFCCC affirmed the principle of common but differentiated responsibilities as an (ill-defined) ground to differentiate between developed and developing States. Thus, owing to their greater financial capacities but also arguably to their historical responsibilities,[18] developed States recognized that they ought to "take the lead in combating climate change and the adverse effects thereof."[19] On these ambivalent grounds, developed States promised to provide support for climate action in developing countries.[20]

---

[14] Paris Agreement, December 12, 2015, in the annex of decision 1/CP.21, "Adoption of the Paris Agreement" (December 12, 2015), art. 9.4. See also United Nations Framework Convention on Climate Change, May 9, 1992, 1771 *UNTS* 107 (hereinafter UNFCCC), art. 4.4.

[15] See Caracas Declaration of the Ministers of Foreign Affairs of the Group of 77 on the Occasion of the Twenty-Fifth Anniversary of the Group (June 21–23, 1989), reproduced in doc. A/44/361, paras. II–34.

[16] See the discussions in Chapter 5, section IV and Chapter 11, section II.

[17] See Chapters 10 and 11.

[18] See decision 1/CP.16, "The Cancún Agreements: outcome of the work of the Ad Hoc Working Group on Long-Term Cooperative Action under the Convention" (December 10–11, 2010), second recital before para. 36.

[19] UNFCCC, *supra* note 14, art. 3.1. See generally Chapter 6, section II.B.

[20] See UNFCCC, *supra* note 14, arts. 4.3 and 11.

C. Self-Interest

Besides the obligation of States to provide assistance and cooperation to the effective protection of human rights and their wrong-based responsibilities, it may also be in the interests of developed States or other actors to support climate action. Climate change is a "common concern of humankind"[21] because anthropogenic emissions in any country affect all countries. The need for "the widest possible cooperation by all countries" was recognized in the Preamble to the UNFCCC. To be effective, the efforts of a State to mitigate climate change do not need to be implemented within its own territory. On the contrary, it appeared that, while developed States had more resources to support climate action, some mitigation projects could be realized at a lower cost in developing countries. From a purely utilitarian perspective, low-hanging fruit – more economical ways to reduce GhG emissions – should be picked wherever it is to be found. As discussed in Chapter 8, the interests of developed States in reaching mitigation outcomes at a reasonable cost led to the creation of flexibility mechanisms, whereby a State can implement its mitigation commitments by supporting projects implemented elsewhere. The same rationale has certainly helped overcome the reluctance of developed States to provide other forms of support for climate action in developed States.

Whereas the two other rationales are essentially normative, being grounded in ethical and legal arguments, this third rationale is a pragmatic one. It places heavy emphasis on climate change mitigation, which ensures global benefits, rather than adaptation.[22] Interested support is also more likely to select activities based on their cost-effectiveness, without necessarily considering the unintended consequences of such activities for local populations or for the environment. Therefore, self-interested support ought to be closely monitored to avoid harmful consequences.

As discussed in the next section, a gap appeared between the text of instruments adopted under the UNFCCC regime and the practice of international support for climate action. Developing States obtained the inclusion of provisions calling for support for climate action in COP decisions which drew from the first two rationales. Yet, the actual provision of international support was most directly influenced by the donors' self-interests. Thus, while COP decisions called for a "balanced allocation between adaptation and mitigation,"[23] developed States have continued to provide considerably more support to mitigation.[24] Likewise, the multilateral funds that were established by the COP were little used, as developed States preferred to retain stronger control on disbursements through multilateral banks or even national agencies.

---

[21] *Ibid.*, preamble, second recital.

[22] Support for adaptation may also be considered as self-interest in a world where economic crises, poverty or political turmoil in one country are likely to spread to the next. Yet, these benefits are less direct, less predictable and arguably less well-understood than those arising from mitigation action.

[23] Decision 1/CP.16, *supra* note 18, para. 95. See also Copenhagen Accord, *supra* note 1, para. 8.

[24] See Barbara K. Buchner *et al.*, *Global Landscape of Climate Finance 2015* (CPI, November 2015), at 2 and *passim*, estimating that around USD 25 billion per year was going to support to adaptation action, as opposed to USD 361 billion in support for mitigation action and USD 4 billion for dual benefits. See also "Summary and recommendations by the Standing Committee on Finance on the 2016 biennial assessment and overview of climate finance flows" in the annex of decision 8/CP.22, "Report of the Standing Committee on Finance" (November 18, 2016), para. 27, estimating that adaptation represents about 25 percent of total public climate finance in developing countries; and generally Erik Haites (ed.), *International Climate Finance* (Routledge, 2013); Bradly J. Condon and Tapen Sinha, *The Role of Climate Change in Global Economic Governance* (Oxford University Press, 2013) 200ff.

II. THE GENERAL FRAMEWORK ON INTERNATIONAL SUPPORT FOR CLIMATE ACTION

International negotiations on climate change have led to the adoption of various provisions to promote international support for climate action. While most emphasis has been placed on financial support, rules and institutions have also been established to promote transfer of technologies as well as capacity building.

A. Financial Support

1. National and Collective Commitments

Although the need for financial support to developing States was recognized from the onset of international negotiations on climate change,[25] its rationale, ambit and modalities remained the object of protracted negotiations. Under the UNFCCC, the responsibility to provide financial support lies on Annex II parties – developed parties except for those from the former Eastern Bloc, which, as they were undergoing a difficult transition to a market economy during the negotiations, were in no position to provide financial resources. Annex II parties committed firstly to provide financial resources for developing States to develop and communicate a national inventory of GhG emissions and a description of steps taken or envisaged to address climate change.[26] The COP interpreted this provision as including activities such as "studies of the possible impacts of climate change, identification of options for implementing adaptation provisions ..., and relevant capacity-building."[27] The costs of such activities were estimated to be relatively low in comparison to the benefits of ensuring universal participation, and Annex II parties committed to meeting the "agreed full costs."[28] This commitment, reiterated in the Kyoto Protocol,[29] was reframed as "capacity building" in the Paris Agreement.[30]

In addition, Annex II parties committed to provide financial support, including for the transfer of technology, "to meet the agreed full incremental costs" of implementing adaptation and mitigation measures in developing States.[31] Annex II parties further committed to "assist the developing country Parties that are particularly vulnerable to the adverse effects of climate change in meeting costs of adaptation to those adverse effects."[32] As a way of promoting the implementation of the latter provision, the Kyoto Protocol dedicated a share of the proceeds of projects under the CDM to being used "to assist developing country Parties that are particularly vulnerable to the adverse effects of climate change to meet the costs of adaptation,"[33] a share which was then defined to 2 percent.[34]

---

[25] See UN General Assembly Resolution 44/228, "United Nations Conference on Environment and Development" (December 22, 1989), paras. 15(j) and 15(k).

[26] UNFCCC, *supra* note 14, art. 4.3. See also *ibid.*, arts. 12.1 and 4.1(a).

[27] Decision 11/CP.1, "Initial guidance on policies, programme priorities and eligibility criteria to the operating entity or entities of the financial mechanism" (April 7, 1995), para. 1(d)(iv), first point.

[28] UNFCCC, *supra* note 14, art. 4.3. See also Daniel Bodansky, "The United Nations Framework Convention on Climate Change: a commentary" (1993) 18:2 *Yale Journal of International Law* 451, at 527.

[29] Kyoto Protocol to the United Nations Framework Convention on Climate Change, December 11, 1997, 2303 *UNTS* 162 (hereinafter Kyoto Protocol), art. 11.2.

[30] See the discussion on capacity building below, section II.C.

[31] UNFCCC, *supra* note 14, art. 4.3, in connection with art. 4.1.

[32] *Ibid.*, art. 4.4.

[33] Kyoto Protocol, *supra* note 29 art. 12.8.

[34] Decision 17/CP.7, "Modalities and procedures for a clean development mechanism, as defined in Article 12 of the Kyoto Protocol" (November 10, 2001), para. 15.

As part of a difficult political deal between developed and developing States, the Copenhagen Accord acknowledged the need for "scaled up, new and additional, predictable and adequate funding"[35] in support of adaptation as well as mitigation. Regarding adaptation, the Copenhagen Accord called upon developed States to "provide adequate, predictable and sustainable financial resources, technology and capacity-building"[36] to developing States. It was recognized that developed States could condition the implementation of their pledge to mitigation action to the availability of relevant international support.[37] Overall, the Copenhagen Accord affirmed the collective commitments of developed States to providing financial support "approaching" USD 30 billion for the period 2010–2012 and USD 100 billion by 2020.[38] The Cancún Agreements later endorsed these national commitments and collective objectives.[39]

The Paris Agreement further highlighted the importance of financial support. It defined, as one of three goals – along with mitigation and adaptation – the objective of "making finance flows consistent with a pathway towards low greenhouse gas emissions and climate-resilient development."[40] Financial support, transfer of technologies and capacity building are plainly integrated, on an equal footing with climate change mitigation, adaptation and transparency, among the different components of the global response to climate change.[41] The pool of potential contributors is also extended as developing States are "encouraged to provide or continue to provide such support voluntarily"[42] – a recognition of growing differences in national circumstances among developing States.

2. The Financial Mechanism of the UNFCCC

Institutional arrangements were also made to convey financial support. In particular, the UNFCCC established a Financial Mechanism "for the provision of financial resources on a grant or concessional basis, including for the transfer of technology."[43] The same Financial Mechanism was endorsed by the Kyoto Protocol and the Paris Agreement.[44] During the negotiations leading to the UNFCCC, a proposal of developing States for a new financial institution was opposed by developed States.[45] Instead, the latter favored an institutional arrangement under the auspices of the International Bank for Reconstruction and Development (the World Bank), where voting power is weighted to reflect each State's participation to the Bank's capital stock.[46] It was eventually agreed in 1992 that the Global Environment Facility (GEF) would be entrusted with the operation of the Financial Mechanism on an "interim basis,"[47] a situation which has persisted for the last quarter of a century.

---

[35] Copenhagen Accord, *supra* note 1, para. 8.
[36] *Ibid.*, para. 3. See also decision 1/CP.13, "Bali Action Plan" (December 14–15, 2013), paras. 1(d) and 1(e).
[37] Copenhagen Accord, *supra* note 1, para. 5.
[38] *Ibid.*, para. 8.
[39] Decision 1/CP.16, *supra* note 18, paras. 95 and 98.
[40] Paris Agreement, *supra* note 14, art. 2.1(c).
[41] See *ibid.*, art. 3.
[42] *Ibid.*, art. 9.2.
[43] UNFCCC, *supra* note 14, art. 11.1.
[44] See Kyoto Protocol, *supra* note 29, art. 11.2; Paris Agreement, *supra* note 14, art. 9.8.
[45] Bodansky, *supra* note 28, at 538.
[46] Thus, as of March 29, 2017, the voting power of the United States represents 16.30 percent of the total, in contrast to 0.05 percent for Tuvalu. See IBRD, "Subscriptions and voting power of member countries" (July 27, 2017).
[47] UNFCCC, *supra* note 14, art. 21.3.

The GEF was established in 1991, on a trial basis, by the World Bank, the United Nations Development Programme (UNDP) and the United Nations Environment Programme (UNEP) to provide financial support to diverse action directed at the protection of the global environment. Following the Earth Summit, it was agreed that the GEF would be "restructured" as a mechanism independent from the World Bank and with its own membership, hence not directly under the control of developed States. Most of the parties to the UNFCCC became "Participants" to the GEF following its restructuration.[48] The restructured GEF was defined as a "mechanism for international cooperation for the purpose of providing new and additional grant and concessional funding to meet the agreed incremental costs of measures to achieve agreed global environmental benefits."[49] This objective of pursuing "global environmental benefits" is certainly consistent with climate change mitigation, but much less so with adaptation to climate change, the benefits of which are generally confined to the local context where the relevant activities are implemented. In addition to climate change, the GEF's focal areas presently include biological diversity, international waters and land degradation (primarily desertification and deforestation), as well as the management of chemicals and waste products.[50]

The institutional structure of the GEF aims to promote a balanced representation of all States. An Assembly of all Participants meets every three years and takes decision by consensus, in particular on the general policies of the GEF.[51] A Council of 32 Participants, with an equal representation of developed and developing countries, is "responsible for developing, adopting and evaluating the operational policies and programs for GEF-financed activities."[52] The Secretariat of the GEF is to work independently, although it is "supported administratively by the World Bank,"[53] which acts as the trustee of the fund.[54] In practice, the GEF appears to be subject to the strong ideological influence of the World Bank, given not only the close proximity of its Washington, D.C. offices, but also the fact that the managers of the GEF are often former World Bank employees.[55]

The GEF coordinates a large range of institutions in channeling financial support for climate action. The three co-founding organizations are recognized as "implementing agencies" which "collaborate with eligible countries in the identification of projects for GEF funding."[56] More specifically, the UNDP focuses on capacity building and technical assistance, the UNEP promotes scientific and technical analysis and environmental management, while the World

---

[48] A total of 140 States became participants in 1994. As of early 2017, the GEF counted 183 participants.

[49] Instrument for the Establishment of the Restructured Global Environmental Facility, March 14–16, 1994, 33 *ILM* 1283, para. 2.

[50] Revised Instrument for the Establishment of the Restructured Global Environmental Facility (as of March 2015) (hereinafter Revised GEF Instrument), para. 2. See also United Nations Convention on Biological Diversity, June 5, 1992, 1760 *UNTS* 79 (hereinafter CBD), art. 39; Stockholm Convention on Persistent Organic Pollutants, May 22, 2001, 2256 *UNTS* 119, art. 14; United Nations Convention to Combat Desertification in those Countries Experiencing Serious Drought and/or Desertification, Particularly in Africa, October 14, 1994, 1954 *UNTS* 3 (hereinafter Convention against Desertification), arts. 20.2(b) and 21; Minamata Convention on Mercury, October 10, 2013, 55 *ILM* 582, arts.13.6, 13.7 and 13.8.

[51] Revised GEF Instrument, *supra* note 50, paras. 13, 14 and 25.

[52] *Ibid.*, para. 15.

[53] *Ibid.*, para. 21.

[54] *Ibid.*, para. 8.

[55] See Joyeeta Gupta, "The global environmental facility in its North-South context" (1995) 4:1 *Environmental Politics* 19.

[56] "Principles of cooperation among the implementation agencies," in Annex D of the Revised GEF Instrument, *supra* note 50, para. 6.

Bank helps in the development and management of investment projects.[57] Beyond these implementing agencies, the GEF seeks to cooperate with other international organizations, in particular multilateral development banks, many of which have been allowed to apply for GEF-financed activities in the same way as implementing agencies.[58]

Despite efforts to restructure the GEF following the Earth Summit, the institution continued to be seen as not sufficiently independent from Western interests, too bureaucratic and not sufficiently efficient in leveraging dedicated multilateral finance. At the Copenhagen Summit in 2009, a consensus appeared on the idea that scaled-up finance required a new, dedicated institutional framework.[59] In addition to the GEF, the Cancún Agreements established the Green Climate Fund (GCF) as another operating entity under the UNFCCC.[60] Although the GCF is also placed under the trusteeship of the World Bank on an interim basis,[61] its headquarters were established in Songdo, in South Korea. The GCF is governed by a board of 24 members, comprising an equal representation of developed and developing States.[62] Unlike the GEF, it possesses a "fully independent" secretariat.[63]

The GCF seeks to "catalyse climate finance" through public and private sources,[64] and to receive financial inputs from "a variety of other sources, public and private, including alternative sources."[65] Thus, it aims to "provide simplified and improved access to funding, including direct access, basing its activities on a country-driven approach."[66] Two separate funding windows are defined for mitigation and adaptation projects.[67] It is intended that a significant portion of financial support for climate action, in particular on adaptation, will be channeled through the GCF.[68] The GCF became operational in 2014. By mid-2017, contributions above USD 10 billion were pledged to the GCF,[69] which started approving projects for funding in 2016. Doubts remain, however, as to whether the GCF will live up to its ambition.[70]

---

[57] *Ibid.*, para. 11.

[58] See Revised GEF Instrument, *supra* note 50, para. 28; and generally Laurence Boisson de Chazournes, "The Global Environment Facility (GEF): a unique and crucial institution" (2005) 14:3 *Review of European, Comparative & International Environmental Law* 193, at 198–199.

[59] Copenhagen Accord, *supra* note 1, para. 10.

[60] Decision 1/CP.16, *supra* note 18, para. 102.

[61] *Ibid.*, para. 107. A permanent trustee should have been appointed at COP23 in November 2017, but no consensus emerged on a nomination by the Board of the Green Climate Fund. See GCF, Report of the Eighteenth Meeting (September 30–October 2, 2017), GCF/B.18/23, para. 58 (Agenda item 15). In its decision B.08/22, the Board of the GCF invited the World Bank "to continue serving as the Interim Trustee until a permanent Trustee is appointed." See GCF, Report of the Eighth Meeting (October 14–17, 2014), GCF/B.08/45, para. 42 (Agenda item 30).

[62] "Governing instrument for the Green Climate Fund," in the annex of decision 3/CP.17, "Launching the Green Climate Fund" (December 11, 2011), para. 9.

[63] *Ibid.*, para. 19.

[64] *Ibid.*, para. 3.

[65] *Ibid.*, para. 30.

[66] *Ibid.*, para. 31.

[67] *Ibid.*, para. 37.

[68] See Copenhagen Accord, *supra* note 1, para. 8; decision 1/CP.16, *supra* note 18, para. 100; and decision 1/CP.18, "Agreed outcome pursuant to the Bali Action Plan" (December 8, 2012), para. 64. See also, generally, Alexander Thompson, "The global regime for climate finance: political and legal challenges" in Cinnamon P. Carlarne, Kevin R. Gray and Richard Tarasofsky (eds.), *The Oxford Handbook of International Climate Change Law* (Oxford University Press, 2016) 137.

[69] According to the Green Climate Fund website, "Resource mobilization," www.greenclimate.fund/partners/contributors/resources-mobilized.

[70] Joëlle de Sépibus, "Green Climate Fund: how attractive is it to donor countries?" (2015) 9:4 *Carbon & Climate Law Review* 298.

3. Other Developments

Besides its Financial Mechanism, the UNFCCC recognizes that finance could be provided through "bilateral, regional or other multilateral channels."[71] Initiatives have been conducted by multilateral development banks. For instance, the World Bank established the Climate Investment Funds in 2008 in order to provide "72 developing and middle-income countries with urgently needed resources to manage the challenges of climate change and reduce their greenhouse gas emissions."[72] In addition to such dedicated funds, many development agencies have recognized the need to mainstream climate-related considerations in all the projects they support. Governments, too, created dedicated agencies for financial support for climate action.[73] Most importantly, financial support has been provided by non-State actors through massive private investments in renewable energies.

Although the myriad of sources and types of financial support makes accountability very difficult, there is clear evidence of a rapid increase in financial support for climate action. In 2014, USD 2.5 billion was channeled through dedicated multilateral funds on climate change managed under the Financial Mechanism of the UNFCCC.[74] Much greater financial support is currently channeled through other multilateral, regional or national institutions – an estimated USD 25.7 billion in climate finance was provided by multilateral development banks alone in 2014.[75] In the same year, the for-profit sector directly invested more than USD 20 billion in relevant sectors of developing States.[76] However, important distinctions exist between diverse financial instruments, ranging from grants to concessional or non-concessional loans and foreign investment. Altogether, there is also a blatant lack of reliable consolidated information on the scope and, even more so, the nature of finance by non-State actors due to the complexity of the landscape of relevant institutions.[77] Moreover, there is virtually no consolidated information available about the quality and effectiveness of climate finance or any concrete information on its actual contribution to prompting additional action on climate change adaptation and mitigation.[78]

B. Transfer of Technology

Transfer of technology is another form of international support for climate action. Technology, here, has a broad meaning; it designates not only goods and equipment, but also knowledge and experience required to produce or use such goods and equipment.[79] Technological development

---

[71] UNFCCC, *supra* note 14, art. 11.5. See also Kyoto Protocol, *supra* note 29, art. 11.3; Paris Agreement, *supra* note 14, art. 9.3.

[72] Climate Investment Funds website, "What we do," www.climateinvestmentfunds.org/about (accessed January 25, 2018).

[73] For instance, China created the Clean Development Mechanism Fund, funded by a tax on CDM project activities. See generally Duan Maosheng, "Clean Development Mechanism Development in China" in Yan Jinyue *et al.* (eds.), *Handbook of Clean Energy Systems* (Wiley, 2015) 3427.

[74] "Summary and recommendations by the Standing Committee on Finance on the 2016 biennial assessment and overview of climate finance flows," *supra* note 24, para. 16.

[75] *Ibid.*, para. 17.

[76] *Ibid.*, para. 18.

[77] See Thompson, *supra* note 68, at 150. See also Buchner *et al.*, *supra* note 24.

[78] Zahar, *supra* note 2, at 111–112.

[79] S.O. Andersen *et al.*, "Summary for policymakers" in B. Metz *et al.* (eds.), *Methodological and Technological Issues in Technology Transfer: A Special Report of IPCC Working Group III* (Cambridge University Press, 2000) 2, at 3.

and deployment is instrumental to climate change mitigation, whether it serves to increase energy efficiency or to explore alternative ways of producing energy. Support for adaptation can also be achieved through the development and deployment of technology ranging from drought-resilient seeds to environmental management systems.[80] Domestic policies in developed States and increasingly in emerging economies have played an important role in developing relevant technology.[81] Yet, the effective deployment of this technology where it is most likely to make a difference requires steps to ensure that useful technology is effectively transferred across jurisdictions and, more specifically, from developed States and emerging economies toward developing States. Obstacles to the transfer of technology include intellectual property rights in the case of proprietary resources,[82] but also a lack of information, human capabilities and capital, as well as various trade and policy barriers.[83] Thus, the transfer of technology is closely related to, yet distinct from financial support and capacity building.

Transfer of technology was one of the demands of developing States from the outset of international negotiations on climate change.[84] In addition to the above-mentioned transfer of technologies as one of the possible actions supported by climate finance,[85] the UNFCCC committed Annex II parties to "take all practical steps to promote, facilitate and finance, as appropriate, the transfer of, or access to, environmentally sound technologies and know-how to other Parties, particularly developing country Parties, to enable them to implement the provisions of the Convention."[86] Similarly, the Kyoto Protocol called on States to consider "the formulation of policies and programmes for the effective transfer of environmentally sound technologies that are publicly owned or in the public domain and the creation of an enabling environment for the private sector."[87] The flexibility mechanisms under the Kyoto Protocol, in particular the CDM and Joint Implementation, have contributed to the diffusion of technologies to emerging economies where project activities were implemented.[88]

As part of the Marrakesh Accords, COP7 established an Expert Group on Technology Transfer and initiated a "Framework for Meaningful and Effective Actions" in order to enhance the transfer of technology through an assessment of needs, exchange of information on available technologies, removing institutional barriers, building capacities and considering relevant institutional

---

[80] See e.g. UNFCCC, *Application of Environmentally Sound Technologies for Adaptation to Climate Change* (May 10, 2006), doc. FCCC/TP/2006/2.

[81] See e.g. O. Edenhofer *et al.*, "Summary for policymakers" in Pichs-Madruga *et al.* (eds.), *Climate Change 2014: Mitigation of Climate Change. Contribution of Working Group III to the Fifth Assessment Report of the Intergovernmental Panel on Climate Change* (Cambridge University Press, 2014) 1, at 29; Ademola A. Adenle, Hossein Azadi and Joseph Arbiol, "Global assessment of technological innovation for climate change adaptation and mitigation in developing world" (2015) 161 *Journal of Environmental Management* 261.

[82] See e.g. Ahmed Abdel-Latif, "Intellectual property rights and the transfer of climate change technologies: issues, challenges, and way forward" (2015) 15:1 *Climate Policy* 103; Varun Rai, Kanye Schultz and Erik Funkhouser, "International low carbon technology transfer: do intellectual property regimes matter?" (2014) 24 *Global Environmental Change* 60; David G. Ockwell *et al.*, "Intellectual property rights and low carbon technology transfer: conflicting discourses of diffusion and development" (2010) 20:4 *Global Environmental Change* 729.

[83] See Andersen *et al.*, *supra* note 79, at 4.

[84] Bodansky, *supra* note 28, at 530.

[85] UNFCCC, *supra* note 14, art. 4.3.

[86] *Ibid.*, art. 4.5. See also UNCED, Rio Declaration on Environment and Development (June 3–14, 1992), available in (1992) 31 *ILM* 874 (hereinafter Rio Declaration), principle 9; Agenda 21, Chapter 34.

[87] Kyoto Protocol, *supra* note 29, art. 10(c).

[88] See e.g. David Popp, "International technology transfer, climate change, and the Clean Development Mechanism" (2011) 5:1 *Review of Environmental Economics & Policy* 131. Flexibility mechanisms are discussed in Chapter 8.

mechanisms.[89] The mandate of this Expert Group was terminated in 2010 when, as announced in the Copenhagen Accord,[90] the Cancún Agreements established a Technology Mechanism composed of an Executive Committee (the Technology Executive Committee) and a Climate Technology Centre and Network.[91] After consultations, it was agreed in 2012 that the Climate Technology Centre and Network would be hosted by the UN Environment Programme, as the leader of a consortium of partner institutions, for an initial term of five years.[92] Efforts had long been made to promote transfer of technology in GEF projects[93] and, following the operationalization of the Technology Mechanism, "linkages" with the Financial Mechanism were encouraged.[94] Lastly, the Paris Agreement included transfer of technologies among the list of priorities in relation to which States are to undertake and communicate ambitious efforts.[95]

C. Capacity Building

A third aspect of international support – capacity building – is a rather ill-defined concept which emerged in the broader field of international development to refer to diverse efforts to empower aid recipients.[96] With regard to climate change, capacity building seeks to enable relevant actors to develop sufficient capacities to plan and implement relevant climate action. Developing relevant knowledge, human resources, public awareness and institutions, for instance, can significantly facilitate future attempts by developing States to promote climate change adaptation and mitigation without depending on international support.[97]

The UNFCCC and the Kyoto Protocol contain a few mentions of capacity building. The UNFCCC gives mandate to the Subsidiary Body for Scientific and Technological Advice to provide advice "on ways and means of supporting endogenous capacity-building in developing countries."[98] As discussed above, Annex II parties committed to meet the agreed full costs of developing

[89] Decision 4/CP.7, "Development and transfer of technologies" (November 10, 2001). An additional set of actions was defined by decision 3/CP.13, "Development and transfer of technologies under the Subsidiary Body for Scientific and Technological Advice" (December 14–15, 2007), para. 2.

[90] Copenhagen Accord, *supra* note 1, para. 11.

[91] Decision 1/CP.16, *supra* note 18, para. 117. See also decision 2/CP.17, "Outcome of the work of the Ad Hoc Working Group on Long-Term Cooperative Action under the Convention" (December 11, 2011), paras. 133–143; and decision 4/CP.17, "Technology Executive Committee – modalities and procedures" (December 9, 2011).

[92] Decision 14/CP.18, "Arrangements to make the Climate Technology Centre and Network fully operational" (December 7, 2012), para. 2.

[93] See in particular GEF, *Elaboration of a Strategic Program to Scale up the Level of Investment in the Transfer of Environmentally Sound Technologies* (November 13, 2008), GEF document GEF/C.34/5.Rev.1.

[94] See e.g. decision 2/CP.17, *supra* note 91, para. 140; 13/CP.18, "Report of the Technology Executive Committee" (December 7, 2012), para. 7; decision 13/CP.21, "Linkages between the Technology Mechanism and the Financial Mechanism of the Convention" (December 13, 2015); and decision 14/CP.22 "Linkages between the Technology Mechanism and the Financial Mechanism of the Convention" (November 17, 2016).

[95] See Paris Agreement, *supra* note 14, arts. 3 and 10.

[96] See, for instance, Beth Walter Honadle, "A capacity-building framework: a search for concept and purpose" (1981) 41:5 *Public Administration Review* 575; Deborah Eade, *Capacity-Building: An Approach to People-Centred Development* (Oxfam, 1997).

[97] For more concrete illustrations of what can be involved by "capacity-building" in relation to climate action, see "List of capacity-building needs of developing country parties" in the annex of decision 10/CP.5, "Capacity-building in developing countries (non-Annex I Parties)" (November 4, 1999).

[98] UNFCCC, *supra* note 14, art. 9.2(d).

the national communication of developing States, thus facilitating processes through which developing States could identify needs and opportunities for future climate action.[99] In the Kyoto Protocol, capacity building is among the actions on which Annex I parties "shall submit information" and that other parties "shall seek to include in their national communications, as appropriate."[100] The parties to the Kyoto Protocol further committed to cooperate in "the strengthening of national capacity-building, in particular human and institutional capacities and the exchange or secondment of personnel to train experts ... in particular for developing countries."[101]

More specific efforts have been made to enhance capacity building in developing States since the adoption of the UNFCCC and the Kyoto Protocol. The initial instructions of the COP to the Financial Mechanism gave priority, among other things, to "enabling activities undertaken by developing country Parties, such as planning and endogenous capacity-building, including institutional strengthening, training, research and education."[102] It rapidly appeared, as confirmed in a decision adopted at COP5, that "capacity-building [was] critical to the effective participation" of developing countries and economies in transition in the negotiations and other processes within the UNFCCC regime.[103] Frameworks for capacity building in developing countries and in countries with an economy in transition, which were adopted as part of the Marrakesh Accords, called upon Annex II parties to provide financial and technical resources in support of country-driven capacity building in these countries, in particular in support for climate change adaptation.[104]

Until 2009, capacity building remained predominantly viewed as a modality of transfer of technology and, to a lesser extent, financial support.[105] The Copenhagen Accord innovated by calling on developed States to offer "financial resources, technology and capacity-building" as three different forms of international support to adaptation action in developing countries.[106] Likewise, the Cancún Agreements encouraged Annex II parties to provide enhanced international support for climate action (including mitigation) on three fronts: finance, technology development and transfer, and capacity building.[107] Moreover, the Cancún Agreements called upon Annex II parties to provide information on the provision of support to capacity building in their biennial reports.[108]

---

[99] See *ibid.*, art. 4.3. See also *supra* note 26 and accompanying text.

[100] Kyoto Protocol, *supra* note 29, art. 10(b)(ii).

[101] *Ibid.*, art. 10(e).

[102] Decision 11/CP.1, *supra* note 27, para. 1(b)(i). See also *ibid.*, para. 1(b)(iv); decision 3/CP.2, "Secretariat activities relating to technical and financial support to Parties" (July 19, 1996), para. 1; decision 2/CP.4, "Additional guidance to the operating entity of the financial mechanism" (November 14, 1998), para. 1(i); and decision 4/CP.4, "Development and transfer of technologies" (November 14, 1998), para. 4.

[103] Decision 10/CP.5, *supra* note 97, recital 5.

[104] See decisions 2/CP.7, "Capacity building in developing countries (non-Annex I Parties)" (November 10, 2001); and 3/CP.7, "Capacity building in countries with economies in transition" (November 10, 2001). See also decisions 2/CP.10, "Capacity-building for developing countries (non-Annex I Parties)" (December 17–18, 2004); and 3/CP.10, "Capacity-building for countries with economies in transition" (December 17–18, 2004).

[105] See e.g. decision 1/CP.13, *supra* note 36, para. 1.

[106] Copenhagen Accord, *supra* note 1, para. 3. By contrast, para. 8 suggests that both technology transfer and capacity building are forms of financial support.

[107] See also, e.g., decision 2/CP.16, *supra* note 18, paras. 95–140.

[108] *Ibid.*, para. 40(a); "UNFCCC biennial reporting guidelines for developed country Parties," in Annex I of decision 2/CP.17, *supra* note 91, para. 23.

Beyond the Cancún Agreements, however, developing States claimed for dedicated institutional arrangements and the adoption of performance indicators on capacity building.[109] The Durban Forum for in-depth discussion on capacity building was established at COP17,[110] followed by the Paris Committee on Capacity-Building at COP21. The Paris Committee has a mandate to "address gaps and needs, both current and emerging, in implementing capacity-building in developing country Parties and further enhancing capacity-building efforts,"[111] in particular through an initial five-year workplan.[112] The Paris Agreement characterizes capacity building as an aspect of "global responses to climate change" on apparently an equal footing with mitigation, adaptation, finance, transfer of technology and transparency.[113] Article 11 commits all parties to enhancing the capacity of developing States through regional, bilateral and multilateral approaches,[114] while Article 13 calls more specifically for support to the "building of transparency-related capacity of developing country Parties."[115] In order to further enhance capacity building, the Paris Agreement calls for States to make "appropriate institutional arrangements" at the first session of the Meeting of the Parties to the Paris Agreement.[116]

Capacity building is not limited to the UNFCCC regime. The UN General Assembly promoted similar efforts by highlighting, as part of the Sustainable Development Goal on climate change, the need to "promote mechanisms for raising capacity for effective climate change-related planning and management in least developed countries and small island developing States."[117] Innumerable agencies have engaged in initiatives to support capacity building or have inserted capacity building components in other activities on climate change.[118] Yet, stakeholders have denounced the ineffectiveness of many "ad-hoc, short-lived, mainly project-based interventions" on capacity building.[119] A more organized institutional framework appears to be needed.

III. REDD+

A large share of GhG emissions originates from land-use change, in particular from the slashing and burning of tropical forests[120] for agricultural production – sometimes, ironically, for the

---

[109] See e.g. IISD, "Summary of the Copenhagen Climate Change Conference: 7–19 December 2009" (2009) 12:459 *Earth Negotiations Bulletin* 1, at 18 (right); UNFCCC SBI, "Submission by the United Republic of Tanzania on behalf of the Group of 77 and China" (September 11, 2009), reproduced in doc. FCCC/SBI/2009/MISC.12/Rev.1 (May 18, 2010) 5, para. 2.

[110] Decision 1/CP.17, "Establishment of an Ad Hoc Working Group on the Durban Platform for Enhanced Action" (December 11, 2011), para. 144. See also IISD, "Summary of the Warsaw Climate Change Conference: 11–23 November 2013" (November 26, 2013) 12:594 *Earth Negotiations Bulletin* 1, at 11–12.

[111] Decision 1/CP.21, *supra* note 14, para. 71. See also decision 2/CP.22 "Paris Committee on Capacity-building" (November 17, 2016).

[112] Decision 1/CP.21, *supra* note 14, para. 73.

[113] Paris Agreement, *supra* note 14, art. 3.

[114] *Ibid.*, art. 11.4.

[115] *Ibid.*, art. 13.15.

[116] *Ibid.*, art. 11.5.

[117] UN General Assembly Resolution 70/1, *supra* note 9, goal 13.b.

[118] See UNFCCC SBI, *Capacity-Building Work of Bodies Established under the Convention and its Kyoto Protocol. Addendum: Compilation of Capacity-Building Activities Undertaken by Bodies Established under the Convention and its Kyoto Protocol* (April 25, 2017), doc. FCCC/SBI/2017/2/Add.1.

[119] Mizan Khan *et al.*, *Capacity Building under the Paris Agreement* (European Capacity Building Initiative, October 2016) 2.

[120] Land use and land-used change were estimated to represent a third of GhG emissions between 1850 and 2000. Yet, it also represents a smaller proportion of recent and current GhG emissions due to a rapid increase in industrial

production of biofuels. Although these practices are often illegal, some developing States lack the resources necessary to protect their forests. Deforestation often occurs in countries which are not otherwise particularly large GhG emitters.[121] Action to protect forests in developing States has progressively become a priority for North-South support, including finance as well as capacity building and technical assistance.[122]

The protection of forests was an opportunity for mutually beneficial international cooperation between developed States willing to provide some resources and developing States willing to use them to protect their forests. Yet, such cooperation needed to be fine-tuned to ensure that support would effectively enhance climate action. Monitoring vast remote forest regions is challenging. Measures would need to be sufficiently well defined to ensure that a State's commitment to protect one section of a forest would not simply divert deforestation to another section or to another forest.[123] Unlike the CDM, support would possibly have to be directed at comprehensive policies rather than at specific project activities. Unlike other forms of international support, international cooperation would be directed specifically at the protection of forests, following pre-determined methodologies. Risks of non-compliance would need to be balanced with the costs of verification.[124]

In 2007, the Bali Action Plan recognized the opportunity for cooperation by calling for "[p]olicy approaches and positive incentives on issues relating to reducing emissions from deforestation and forest degradation in developing countries."[125] Another decision adopted at COP13 encouraged parties to "further strengthen and support ongoing efforts to reduce emissions from deforestation and forest degradation on a voluntary basis."[126] The acronym "REDD" was retained to refer to "Reducing Emissions from Deforestation and forest Degradation in developing countries." In addition, the Bali Action Plan highlighted the need to consider "the role of conservation, sustainable management of forests and enhancement of forest carbon stocks in developing countries,"[127] which are encapsulated in the "+" of "REDD+."

In the following years, the COP adopted slightly more detailed methodological guidance and discussed further steps to promote REDD+.[128] The Cancún Agreements, in particular, announced

sources. See generally R.T. Waton et al., Land Use, Land-Use Change and Forestry: A Special Report of the Intergovernmental Panel on Climate Change (Cambridge University Press, 2000); R.K. Pachauri et al., Climate Change 2014: Synthesis Report. Contribution of Working Groups I, II and III to the Fifth Assessment Report of the Intergovernmental Panel on Climate Change (IPCC, 2015) 45.

[121] Thus, Nigeria, Uganda and Sudan were among the countries which lost the highest proportion of land covered by forests from 1990 to 2015, according to the World Bank's data, "Forest Area (% of land area)" (accessed July 13, 2017).

[122] See e.g. Paris Agreement, supra note 14, art. 5; decision 10/CP.19, "Coordination of support for the implementation of activities in relation to mitigation actions in the forest sector by developing countries, including institutional arrangements" (November 22, 2013), para. 3(d).

[123] See generally Michael Dutschke, "Key Issues in REDD+ verification" (CIFO occasional paper 88, 2013).

[124] Lee J. Alston and Krister Andersson, "Reducing greenhouse gas emissions by forest protection: the transaction costs of implementing REDD" (2011) 2:2 Climate Law 281.

[125] Decision 1/CP.13, supra note 36, para. 1(b)(iii).

[126] Decision 2/CP.13, "Reducing emissions from deforestation in developing countries: approaches to stimulate action" (December 14–15, 2007), para. 1.

[127] Decision 1/CP.13, supra note 36, para. 1(b)(iii).

[128] See decision 4/CP.15, "Methodological guidance for activities relating to reducing emissions from deforestation and forest degradation and the role of conservation, sustainable management of forests and enhancement of forest carbon stocks in developing countries"; decision 1/CP.16, supra note 18, paras. 68–79 and Appendixes I and II;

the coordination of activities "in phases, beginning with the development of national strategies ... followed by the implementation of national policies ... and evolving into results-based actions that should be fully measured, reported and verified."[129] The following year, COP17 agreed on the principle that result-based finance could come "from a wide variety of sources, public and private, bilateral and multilateral, including alternative sources."[130] It also decided that such support could be provided through market-based approaches (such as the CDM) or through alternative approaches.[131] Yet, the progress of negotiations, involving relatively complex modalities of application, was slow; it was not the priority of intense negotiations centered, largely, on framing the post-Kyoto mitigation regime. Fifty States initiated parallel negotiations in 2010 through an Agreement on Financing and Quick-Start Measures to Protect Rainforests.[132] This put additional pressure on the COP to achieve rapid progress.

In 2013, COP19 finally adopted a comprehensive agreement, the Warsaw Framework for REDD+, through a set of seven decisions.[133] The Warsaw Framework facilitates the provision of financial support to developing States which achieve to protect their forests. In order to participate, a developing State must set up a national forest monitoring system applying the most recent IPCC guidance and guidelines to account for GhG emissions and removals from forests.[134] It must also communicate an estimation of its baseline (a national reference mission level and/or forest reference level) to the UNFCCC Secretariat, which is to organize a technical assessment of this communication by a team of experts.[135] The mitigation outcome of relevant national policies deemed appropriate by the developing party is then to be reported through biennial update reports along with any other climate action by that party.[136] Finally, the Warsaw Framework "recogniz[es] the need for adequate and predictable support" for the implementation of forest-related activities in developing States and promotes consultations between interested parties.[137] Article 5.2 of the Paris Agreement endorses the essential features of the Warsaw Framework.[138]

decision 2/CP.17, *supra* note 91, paras. 63–73; and decision 1/CP.18, *supra* note 68. See generally Christina Voigt and Felipe Ferreira, "The Warsaw Framework for REDD+: implications for national implementation and result-based finance" in Christina Voigt (ed.), *Research Handbook on REDD-Plus and International Law* (Edward Elgar, 2016) 30; Louisa Denier *et al.*, *The Little Book of Legal Frameworks for REDD+: How Policy and Legislation Can Create an Enabling Environment for REDD+* (Global Canopy Programme, 2014); special issue in (2015) 9:2 *Carbon & Climate Law Review* 99.

[129] Decision 1/CP.16, *supra* note 18,para. 73.

[130] Decision 2/CP.17, *supra* note 91, para. 65.

[131] *Ibid.*, paras. 66 and 67.

[132] See Chapter 4, section III.

[133] See decisions 9–15/CP.19.

[134] Decision 11/CP.19, "Modalities for national forest monitoring systems" (November 22, 2013), para. 2. This monitoring should, according to a previous decision, "use a combination of remote sensing and ground-based forest carbon inventory approaches for estimating, as appropriate, anthropogenic forest-related greenhouse gas emissions by sources and removals by sinks, forest carbon stocks and forest area changes." See decision 4/CP.15, *supra* note 128, para. 1(d)(i).

[135] Decision 13/CP.19, "Guidelines and procedures for the technical assessment of submissions from Parties on proposed forest reference emission levels and/or forest reference levels" (November 22, 2013).

[136] Decision 14/CP.19, "Modalities for measuring, reporting and verifying" (November 22, 2013). See also Chapter 13, section II.B.2.

[137] Decision 10/CP.19, *supra* note 122, recital 4.

[138] See Paris Agreement, *supra* note 14, art. 5.2. See also decision 1/CP.21, *supra* note 14, para. 54.

REDD+ finance can either be constituted by *ex ante* grants or by *ex post* performance-based payments. *Ex post* payment could in principle be connected to international market mechanisms such as the CDM; this, however, would require a fair, reliable and quantifiable assessment of the actual impact of a policy on overall forest coverage, which remains a challenging task.[139] At the moment, it is estimated that about USD 800 million are channeled annually in support for the REDD+ project, mostly through *ex ante* public grants from about 20 donor States to about 80 developing States.[140] Dedicated multilateral funds have been created to channel financial support to REDD+ projects. In particular, the UN-REDD program was created by the Food and Agriculture Organization, the UN Development Programme and the UN Environment Programme. However, it remains difficult to assess the effective climate benefits of many projects supported under the REDD+ program,[141] which often takes place in countries with poor enforcement capacities.[142]

Forest policies should not only strive to promote climate change mitigation; the interests of communities whose livelihood depends on forests, including some indigenous communities, must not be overlooked. To ensure that climate change mitigation does not come at a cost to communities which are often politically underrepresented, the COP has adopted a number of safeguards. In particular, activities falling within the scope of REDD+ should be consistent with the "national development needs and goals" of the developing State where they are implemented and should "promote sustainable management of forests."[143] They should also promote "respect for the knowledge and rights of indigenous peoples and members of local communities" and "the full and effective participation of relevant stakeholders."[144] Developing States are encouraged to report information on compliance with these safeguards and other non-carbon benefits.[145] It is uncertain whether these provisions really make a difference on the ground.

---

[139] See Chapter 8, section II.

[140] Marigold Norman and Smita Nakhooda, *The State of REDD+ Finance* (Center for Global Development Working Paper No. 378, May 2015). See also Marigold Norman *et al.*, "*Climate finance thematic briefing: REDD+ finance*" (Henrich Böll Stiftung, December 2015).

[141] See Astrid B. Bos *et al.*, "Comparing methods for assessing the effectiveness of subnational REDD+ initiatives" (2017) 12:7 *Environmental Research Letters* 1; S. Naeem *et al.*, "Get the science right when paying for nature's services" (2015) 347:6227 *Science* 1206.

[142] See R.M. Ochieng *et al.*, "Institutional effectiveness of REDD+ MRV: countries progress in implementing technical guidelines and good governance requirements" (2016) 61 *Environmental Science & Policy* 42; Richard S. Mbatu, "Domestic and international forest regime nexus in Cameroon: an assessment of the effectiveness of REDD + policy design strategy in the context of the climate change regime" (2015) 52 *Forest Policy & Economics* 46.

[143] Decision 1/CP.16, *supra* note 18, Appendix I, para. 1(f) and (k).

[144] *Ibid.*, Appendix I, para. 2(c) and (d).

[145] See decision 12/CP.17, "Guidance on systems for providing information on how safeguards are addressed and respected and modalities relating to forest reference emission levels and forest reference levels as referred to in decision 1/CP.16" (December 9, 2011); decision 11/CP.19, *supra* note 134, para. 5; decision 12/CP.19, "The timing and the frequency of presentations of the summary of information on how all the safeguards referred to in decision 1/CP.16, appendix I, are being addressed and respected" (November 22, 2013); decision 17/CP.21, "Further guidance on ensuring transparency, consistency, comprehensiveness and effectiveness when informing on how all the safeguards referred to in decision 1/CP.16, appendix I, are being addressed and respected" (December 10, 2015); and decision 18/CP.21, "Methodological issues related to non-carbon benefits resulting from the implementation of the activities referred to in decision 1/CP.16, paragraph 70" (December 10, 2015).

IV. AN EMERGING VISION OF INTERNATIONAL SUPPORT FOR CLIMATE ACTION

Diverse documents have been adopted, in particular under the UNFCCC regime, to encourage and to guide international support for climate action. While such support remains generally voluntary, a vision of what international support for climate action ought to be has started to emerge through language repeated multiple times in treaties, COP decisions and other relevant statements. This slowly emerging vision of international support for climate action reveals not only a tension between the claims of some of the most vulnerable nations and their rejection by developed States, but also echoes two different visions of international support for climate action, as either guided by conceptions of justice and fairness or by national interests alone.

A. Novelty and Additionality

A first element of the emerging vision of international support for climate action is that such support, in particular financial support, should be "new and additional."[146] The idea that funding for environmental protection in developing States should be "new and additional" had emerged even before the Earth Summit.[147] Similar provisions were inserted into the Convention on Biodiversity as well as, two years later, the Convention against Desertification.[148] The general idea was that "money to implement the Convention should not be diverted from existing development aid."[149] To transform this general idea into a concrete legal requirement would have required a clear methodology to assess not just the novelty of a fund, but also the nature of particular expenses as "additional to what donor countries would have allocated anyway."[150] This would have required the determination of a baseline – what parties would have been expected to allocate anyway. Yet, no methodology to determine such a baseline could be agreed upon and, instead, the COP left it to each individual developed party, when reporting on its provision of financial support, to "indicate clearly how they have determined resources as being 'new and additional.'"[151]

---

[146] UNFCCC, *supra* note 14, art. 4.3. See also Kyoto Protocol, *supra* note 29, art. 11.2(a); Copenhagen Accord, *supra* note 1, para. 8; decision 1/CP.16, *supra* note 18, paras. 18, 95 and 97. See generally Charlotte Streck, "Ensuring new finance and real emission reduction: a critical review of additionality concept" (2011) 5:2 *Carbon & Climate Law Review* 158.

[147] See UN General Assembly Resolution 44/207, "Protection of global climate for present and future generations of mankind" (December 22, 1989), para. 14; UN General Assembly Resolution 44/228, "United Nations Conference on Environment and Development" (December 22, 1989), para. 15(j); Amendment to the Montreal Protocol on Substances that Deplete the Ozone Layer adopted in London on June 29, 1990, 1598 *UNTS* 469, art. 1T (calling for contributions that would "be additional to other financial transfers").

[148] See respectively CBD, *supra* note 50, art. 20.2; and Convention against Desertification, *supra* note 50, art. 20.2(b).

[149] Bodansky, *supra* note 28, at 526.

[150] Zahar, *supra* note 2, at 26.

[151] See "Revised Guidelines for the Preparation of National Communications by Parties included in Annex I to the Convention," in the annex of decision 9/CP.2, "Communication from Parties included in Annex I to the Convention: Guidelines, schedule and process for consideration" (July 19, 1996), para. 42; "Guidelines for the preparation of national communications by Parties included in Annex I to the Convention," in the annex of decision 4/CP.5, "Guidelines for the preparation of national communications by Parties included in Annex I to the Convention, Part II: UNFCCC reporting guidelines on national communications" (November 4, 1999), at para. 51; "Guidelines for the preparation of the information required under Article 7 of the Kyoto Protocol," in the annex of decision 15/CMP.1, "Guidelines for the preparation of the information required under Article 7 of the Kyoto Protocol" (November 30, 2005), para. 41; "UNFCCC biennial reporting guidelines for developed country Parties," *supra* note 108, para. 18(f).

Through time and practice, consideration for the "new and additional" nature of climate finance in national reports has progressively boiled down to a demonstration that the provision of climate finance had increased over time.[152] The Paris Agreement took stock of this evolution: there, the requirement was simply that financial support should "represent a progression beyond previous efforts"[153] and be "scaled-up."[154] This, however, tends to set aside the requirement of additionality, with a persistent risk that development aid will be displaced or requalified as climate finance.

B. Adequateness

A second element of an emerging vision of financial support is that such support should be of an adequate scope, although the specific meaning of the term remains ill-determined. The UNFCCC, the Kyoto Protocol and multiple COP decisions have directed Annex II parties to meet the "agreed full costs" of reporting activities and the "agreed full incremental costs" of implementation measures in developing States.[155] "Full," here, is reminiscent of the obligation of a State to provide "full" reparation for the injury caused by its internationally wrongful acts under the general international law of State responsibility.[156] In particular circumstances, however, States have agreed to a more flexible determination of "appropriate" reparation.[157] Likewise, any (full) financial support for climate action is conditioned by an ad hoc agreement of developed States.[158] While efforts have been made to enhance support to reach a level States consider to be adequate, it has been widely accepted that such support would not necessarily represent the totality of the costs faced by developing States.[159]

Climate change action, whether related to mitigation or adaptation, is often mainstreamed through unspecific policies, programs and projects. For instance, consideration should be given to climate change in a national development strategy. In such cases, there could be no claim made by developing States for international support to cover all the costs of such broad activities. Rather, international support for climate action should be limited to the incremental costs: the additional costs generated by the inclusion of climate change considerations in these activities. Thus, international support could, for instance, be directed to the incremental costs of building a more efficient power plant (mitigation) and to carrying out a development project able to withstand extreme weather events which are becoming more likely (adaptation). Limiting support to incremental efforts is critical to maximizing the impact of the scarce resources made available by donors.

---

[152] See, for instance, United States, *Second Biennial Report under the UNFCCC* (December 31, 2015), at 46.

[153] Paris Agreement, *supra* note 14, art. 9.3.

[154] *Ibid.*, art. 9.4. See also Copenhagen Accord, *supra* note 1, para. 8.

[155] UNFCCC, *supra* note 14, art. 4.3; Kyoto Protocol, *supra* note 29, art. 11.2; "Governing instrument for the Green Climate Fund," *supra* note 62, para. 35.

[156] ILC, *Draft Articles on Responsibility of States for Internationally Wrongful Acts with Commentaries*, in (2001) *Yearbook of the International Law Commission*, vol. II, part two (hereinafter *Articles on State Responsibility*), art. 31.1.

[157] See Chapter 5, section IV.B. Remedial obligations would not necessarily require that developed States meet the *full* costs of mitigation in developing States.

[158] See UNFCCC, *supra* note 14, art. 4.3.

[159] See, for instance, *ibid.*, art. 4.4.

Concretely, however, identifying incremental costs can turn out to be very challenging, here again, due to a lack of a well-defined baseline.[160] The COP recognized this difficulty and suggested a flexible and pragmatic approach which should be followed through a transparent process.[161] The COP relied on the GEF to develop technical guidance in this sense.[162] The GEF initially developed complex quantitative methodologies to determine baseline costs from incremental costs.[163] An internal evaluation carried out in 2006 found this approach to be "clearly unrealistic" and even counterproductive, creating additional hurdles for developing States seeking access to financial support.[164] In June 2007, following the recommendation of its Evaluation Office, the GEF adopted a new policy putting much more emphasis on a qualitative approach of "incremental reasoning."[165]

While the Paris Agreement placed far less emphasis on the requirement of incrementality, practical efforts were made toward leveraging greater financial resources for agreed-upon projects. Following the evolving policies of the GEF, States came to agree that a strictly quantitative approach to incrementality was not possible, especially as emphasis was increasingly placed on the unfulfilled need for support for action on climate change adaptation, which may be particularly difficult to distinguish from development.[166] On the other hand, the adoption of ambitious collective mobilization goals for 2020 and beyond seeks to come to terms with the commitment of comprehensive financial support.[167] The objective of the Paris Agreement to "mak[e] finance flows consistent with a pathway towards low greenhouse gas emissions and climate-resilient development"[168] reaffirms the requirement for greater financial support for climate action.

C. Efficiency

A third element of an emerging vision of international support for climate action is that such support should be efficient. This requirement emerged at the outset of international climate change negotiations as a recognition of the need to make the best use possible of whatever support was to be provided. For instance, the UNFCCC requires that the implementation of financial commitments "take into account the need for adequacy and predictability in the flow of funds."[169] The predictability of support allows receiving States to make the best use of it. As to adequacy, it relates here not only to the overall amount, but also possibly to the necessary

---

[160] See Bodansky, *supra* note 28, at 526; Luis Gomez-Echeverri, "The changing geopolitics of climate change finance" (2013) 13:5 *Climate Policy* 632, at 635.

[161] See decision 11/CP.1, *supra* note 27, para. 1(e); decision 11/CP.2, "Guidance to the Global Environment Facility" (July 19, 1996), para. 1(b); and decision 2/CP.4, *supra* note 102, recital 5.

[162] See decision 11/CP.2, *supra* note 161, para. 3; decision 2/CP.4, *supra* note 102, para. 3(c); and decision 5/CP.8, "Review of the financial mechanism" (November 1, 2002), para.4(c).

[163] See e.g. decision 11/CP.2, *supra* note 161, recital 6; decision 2/CP.4, *supra* note 102, recitals 5 and 6; decision 7/CP.13, "Additional guidance to the Global Environment Facility" (December 14–15, 2007), para. 1(c).

[164] Global Environmental Facility, *Evaluation of Incremental Cost Assessment* (November 2, 2006), GEF document GEF/MEC/C.30/2, para. 43.

[165] GEF, *Operational Guidelines for the Application of the Incremental Cost Principle* (GEF Policy Paper, June 13, 2007), para. 6(d). See also GEF, *Evaluation of Incremental Cost Assessment, supra* note 164, para. 46.

[166] See e.g. Bodansky, *supra* note 28, at 528; Zahar, *supra* note 2, at 27.

[167] See Paris Agreement, *supra* note 14, art. 9.3; and decision 1/CP.21, *supra* note 14, para. 53.

[168] Paris Agreement, *supra* note 14, art. 2.1(c).

[169] UNFCCC, *supra* note 14, art. 4.3. See also Kyoto Protocol, *supra* note 29, art. 11.2.

balance between different receiving States and between different functions (adaptation and mitigation).

Efficiency is assessed in relation to a given objective. Yet, three distinct rationales exist to justify international support for climate action.[170] Accordingly, efficiency can alternatively be assessed in relation to the ability of climate finance to assist the populations in need, to convey adequate reparation for the harms caused by excessive GhG emissions or to reduce global GhG emissions. In practice, while donors have placed emphasis on support for climate change mitigation with global benefits, developing States have insisted throughout climate change negotiations on the need for assistance and, sometimes, on the duty of responsible States to pay adequate reparation. This has led to a tension between the prevalence of assistance in formal decisions and the influence of self-interests in the practice of international support for climate action. It certainly did not help that the GEF is under the trusteeship of the World Bank, where weighted votes gave developed States a much stronger level of influence in the determination of relevant modalities.

Efficient support should target countries and activities which need it the most. The UNFCCC clearly recognized that developed States should provide support to developing States in application of the principle of common but differentiated responsibilities.[171] It was also understood that the needs of particular categories of developing States should command greater consideration.[172] Small island States and other States particularly prone to natural disasters, sea-level rise, drought and desertification were also highlighted, along with the least developed States, as requiring "full consideration."[173] These States need financial support for their adaptation programs the most urgently, given the severity of the adverse impacts of climate change that they are facing and their lack of adaptation capacity. In practice, however, most of the financial support for climate action was directed toward mitigation rather than adaptation, and to emerging economies rather than the least developed or most vulnerable States.[174]

Attempts have been made at encouraging a rebalancing in the provision of financial support in favor of adaptation and of the most vulnerable States. In implementing the Kyoto Protocol, the Marrakesh Accord dedicated 2 percent of the proceeds of the CDM to adaptation,[175] channeled through the Adaptation Fund toward the developing country parties that are particularly vulnerable to the adverse effects of climate change.[176] The COP has also shown particular concern for the financial situation of the least developed States. In 2000, it encouraged consideration for the establishment of a debt relief mechanism through the United Nations Conference on the Least Developed Countries.[177] The following year, the Marrakesh Accords established the Least Developed Countries Fund in order to implement a work program on

[170] See above, section I of this chapter.
[171] UNFCCC, *supra* note 14, art. 4.4.
[172] *Ibid.*, art. 3.2.
[173] *Ibid.*, arts. 4.8 and 4.9.
[174] See e.g. Buchner *et al.*, *supra* note 24, at 2 and *passim*.
[175] Decision 17/CP.7, *supra* note 34, para. 15.
[176] See decision 28/CMP.1, "Initial guidance to an entity entrusted with the operation of the financial mechanism of the Convention, for the operation of the Adaptation Fund" (December 9–10, 2005). For a rather optimistic account, see Britta Horstmann and Achala Chandani Abeysinghe, "The Adaptation Fund of the Kyoto Protocol: a model for financing adaptation to climate change?" (2011) 2:3 *Climate Law* 415.
[177] See resolution 2/CP.6, "Input to the Third United Nations Conference on the Least Developed Countries" (November 25, 2000), para. 2.

least developed States with a view, in particular, to facilitating the development of reports on the adaptation needs of these States, the national adaptation programs of actions.[178] Through the Copenhagen Accord, the Cancún Agreements and eventually the Paris Agreement, States agreed to the aspirational goal of a "balanced allocation between adaptation and mitigation."[179] Consistently, the GCF was requested to balance funding between adaptation and mitigation activities.[180]

Efficient support should also be consistent with national priorities and orientations. In parallel to the progressive recognition that development aid should defer to the priorities defined by aid-receiving countries,[181] States agreed on the need for climate finance to support country-driven policies in developing States. At its first session, the COP stated that "[p]rojects funded through the financial mechanism should be country-driven and in conformity with, and support of, the national development priorities of each country."[182] Likewise, the Paris Agreement emphasized the need for climate finance to "tak[e] into account the needs and priorities of developing country Parties."[183] Consistently, the GCF adopted measures to promote and enhance country ownership and "drivenness" in supported activities.[184] Increased emphasis on capacity building aims to promote such ownership of climate action in developing States.

D. Burden Sharing

A fourth element of the emerging vision of international support for climate action is that of burden sharing. The UNFCCC and the Kyoto Protocol recognize "the importance of appropriate burden sharing among the developed country Parties."[185] The COP agreed repeatedly that "appropriate modalities for burden sharing need to be developed,"[186] but developed States rejected any quantified financial commitment. Instead, financial disbursements remained generally contingent on ad hoc agreements. Little could be done to realize the aspiration of systematic burden sharing among developed States. A tension appears in the UNFCCC between the collective commitment of developed States to provide adequate support to developing States and the voluntary basis on which such support is actually provided. As noted above, developed States insisted on specifying that financial support would be limited to the "*agreed* full costs" and "*agreed* full incremental costs."[187] Unsurprisingly, a gap rapidly appeared between what virtually any reasonable observer would consider as adequate support and what was actually provided to developing States.

---

[178] See decision 5/CP.7, "Implementation of Article 4, paragraphs 8 and 9, of the Convention" (November 10, 2001), paras. 11–17; and decision 7/CP.17, "Work programme on loss and damage" (December 9, 2011), para. 6.

[179] Copenhagen Accord, *supra* note 1, para. 8; decision 1/CP.16, *supra* note 18, para. 95; Paris Agreement, *supra* note 14, art. 9.1.

[180] Decision 3/CP.17, *supra* note 62, para. 8.

[181] See, for instance, Paris Declaration on Aid Effectiveness (2005) and Accra Agenda for Action (2008).

[182] Decision 11/CP.1, *supra* note 27, para. 1(a)(ii).

[183] Paris Agreement, *supra* note 14, art. 9.3.

[184] GCF, *Guidelines for Enhanced Country Ownership and Country Drivenness* (March 31, 2017), doc. GCF/B.16/06.

[185] UNFCCC, *supra* note 14, art. 4.3; Kyoto Protocol, *supra* note 29, art. 11.2.

[186] See "Core elements for the implementation of the Buenos Aires Plan of Action," in the annex of decision 5/CP.6, "The Bonn Agreements on the implementation of the Buenos Aires Plan of Action" (July 25, 2001); and decision 7/CP.7, "Funding under the Convention" (November 10, 2001), para.1(d).

[187] UNFCCC, *supra* note 14, art. 4.3. See also *Kyoto Protocol, supra* note 29, art. 11.2. See generally the discussions in Bodansky, *supra* note 28, at 527.

One potential way to bridge the gap was to encourage additional sources of support beyond Annex II parties. Thus, the Bali Action Plan called for "mobilization of public- and private-sector funding and investment, including facilitation of climate-friendly investment choices."[188] The Paris Agreement noted that, beyond developed States, "[o]ther Parties are encouraged to provide or continue to provide [financial] support voluntarily."[189] Co-financing was strongly encouraged, and sometimes even required, not only as a way to enhance the effectiveness of financial support but also in order to "strengthen partnerships with recipient country governments, multilateral and bilateral financing entities, the private sector, and civil society."[190] The UNFCCC regime has thus attempted to catalyse efforts by non-parties.[191]

Efforts have been made to promote transparency in the provision of international support for climate action, thus making it possible to name and shame the States most reluctant to contribute. Annex II parties to the UNFCCC are thus required to "incorporate details of measures taken in accordance with" their financial commitment as part of their national communications.[192] The Cancún Agreements established a Standing Committee to oversee financial support, in particular through "measurement, reporting and verification of support provided to developing country Parties."[193] The following year, Annex II parties were directed to detail the provision of financial, technological and capacity-building support to non-Annex I parties as part of a new biennial reporting requirement.[194] Likewise, the Paris Agreement requires developed parties and encourages others to provide "transparent and consistent information on support for developing country Parties provided and mobilized through public interventions."[195]

V. CONCLUSION

International support for climate action remains one of the most controversial aspects of the international law on climate change. The outcome of protracted negotiations has fallen short of the great expectations of some for systematic assistance to developing States or *a fortiori* for any form of reparation. Instead, international support has largely focused on climate change mitigation in emerging and newly industrialized economies. With the Copenhagen Accord and the Cancún Agreements, however, a new momentum for international support has been created. The collective objective of developed States of mobilizing USD 100 billion a year by 2020 in support of climate action in developing States is ambitious, but its success will depend on the nature of this finance.

[188] Decision 1/CP.13, *supra* note 36, para. 1(e)(v).
[189] Paris Agreement, *supra* note 14, art. 9.2.
[190] GEF, *Co-Financing Policy* (June 30, 2014), doc. FI/PL/01, para. 2(b).
[191] Decision 1/CP.21, *supra* note 14, para. 118.
[192] UNFCCC, *supra* note 14, art. 12.3.
[193] Decision 1/CP.16, *supra* note 18, para. 112.
[194] "UNFCCC biennial reporting guidelines for developed country Parties," *supra* note 108, para. 13.
[195] Paris Agreement, *supra* note 14, art. 9.7.

13

Ambition and Compliance

Despite protracted international negotiations over the last quarter of a century, it is far from clear that international cooperation will avert dangerous interference with the climate system. Scientific and political discussions on international responses to climate change often convey a sense of frustration. There is no doubt that "political and sectoral interests have contributed to delay collective efforts."[1] So far, the INDCs communicated by States in the run-up to the Paris Agreement appear inconsistent with the collective target, affirmed in the same Agreement, of holding the increase in the global average temperature to "well below 2°C above pre-industrial levels" – let alone its aspirational objective of holding this temperature within 1.5 degrees Celsius.[2] And even if they were achieved, these objectives would not prevent the infliction of serious harm on populations around the world, most obviously those living in low-lying islands or coastal regions.

As discussed in Chapter 5, the principles of general international law contain norms relevant to climate change. Under the no-harm principle, States must refrain from causing significant harm to others through excessive GhG emissions. Under the law of State responsibility, States which have breached this obligation must pay adequate reparation. The current climate crisis is not caused by a lack of international legal obligations; rather, it results from a lack of compliance with these obligations. Accordingly, promoting State compliance with general international law is key to addressing climate change.

This chapter suggests that the UNFCCC regime as a whole can be understood as an attempt to promote the compliance of States with their obligations under general international law, in particular obligations stemming from the no-harm principle and the law of State responsibility.[3] The UNFCCC regime defines a series of intermediate steps and institutional frameworks to promote compliance with general international law. National commitments establish substantive obligations, in particular on the limitation of GhG emissions, whose ambition increases over time, thus gradually pushing States toward compliance with general international law. Procedural rules on reporting and institutional developments on verification incentivize States

---

[1] Sir Robert Watson *et al.*, *The Truth about Climate Change* (FEU-US, September 2016) 3.

[2] See Paris Agreement, December 12, 2015, in the annex of decision 1/CP.21, "Adoption of the Paris Agreement" (December 12, 2015), art. 2.1(a). See also Watson *et al.*, *supra* note 1, at 7.

[3] This chapter builds in part upon Benoit Mayer, "Construing international climate change law as a compliance regime" (2018) 7:1 *Transnational Environmental Law* 115. A similar analysis could be developed in relation to other treaty-based rules such as the rules developed under the ICAO and the IMO; see Chapter 4, section II.

to fulfill their national commitments while also pushing them to enhance the ambition of these commitments. In this dynamic system, compliance with procedural rules encourages compliance with substantive rules, which, in turn, advances compliance with the norms of general international law.

This chapter first explores the gap in compliance with general international law, in particular with regard to emissions and reparations. It then retraces efforts under the UNFCCC regime to bridge this compliance gap by breaking it down into three intermediate gaps: a gap in State conduct, a gap in national commitments and a gap in collective objectives.

I. THE GENERAL ARCHITECTURE OF INTERNATIONAL CLIMATE AGREEMENTS

Under general international law, each State has sovereign rights as well as the corollary obligation to respect the rights of other States. As discussed before,[4] this implies that each State must refrain from causing, or allowing activities under its jurisdiction from causing, significant harm to the territories of other States and that, if a State breaches this obligation, it must pay adequate reparation to those affected by its fault. There remains significant indeterminacy in the modalities of relevant general international law obligations. Nevertheless, current efforts are arguably not preventing "significant" harm from affecting many countries, in particular small island developing States and low-lying coastal countries. Likewise, very little effort has been made by the States responsible for excessive GhG emissions to pay anything close to adequate reparation to those injured as a consequence. In other words, there is a wide gap between the obligations of States under general international law and their actual conduct.

It is possible to distinguish two parallel aspects of this compliance gap. The *compliance gap on emissions* is constituted by the failure of States to comply with their obligation to avoid and prevent activities under their jurisdiction from causing serious harm to the environment of other States. International action on climate change mitigation seeks to overcome this gap in compliance by fostering national commitments to gradually limit and reduce GhG emissions. The *compliance gap on reparations* is constituted by the failure of States to make adequate reparation for the harm caused by the internationally wrongful act constituted by excessive GhG emissions. Calls to address this gap in compliance led to international cooperation on adaptation, calls for international support and, more recently, discussions on possible approaches to address loss and damage.

Climate agreements can be interpreted as successive attempts to bridge the compliance gap on emissions and, to a lesser extent, on reparations. This bridge rests on two pillars. Firstly, each agreement defines a collective objective, at least on climate change mitigation and sometimes also on adaptation and support. Secondly, each agreement endorses national commitments which contribute to the realization of these objectives. Thus, three intermediary gaps appear under each international climate agreement (see Figure 13.1). Firstly, a gap appears between general international law and the collective objectives defined by climate agreements. For instance, the collective objective of the Kyoto Protocol of reducing the GhG emissions of developed States by 5 percent is probably not sufficient to fulfill their obligation

---

[4] See Chapter 5, section I.

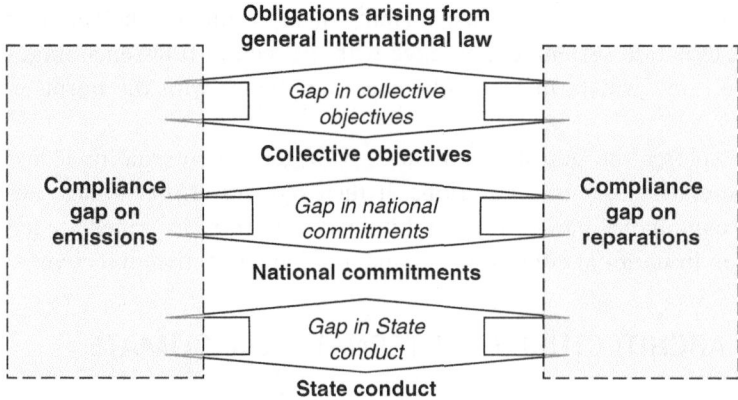

Figure 13.1 Breakdown of the compliance gap

under general international law. Secondly, a gap may appear between collective objectives and national commitments. Thus, INDCs are insufficiently ambitious to achieve the 2- or 1.5-degree-Celsius target endorsed by the Paris Agreement. Thirdly, a gap frequently appears between national commitments and actual State conduct. This gap in state conduct may result from non-participation (whether non-ratification or subsequent withdrawal) or from lack of implementation.

Beyond this general framework, each international climate agreement addresses the compliance gap in a different way. The UNFCCC defines a collective mitigation ambition ("stabilization of greenhouse gas concentrations in the atmosphere at a level that would prevent dangerous anthropogenic interference with the climate system")[5] and national commitments.[6] Yet, because this collective objective and these national commitments are couched in vague provisions, the UNFCCC does not allow for a clear and objective assessment of whether national commitments have been implemented and whether the collective objective has been achieved.

The Kyoto Protocol aimed precisely at defining more specific national commitments. Developed States Parties committed to achieve their quantified emission limitation and reduction commitment (QELRC) over the 2008–2012 period. Taken as a whole, these commitments would have achieved a reduction in the GhG emissions of developed States by 5 percent below 1990 levels during the commitment period.[7] A gap in State conduct appeared with the decision of the United States not to ratify the Kyoto Protocol and, later, the notification that Canada would withdraw from the Protocol before the expiration of the commitment period. In addition, there was certainly a gap in collective objectives between the objective of reducing the overall GhG emissions of developed States by 5 percent during the commitment period and the obligations of developed as well as developing States under general international law. This gap in collective objectives became all the more obvious as GhG emissions from emerging economies increased much faster than emissions in developed States decreased. The approach of the Kyoto Protocol

---

[5] United Nations Framework Convention on Climate Change, May 9, 1992, 1771 *UNTS* 107 (hereinafter UNFCCC), art. 2.

[6] *Ibid.*, art. 4.

[7] Kyoto Protocol to the United Nations Framework Convention on Climate Change, December 11, 1997, 2303 *UNTS* 162 (hereinafter Kyoto Protocol), art. 3.1.

did, however, exclude any gap in national commitments, since the collective objective on mitigation was due to be met by the sum of national commitments initially negotiated.

By contrast, the Paris Agreement defines more demanding collective objectives but more flexible national commitments, leaving room – initially at least – to a wide gap in national commitments. Collective objectives regard not only climate change mitigation, but also adaptation and finance, although the latter objectives were defined in vaguer terms.[8] National commitments are to be nationally determined and communicated.[9] As in the case of the Kyoto Protocol, a gap in State conduct may appear, for instance, if a State fails to implement its commitment or even, as the United States could do, withdraws from the Agreement. Furthermore, there is arguably a gap in collective objectives between the objectives of the Paris Agreement and general international law, as, for instance, even at 1.5-degree-Celsius warming would cause great harm to many nations around the world. Overall, however, it is the gap in national commitments which represents the greatest initial challenge to the effectiveness of the Paris Agreement: as of early 2018, the INDCs communicated by parties appeared very unlikely to be consistent with the objective of holding the increase in global average temperature within 2 degrees Celsius.[10]

The success of an agreement on climate change can be assessed by its ability to reduce simultaneously the gaps in collective objective, national ambition and State conduct, thus genuinely promoting compliance with States' obligations under general international law. From this perspective, the Kyoto Protocol achieved few results because of the limited scope and low ambition of its collective objective; it was further affected by the non-participation of the United States and, to a lesser extent, by Canada's withdrawal. By contrast, the Paris Agreement could rapidly achieve quasi-universal participation[11] while defining ambitious targets, but this may come at the cost of a wide gap between national commitments (NDCs) and the collective objective of the Agreement. Accordingly, the success of the Paris Agreement will depend on the ability of relevant institutions not only to promote effective implementation while maintaining participation, but also to foster the willingness of States to significantly enhance their commitments.

The next three sections explore how the successive agreements have attempted to reduce the gaps in State conduct, national commitments and collective objectives.

II. ADDRESSING THE GAP IN STATE CONDUCT

The gap in State conduct is one of the components of the gap in compliance with general international law obligations. This gap appears whenever a State's actual conduct differs from what would be expected from it under an international climate agreement, whether the State is party to this agreement or not. Thus, a gap in State conduct can result from the *non-participation* of a State in the particular agreement. Alternatively, a gap in State conduct can also result from *non-compliance* by a party to an agreement to the terms of this agreement. In this sense, "compliance"

[8] Paris Agreement, *supra* note 2, art. 2.1.
[9] *Ibid.*, art. 3.
[10] See e.g. UNEP, *The Emissions Gap Report 2017: A UN Environment Synthesis Report* (November 2017).
[11] The withdrawal of the United States would only be temporary, it appears, given the political support that participation to the Paris Agreement has achieved among American voters. See Scott Clement and Brady Dennis, "Post-ABC poll: nearly 6 in 10 oppose Trump scrapping Paris Agreement" *Washington Post* (June 5, 2017).

is understood in a narrow sense as compliance with a particular climate agreement, rather than in a broader sense as compliance with general international law. States have generally complied with national commitments established in clear and specific terms, such as the QELRCs under the Kyoto Protocol.[12] Yet, they may not have complied with vaguer commitments such as those contained in the UNFCCC,[13] where an objective assessment is more challenging precisely because of the broad language in which these provisions are couched.

This section documents a number of efforts that have been made to avoid or reduce gaps in State conduct, and often, more specifically, to ensure compliance with the national commitments of the parties to international climate agreements. In particular, peer-review or "naming-and-shaming" mechanisms have been used to increase the political costs of non-compliance with national commitments. Because "naming" non-compliance requires an assessment of relatively complex data, sophisticated rules and institutional arrangements have sought to promote consistent and transparent reporting as well as independent verification.

A. Socio-political Mechanisms to Promote Cooperative State Conduct

International law is rather poorly equipped to narrow the gap in State conduct. States are free to consent to international agreements or not. Once they are party to an agreement, enforcement mechanisms at the international level do not bear comparison with domestic courts and tribunals.[14] Therefore, it may come as a surprise, to paraphrase Louis Henkin, that most States have participated in most climate agreements and have implemented most of their commitments most of the time.[15] Only a contextualized analysis of legal tools can explain this "miracle."

National interests are an obvious determinant of States' conduct. Efforts have been made to present participation to climate agreements and compliance with State obligations under these agreements as States' own interests. For instance, the conditional provision of international support incentivized the participation of developing States. Yet, an effective international cooperation against climate change requires States to adopt conducts which, sometimes, go against their immediate national interests. Even then, compliance with international obligations is encouraged by a set of social and political processes that the law triggers and reinforces. Governments are made of people whose agenda is not confined to promoting national interests; some members of national governments are sensitive to the need for a State to comply with its international obligations. Likewise, NGOs and public opinions may also incentivize compliance with international climate agreements.[16] In certain political systems, a politician denying climate change may lose votes, and a government reluctant to mitigate climate change might not be re-elected. A government which blatantly fails to honor

---

[12] Every developed State which remained a party to the Kyoto Protocol until the end of the initial commitment period complied with its emission limitation and reduction commitment, while Canada preferred to withdraw to avoid a finding of non-compliance. See Igor Shishlov, Romain Morel and Valentin Bellassen, "Compliance of the parties to the Kyoto Protocol in the first commitment period" (2016) 16:6 *Climate Policy* 768.

[13] See UNFCCC, *supra* note 5.

[14] See Chapter 14, section I.

[15] Louis Henkin, *How Nations Behave*, 2nd edn (Columbia University Press, 1979) 47.

[16] See e.g. Martha Finnemore and Kathryn Sikkink, "International norm dynamics and political change" (1998) 52:4 *International Organization* 887; Thomas M. Franck, *The Power of Legitimacy among Nations* (Oxford University Press, 1990).

its promises will encounter criticisms at the international level if not also at the domestic level. The decision of US President Trump to withdraw from the Paris Agreement in the name of national interests led to many criticisms, internally from all political boards and internationally from various States. Over time, compliance with international law becomes a habit rather than a constraint.[17]

The development of international law can reinforce these social and political processes. Compliance with an international obligation is generally more likely when "[t]he parties have agreed in advance to the standards by which the conduct is to be judged."[18] This clarity in national commitments makes non-compliance more obvious and hence more politically costly. Precisely, the rules adopted by international climate agreements provide the clarity that norms of general international law often lack, thus favoring compliance. As discussed below, sophisticated rules have been developed to increase transparency in climate action.

B. Transparency

Naming and shaming is likely to become more effective, as a tool to promote compliance, if a State's compliance with its obligations is assessed through an objective and public process.[19] Such an assessment is difficult in relation to norms of general international law due to a lack of well-defined touchstones, but it is easier in relation to the specific rules set up by international climate agreements. In other words, the most effective national commitments are those which allow compliance to be verified in a public and objective process.[20]

To ensure that compliance can readily be assessed, a key aspect of international cooperation on climate change is to ensure that the necessary information is accessible from consistent and reliable sources. This raises unique technical difficulties in the context of climate change mitigation because the sources and sinks of GhGs are multiple and diffuse. Despite considerable improvements over the last few decades, developing a complete inventory of the net GhG emissions occurring within the territory of any given State remains challenging. In particular, GhG emissions from land-use change and forestry are still often excluded from the reports of many States, including developed ones, because of a lack of accurate accounting methodology. Many developing States face a great challenge to estimate GhG emissions in many other sectors as well. A global system of satellites is slowly being developed to measure GhG emissions from the sky; this, in the medium term, may provide a source for global, independent and reliable information to complement national reports.[21]

[17] Harold Hongju Koh, "Why do nations obey international law?" (1997) 106:8 *Yale Law Journal* 2599, at 2646.

[18] Abram Chayes and Antonia Handler Chayes, *The New Sovereignty Compliance with International Regulatory Agreements* (Harvard University Press, 1995) 120.

[19] See e.g. Sylvia I. Karlsson-Vinkhuyzen *et al.*, "Entry into force and then? The Paris Agreement and state accountability" *Climate Policy* (forthcoming).

[20] See in particular Kyoto Protocol, *supra* note 7, art. 3.1; and compare with UNFCCC, *supra* note 5, art. 4.2.

[21] See decision 19/CP.22, "Implementation of the global observing system for climate" (November 17, 2016).

Thus, great efforts have been made from the outset of the UNFCCC regime to develop tools for assessing national GhG emissions and their evolution over time as a necessary preliminary step to assessing the compliance of States with their national commitments.[22] Intricate technical methodologies have been adopted to develop reliable and consistent GhG accountancy. The following highlights the key features of technical frameworks for the measurement, reporting and verification (MRV) of climate change mitigation.

1. Measurement

Technical guidance had to be adopted if States were to report on GhG emissions taking place within their territory using consistent assumptions and similar methodologies, thus producing comparable data. The UNFCCC gave the COP a mandate to develop and refine "comparable methodologies ... for preparing inventories of greenhouse gas emissions by sources and removals by sinks, and for evaluating the effectiveness of measures to limit the emissions and enhance the removals of these gases."[23] The technicality of the work, however, made it particularly challenging for the COP, which largely relied on the work of the IPCC.

The IPCC adopted the three-volume Guidelines for National Greenhouse Gas Inventories in 1994. In order to build on new research and necessary refinements, this document was revised in 1996 and again in 2006, when it was expanded to five volumes;[24] a further update is announced for 2019.[25] In addition, the IPCC adopted more advanced guidelines to specify the methodologies for taking uncertainty into account[26] and for accounting GhG emissions and sinks from land-use change and forestry.[27] Beyond climate change mitigation, it also developed Technical Guidelines for Assessing Climate Change Impacts and Adaptations.[28]

The first sessions of the COP endorsed the use of the Guidelines for National Greenhouse Gas Inventories as default methodologies for estimating, reporting and verifying inventory data.[29] Through the subsequent endorsement of the revised Guidelines and "any future elaboration of these guidelines, or parts of them,"[30] the COP, serving as the meeting of the parties to the Kyoto Protocol, delegated some of its technical role to the IPCC. Consistently, the Paris Agreement requires that each party assesses its GhG emissions "using good practice methodologies accepted by the Intergovernmental Panel on Climate Change and agreed upon by the Conference of the Paris serving as the meeting of the Parties to this Agreement."[31]

---

[22] See e.g. M.J. Grubb, D.G. Victor and C.W. Hope, "Pragmatics in the greenhouse" (1991) 354:6352 *Nature* 348, at 349; Daniel Bodansky, "The United Nations Framework Convention on Climate Change: a commentary" (1993) 18:2 *Yale Journal of International Law* 451, at 546.

[23] UNFCCC, *supra* note 5, art. 7.2 (d).

[24] Simon Eggleston *et al.*, *2006 IPCC Guidelines for National Greenhouse Gas Inventories* (IGES, 2006).

[25] IPCC, "Update of methodologies on national greenhouse gas inventories" (April 11–13, 2016), doc. IPCC-XLIII/Doc. 6, Corr.1 (22.III.2016), para. 19.

[26] IPCC, *Good Practice Guidance and Uncertainty Management in National Greenhouse Gas Inventories* (2000).

[27] IPCC, *Good Practice Guidance for Land Use, Land-Use Change and Forestry* (2003).

[28] IPCC, *Technical Guidelines for Assessing Climate Change Impacts and Adaptations* (1995).

[29] See e.g. decision 4/CP.1, "Methodological issues" (April 7, 1995), para. 1; and "Revised Guidelines for the Preparation of National Communications by Parties included in Annex I to the Convention," in the annex of decision 9/CP.2, "Communication from parties included in Annex I to the Convention: guidelines, schedule and process for consideration" (July 19, 1996), para. 14.

[30] Decision 16/CMP.1, "Land use, land-use change and forestry" (December 9–10, 2005), para. 3. See also Kyoto Protocol, *supra* note 7, art. 5.2.

[31] Paris Agreement, *supra* note 2, art. 13.7(a).

2. Reporting

Each State is to report GhG emissions originating from within its territory and on the national responses to climate change to the UNFCCC Secretariat on a regular basis. More specifically, the UNFCCC requires each party to communicate a national inventory of its GhG emissions, a general description of the steps taken or envisaged, and any other relevant information, including, if applicable for developing States, considerations for the need for support.[32] Annex I parties must also include in their report a specific description of the measures of implementation adopted and their intended impact on GhG emissions.[33] Annex II parties must specify the measures taken to provide financial and technological support to developing States.[34]

In addition, the Kyoto Protocol requires Annex I parties to include additional information to their GhG inventories and national communications under the UNFCCC as defined by a decision of the parties.[35] Subsequent decisions established registry systems, including an international transaction log, where the transactions carried out under the flexibility mechanisms are recorded.[36] The international transaction log facilitates the verification of national accounts communicated as part of the annual GhG inventory.[37]

The frequency of these reports differs for developed and developing States. As decided by the COP, Annex I parties are required to communicate their GhG inventories on April 15 of each year[38] and their complete national communication on January 1 every four years (e.g. 2014, 2018).[39] In contrast, no firm deadline was initially agreed upon for the regular communications of non-Annex I parties because it was agreed that reporting by these parties would be contingent upon financial support from Annex II parties.[40]

The Copenhagen Accord and the Cancún Agreements require more frequent reporting to verify the implementation of developed States' quantified economy-wide emission reduction

---

[32] UNFCCC, *supra* note 5, art. 12.1. See also Alexander Zahar, *International Climate Change Law and State Compliance* (Routledge, 2015) 31–39.

[33] UNFCCC, *supra* note 5, art. 12.2.

[34] *Ibid.*, art. 12.3.

[35] Kyoto Protocol, *supra* note 7, art. 7.

[36] See decisions 12/CMP.1, "Guidance relating to registry systems under Article 7, paragraph 4, of the Kyoto Protocol" (December 9–10, 2005); 13/CMP.1, "Modalities for the accounting of assigned amounts under Article 7, paragraph 4, of the Kyoto Protocol" (November 30, 2005); and 14/CMP.1, "Standard electronic format for reporting Kyoto Protocol units" (November 30, 2005).

[37] Decision 15/CMP.1, "Guidelines for the preparation of the information required under Article 7 of the Kyoto Protocol" (November 30, 2005).

[38] Decision 3/CP.1, "Preparation and submission of national communications from the parties included in Annex I to the Convention" (April 7, 1995), para. 2(b).

[39] Decision 2/CP.17, "Outcome of the work of the Ad Hoc Working Group on Long-Term Cooperative Action under the Convention" (December 11, 2011), para. 14. Before this decision and following the first national communication required by art. 12.5 of the Convention, the deadline of each national communication had been set by the COP on April 15, 1997 (second national communication), November 30, 2001 (third), January 1, 2006 (fourth), January 1, 2010 (fifth) and January 1, 2014 (sixth). See respectively decision 3/CP.1, *supra* note 38, para. 2(a); decision 11/CP.4, "National communications from parties included in Annex I to the Convention" (November 14, 1998), para. 2(a); decision 4/CP.8, "National communications from parties included in Annex I to the Convention" (November 1, 2002), para. 3; decision 10/CP.13, "Compilation and synthesis of fourth national communications" (December 14–15, 2007), para. 2; and decision 9/CP.16, "National communications from parties included in Annex I to the Convention" (December 10–11, 2010), para. 5.

[40] UNFCCC, *supra* note 5, art. 4.7. See also Chapter 12, section II.A(1). The initial communication was required within three years of the entry into force of the Convention with respect to the particular party, subject to funding. See UNFCCC, *supra* note 5, art. 12.5. See also decision 8/CP.11, "Submission of second and, where appropriate, third national communications from parties not included in Annex I to the Convention" (December 9–10, 2005), para. 2.

Table 13.1 Overview of national commitments on reporting applicable in 2018

Parties concerned	Original legal basis	Report	Frequency (as of 2018)	Content (highlights)
Annex I	UNFCCC art. 12.1–3	GhG inventory	Annual	Estimates of GhG emissions and sinks
	Kyoto Protocol art. 7	National communications	Every four years	Implementation of UNFCCC commitments and Kyoto Protocol commitments
	Cancún Agreements	Biennial Report	Every two years	Implementation of Cancún pledges
Non-Annex I	UNFCCC art. 12.1	National communications	Every four years	Estimates of GhG emissions and sinks; implementation of the UNFCCC commitments
	Cancún Agreements	Biennial Update Report	Every two years	Implementation of Cancún pledges

targets and developing States' nationally appropriate mitigation actions,[41] thus adding another layer of complexity to national commitments on reporting (see Table 13.1). Annex I parties committed to provide biennial reports, on January 1 of every even year, containing information on GhG emissions, progress toward achievement of their quantified economy-wide emission reduction target and the provision of support to developing States.[42] Non-Annex I parties committed, contingent upon financial support, to make national communications (including a GhG inventory) every four years and to produce Biennial Update Reports containing information on the implementation of their nationally appropriate mitigation action.[43] The scope of national communications was also extended more clearly beyond mitigation to include international support as well as adaptation.[44]

The Paris Agreement sets up "an enhanced transparency framework for action and support" building on acquired experience.[45] The purpose of this framework is to provide "a clear understanding of climate action in the light of the objective of the Convention"[46] as well as "clarity on support provided and received by relevant individual Parties."[47] The modalities of implementation decided through negotiations conducted by the Ad Hoc Working Group on the Paris Agreement will determine, among other things, the frequency of such reports and their relation to pre-existing reporting obligations.

---

[41] See in particular Copenhagen Accord, in the annex of decision 2/CP.15 (December 18–19, 2009), paras. 4 and 5.
[42] See *ibid.*, para. 40(a).
[43] *Ibid.*, para. 60(b) and (c).
[44] Decision 1/CP.16, "The Cancún Agreements: outcome of the work of the Ad Hoc Working Group on Long-Term Cooperative Action under the Convention" (December 10–11, 2010), paras. 40(c), 41 and 42.
[45] Paris Agreement, *supra* note 2, art. 13.1.
[46] *Ibid.*, art. 13.5.
[47] *Ibid.*, art. 13.6.

3. Verification

Verification processes are essential to ensuring that reliable and consistent information is reported by the parties. Verification processes can be understood as processes to ensure compliance with national obligations on reporting, an essential step toward ensuring compliance with more substantive national commitments on climate change mitigation, adaptation and support, which, in turn, promote compliance with the norms of general international law.

Although the question of verification was amply discussed during the negotiations which led to the adoption of the UNFCCC,[48] little could be agreed upon at this stage. The Convention only gives the COP the power to "[a]ssess, on the basis of all information made available to it in accordance with the provisions of the Convention, the implementation of the Convention by the Parties."[49] As recommended by the Intergovernmental Negotiating Committee for a Framework Convention on Climate Change (INC/FCC), the first session of the COP adopted the "statement of purpose"[50] of a review process. Accordingly, this review process would largely be limited to verifying the information contained in national communications, thus ensuring "that the Conference of the Parties has accurate, consistent and relevant information at its disposal"[51] to assess whether each party complies with its substantive commitments. The review, carried out by an expert review team (ERT), was to be conducted "in a facilitative, non-confrontational, open and transparent manner."[52] A similar process was later established to review the GhG inventories of Annex I parties.[53] These review processes have involved initial checks and summary reviews, but also in-depth desk studies, centralized reviews (i.e. discussions by experts) and in-country visits.

The Kyoto Protocol relied on the same processes for the verification of national reports. The steps taken by Annex I parties to implement their QELRC were considered as part of the review of national communications under the Convention.[54] Likewise, the supplementary data on GhG emissions necessary for assessing compliance with these States' QELRCs was reviewed as part of the review of GhG inventories.[55]

More stringent verification processes were established under the Cancún Agreements, extending for the first time to non-Annex I parties. The Cancún Agreements established an International Assessment Review (IAR) process for Annex I parties and an International Consultations and Analysis (ICA) process applicable to non-Annex I parties.[56] The relevant modalities and procedures for these two processes were adopted the following year at COP17.[57] Accordingly, the IAR of an

---

[48] Bodansky, *supra* note 22, at 546–547.

[49] UNFCCC, *supra* note 5, art. 7.2 (e).

[50] Decision 2/CP.1, "Review of first communications from the parties included in Annex I to the Convention" (April 7, 1995), para. 1.

[51] "Purpose of the review of the first communications from Annex I parties," in Annex I of decision 2/CP.1, *supra* note 50.

[52] *Ibid.*

[53] Decision 3/CP.5, "Guidelines for the preparation of national communications by parties included in Annex I to the Convention, Part I: UNFCCC reporting guidelines on annual inventories" (November 4, 1999) and UNFCCC, *Guidelines on Reporting and Review* (February 16, 2000), doc. FCCC/CP/1999/7. See also decision 19/CP.8, "UNFCCC guidelines for the technical review of greenhouse gas inventories from parties included in Annex I to the Convention" (November 1, 2002).

[54] Kyoto Protocol, *supra* note 7, arts. 8.1 and 7.2.

[55] *Ibid.*, arts. 8.1 and 7.1.

[56] See decision 1/CP.16, *supra* note 42, paras. 44 and 63.

[57] See Annex II and Annex IV of decision 2/CP.17, *supra* note 39.

Annex I party consists in a technical review by an ERT of the biennial report, GhG inventory and national communications of this party, followed by a multilateral assessment of the implementation of quantified economy-wide emission reduction targets conducted under the SBI, involving written and oral questions by any other parties.[58] Only slightly less demanding, the ICA process of a non-Annex I party involves a technical review by ERTs of the biennial update report followed by a "facilitative sharing of views" through a workshop.[59] In both cases, this multilateral verification processes is likely to put greater public pressure on States which do not fulfill their national commitments on reporting and thus facilitate naming-and-shaming processes.[60] Further guidelines called for the process to be "facilitative, non-confrontational, open and transparent" and to "promote consistency, comparability and transparency in the review of information reported."[61]

The "enhanced transparency framework for action and support"[62] announced by the Paris Agreement will likely build on the experience of IAR and ICA. Each party committed to "participate in a facilitative, multilateral consideration of progress" with respect to its NDC.[63] The first session of the COP serving as the Meeting of the Parties to the Paris Agreement shall "adopt common modalities, procedures and guidelines, as appropriate, for the transparency of action and support."[64] The work of the Ad Hoc Working Group on the Paris Agreement, preparing a decision on these modalities for adoption by the end of 2018,[65] will seek to push further toward processes which cause public embarrassment to parties failing to comply with their commitments.

C. Reviewing Compliance

Institutional mechanisms have been established to assess the compliance of States with substantive national commitments under successive climate agreements, including commitments on climate change mitigation.[66] In contrast to verification processes, reviews of compliance go a step further by drawing conclusions on the compliance of States with their national commitments, including substantial national commitments on climate change mitigation.[67]

---

[58] Decision 2/CP.17, *supra* note 39, para. 23, and Annex II, part IV.

[59] Annex IV of decision 2/CP.17, *supra* note 39, para. 3(b).

[60] Antto Vihma, "Analyzing soft law and hard law in climate change" in Erkki J. Hollo, Kati Kulovesi and Michael Mehling (eds.), *Climate Change and the Law* (Springer, 2013) 143, at 161.

[61] "Guidelines for the technical review of information reported under the Convention related to greenhouse gas inventories, biennial reports and national communications by parties included in Annex I to the Convention," in the annex of decision 23/CP.19, "Work programme on the revision of the guidelines for the review of biennial reports and national communications, including national inventory reviews, for developed country parties" (November 22, 2013), paras. 5(a) and 6. See also the annex of decision 13/CP.20, "Guidelines for the technical review of information reported under the Convention related to greenhouse gas inventories, biennial reports and national communications by parties included in Annex I to the Convention" (December 12, 2014).

[62] Paris Agreement, *supra* note 2, art. 13.1.

[63] *Ibid.*, art. 13.11.

[64] *Ibid.*, art. 13.13.

[65] Decision 1/CP.21, *supra* note 2, para. 91.

[66] These mechanisms have taken inspiration in pre-existing compliance systems under other multilateral environmental agreements. See Meinhard Doelle, "Early experience with the Kyoto system: possible lessons from MEA compliance system design" (2010) 1:2 *Climate Law* 237.

[67] In practice, the distinction between documenting compliance and facilitating it is not always clear. See, for instance, Sebastian Oberthür, "Compliance under the evolving climate change regime" in Kevin R. Gray, Cinnamon Piñon Carlarne and Richard Tarasofsky (eds.), *The Oxford Handbook of International Climate Change Law* (Oxford University Press, 2016) 120, dealing with IAR and ICA as compliance rather than verification mechanisms.

In practice, reviews of compliance have often focused on technical issues of compliance with reporting obligations which had been identified but not solved at the stage of verification.

The need to develop a review mechanism was recognized in the UNFCCC. Article 7 makes it one of the functions of the COP to "keep under regular review the implementation of the Convention," including through "assess[ing] ... the implementation of the Convention by the Parties."[68] More specifically, Article 13 provides that the first session of the COP shall "consider the establishment of a multilateral consultative process, available to Parties on their request, for the resolution of questions regarding the implementation of the Convention."[69] This formulation left many options on the table, including a possible standing by non-State actors and the possibility of penalties for non-compliance.[70]

The ensuing consultations reflected an understanding that, although multilateral consultative process established in other multilateral environmental agreements "usually lack[ed] real enforcement power, their strength was manifested through peer pressure and regular accountability by Parties in an intergovernmental process."[71] A decision adopted at COP 4 contained the draft terms of reference of a standing "multilateral consultative committee,"[72] with only questions of composition and voting procedures remaining to be decided. Yet, priority was given to the drafting of a protocol with more specific obligations rather than to enhance the implementation of vague commitments contained in UNFCCC. Following COP4, negotiations on a multilateral consultative process froze and this committee was never established. Instead of being first tested under the UNFCCC, a mechanism to review compliance would be established directly under the Kyoto Protocol.[73]

Article 18 of the Kyoto Protocol called upon the COP serving as the meeting of the parties to this Protocol to "approve appropriate and effective procedures and mechanisms to determine and to address cases of non-compliance with the provisions of this Protocol."[74] This should include "an indicative list of consequences," although it was agreed that "[a]ny procedures and mechanisms ... entailing binding consequences shall be adopted by means of an amendment to this Protocol."[75] Accordingly, a decision establishing a "Compliance Committee" formed part of the 2001 Marrakesh Agreements on the implementation of the Kyoto Protocol, later confirmed by the first session of the Meeting of the Parties (CMP1).[76]

The Compliance Committee is a body of 20 experts elected by the Meeting of the Parties, but serving in their individual capacities. It is composed of two branches: a Facilitative Branch and an Enforcement Branch. "Questions of implementation" initiating cases before the Compliance Committee can be submitted either by a party or by an ERT upon verification of national

---

[68] UNFCCC, *supra* note 5, arts. 7.2 and 7.2(e).

[69] *Ibid.*, art. 13.

[70] Bodansky, *supra* note 22, at 548.

[71] INC/FCC, *Consideration of the Establishment of a Multilateral Consultative Process for the Resolution of Questions Regarding Implementation (Article 13)* (July 26, 1994), doc. A/AC.237/59, para.11.

[72] Decision 10/CP.4, "Multilateral consultative process" (November 6, 1998).

[73] See decision 5/CP.2, "Linkage between the Ad Hoc Group on Article 13 and the Ad Hoc Group on the Berlin Mandate" (July 19, 1996); Kyoto Protocol, *supra* note 7, art. 16.

[74] Kyoto Protocol, *supra* note 7, art. 18.

[75] *Ibid.*

[76] See respectively decision 24/CP.7, "Procedures and mechanisms relating to compliance under the Kyoto Protocol" (November 10, 2001); and decision 27/CMP.1, "Procedures and mechanisms relating to compliance under the Kyoto Protocol" (November 30, 2005).

reports.[77] The Bureau of the Compliance Committee allocates questions of implementation to either of the two Branches as it deems appropriate in accordance with the mandates of each Branch.[78]

The Facilitative Branch deals pre-emptively with potential non-compliance of any commitment under the Kyoto Protocol by providing advice, facilitating support and formulating recommendations.[79] By contrast, the Enforcement Branch is "responsible for determining whether a Party included in Annex I is not in compliance" with its mitigation commitments, its reporting commitments, or the provisions establishing flexibility mechanisms.[80] The Enforcement Branch has greater powers, which extend to the adoption of a "declaration of non-compliance."[81] This Branch can request the party to report on the remedial measures it intends to take and to submit progress reports toward the implementation of these remedial measures.[82] The Enforcement Branch can also suspend the eligibility of a party to use the flexibility mechanisms or deduct 1.3 times the excess emissions of a party from its assigned amount for a following commitment period.[83] Appeal against decisions of the Enforcement Branch regarding compliance with an emission limitation and reduction commitment can be made before the COP serving as the meeting of the parties to the Protocol, but only on the very narrow ground of a denial of due process.[84]

In 2006, South Africa submitted a question of implementation on behalf of the G77 and China regarding the delayed submission of national communications and GhG inventories by several Annex I parties.[85] The question was allocated to the Facilitative Branch, whose ten members could not make a decision on whether this question of implementation had been properly submitted, considering that it had not been filed on behalf of "any Party," but of multiple parties in concert.[86] Consequently, no action was taken. No other question of implementation was subsequently allocated to the Facilitative Branch. To try to remain relevant, the Facilitative Branch attempted to take the initiative of offering its services to countries with foreseeable compliance problems, such as Canada. Yet, it had few services to offer and Canada showed little interest in cooperating.[87]

The Enforcement Branch has been more active, although, with the sole exception of Canada, its practice has also focused on reporting obligations rather than on substantive national commitments. The Enforcement Branch has been allocated nine questions of implementation concerning eight different parties, with two questions concerning Ukraine (see Table 13.2).

---

[77] Annex of decision 27/CMP.1, *supra* note 76, section VI.1.

[78] *Ibid.*, section VII.1.

[79] *Ibid.*, sections IV and XIV.

[80] *Ibid.*, section V.4.

[81] *Ibid.*, section XV.1.

[82] *Ibid.*, sections XV.2–3.

[83] *Ibid.*, sections XV.4–5.

[84] *Ibid.*, section XI.1. See also UNFCCC, *Procedural Requirements and the Scope and Content of Applicable Law for the Consideration of Appeals under Decision 27/CMP.1* (September 15, 2011), doc. FCCC/TP/2011/6.

[85] Facilitative Branch of the Compliance Committee, letter of the Chairman of the Group of 77 and China (May 26, 2006), doc. CC-2006-1-1/FB.

[86] Facilitative Branch of the Compliance Committee, "Report to the Compliance Committee on the Deliberations in the Facilitative Branch Relating to the Submission Entitled 'Compliance with Article 3.1 of the Kyoto Protocol'" (2006) doc. CC-2006-11-3/FB. See also Meinhard Doelle, "Compliance and enforcement in the climate change regime" in Hollo *et al.* (eds.), *supra* note 60, 165, at 171.

[87] See Zahar, *supra* note 32, section 3.2.2; Oberthür, *supra* note 67, at 125.

Table 13.2 Overview of the cases considered by the Enforcement Branch of the Compliance Committee

Party/ date of referral	Consequences	End of the proceedings
Greece December 28, 2007	• Finding of non-compliance • Request for remedial measures • Suspension of eligibility for flexibility mechanisms	No longer an issue of implementation November 13, 2008
Canada April 11, 2008	None	No finding of non-compliance June 15, 2008
Croatia August 26, 2009	• Finding of non-compliance • Request for remedial measures • Suspension of eligibility for flexibility mechanisms	No longer an issue of implementation February 8, 2012
Bulgaria March 9, 2010	• Finding of non-compliance • Request for remedial measures • Suspension of eligibility for flexibility mechanisms	No longer an issue of implementation November 13, 2008
Romania May 11, 2011	• Finding of non-compliance • Request for remedial measures • Suspension of eligibility for flexibility mechanisms	No longer an issue of implementation July 13, 2012
Ukraine June 3, 2011	• Finding of non-compliance • Request for remedial measures • Suspension of eligibility for flexibility mechanisms	No longer an issue of implementation March 9, 2013
Lithuania September 7, 2011	• Finding of non-compliance • Request for remedial measures • Suspension of eligibility for flexibility mechanisms	No longer an issue of implementation October 24, 2012
Slovakia May 8, 2012	• Finding of non-compliance • Request for remedial measures	No longer an issue of implementation July 4, 2013
Ukraine April 8 ,2016	• Finding of non-compliance • Request for remedial measures	No longer an issue of implementation September 6, 2017

Often, these questions regarded rather complex and technical questions of accounting and reporting of GhG emissions. Every question of implementation originated from the reports of ERTs: as is often the case in multilateral agreements, no single party was sufficiently directly affected by non-compliance by another party to be ready to take the relatively "hostile" step of initiating a review of compliance against a counterpart. Seven of the nine questions of implementation regard a country with an economy in transition. Besides a question regarding Canada which led to no substantive decision, every other question of implementation led to a finding of non-compliance and requests for remedial measures, often along with a temporary suspension of the party's eligibility to participate in flexibility mechanisms. One party, Croatia, appealed against a finding of non-compliance before the COP, but then withdrew its appeal.[88] Eventually, all eight cases were solved by a determination of the Enforcement Branch that there was no longer an issue of implementation.

The practice of the Compliance Committee has largely been confined to procedural matters. The only question of implementation regarding compliance with a substantive commitment related to Canada's foreseeable inability to achieve its QELRC. However, the Compliance Committee appeared unable to deal with this looming case of non-compliance with emission limitation and reduction commitments. On the one hand, the Enforcement Branch can only deal with ascertained cases of non-compliance after the expiration of a commitment period. Before the expiration of a commitment period, one cannot strictly rule out the possibility of a party purchasing emission units from flexibility mechanisms to balance domestic emissions. On the other hand, the Facilitative Branch appeared ill-equipped, in the absence of a more specific mandate, to pre-emptively address compliance. Canada's withdrawal from the Kyoto Protocol with effect two weeks before the expiration of the first Commitment Period prevented it from being found formally in breach of its emission limitation and reduction commitment.[89] This illustrates the inherent limitations of international mechanisms – their ability to correct small technical issues of compliance, but impotence when facing the downright unwillingness of a sovereign State to cooperate.

The Paris Agreement established a new "mechanism to facilitate implementation of and promote compliance with"[90] its provisions. Like the Kyoto Protocol's Compliance Committee, the Paris Agreement mechanism is to be "expert-based and facilitative in nature" and to function "in a manner that is transparent, non-adversarial and non-punitive."[91] This committee under the Paris Agreement should be composed of 12 members elected by the COP Serving as the Meeting of the Parties to the Paris Agreement.[92] The other modalities and procedures of this committee are to be adopted at the first session of the COP serving as the meeting of the parties to the Paris Agreement, which is to be concluded by the end of 2018.[93] Current negotiations regard, for instance, the nature of the mechanism, the scope of the review of compliance and the triggers of such reviews.

Verification and review mechanisms ultimately depend on the willingness of national governments to implement their commitments. Per se, neither the reports of ERTs nor the issuance of a declaration of non-compliance by the Enforcement Branch of the Compliance Committee could force a reluctant government to change its course of action. Suspension of the right of a State to participate in flexibility mechanisms may constitute an economic incentive, albeit limited. Instead of "enforcing" national commitments, verification and review mechanisms seek to feed various national and transnational socio-political processes which push national governments to comply with international obligations and shame those which do not. From this perspective, the publicizing of verification and compliance processes is an

---

[88] See decision 14/CMP.7, "Appeal by Croatia against a final decision of the enforcement branch of the Compliance Committee in relation to the implementation of decision 7/CP.12" (December 9, 2011).

[89] UNFCCC Compliance Committee, Facilitative Branch, "Report on the Meeting" UN Doc CC/FB/12/2012/3 (November 9, 2012).

[90] Paris Agreement, *supra* note 2, art. 15.1.

[91] *Ibid.*, art. 15.2.

[92] Decision 1/CP.21, *supra* note 2, para. 102.

[93] See Paris Agreement, *supra* note 2, art. 15.3; decision 1/CP.21, *supra* note 2, para. 103.

essential step, as the COP recognized, to "mobilize opinion in support of the implementation of the Convention."[94] The UNFCCC Secretariat has engaged in efforts to facilitate an understanding of these somewhat opaque processes through which compliance with national commitments are assessed, if not by the general public, then at least by a broader interested audience.[95]

III. ADDRESSING THE GAP IN NATIONAL COMMITMENTS

As described above, the gap between the conduct of States and their obligations under general international law (the compliance gap) is composed of a gap in State conduct, a gap in national commitments and a gap in collective objectives. The previous section discussed methods used to reduce the gap in State conduct, including by promoting compliance (in a narrower sense) with national commitments. While reducing the gap in State conduct is necessary, it is not suffi-cient. Little would be achieved if the gap in State conduct were to be bridged solely by reducing the ambition of national commitments. It is only if agreed-upon commitments are sufficiently ambitious that they may contribute to bringing States into compliance with their general inter-national law obligations. The present section explores the efforts made to address the gap in national commitments, while the next section turns to the gap in collective objectives.

The gap in national commitments relates to inconsistencies between what national governments commit to do individually and what, in their own assessment, should be achieved collectively. It may appear counterintuitive that such inconsistencies exist at all within any international climate agreement. Yet, as will be discussed below, a gap in national commitments appears in most such agreements, most clearly with regard to climate change mitigation. By formulating a collective objective whose implementation requires more action than they are ready to commit at that particular time, States recognize the need to pursue fur-ther negotiations in order to enhance national commitments.[96] Thus, the gap between national commitments and collective objectives reflects the nature of the UNFCCC regime as, essen-tially, a series of steps in a gradual process to foster compliance with general international law.

A. The UNFCCC

Despite the vagueness of the relevant provisions, a gap in national commitments could arguably be distinguished in the UNFCCC. On the one hand, the collective objective of a "stabilization of greenhouse gas concentrations in the atmosphere at a level that would prevent dangerous anthropogenic interference with the climate system"[97] offers an ill-defined touchstone to assess the consistency of national commitments. On the other hand, these national commitments are themselves defined in general terms. Yet, it appears unlikely that commitments as unspecific as the commitment of developed States to "adopt national policies and take corresponding

---

[94] Decision 7/CP.1, "The report on implementation" (April 7, 1995), recital 3.
[95] See e.g. *ibid.* Complementary initiatives have also been carried out by NGOs, for instance by the Climate Policy Implementation Tracking Framework of the World Resource Institute. See Priya Barua, Taryn Fransen and Davida Wood, *Climate Policy Implementation Tracking Framework* (World Resource Institute Working Paper, November 2014).
[96] See e.g. UNFCCC, *supra* note 5, art. 4.2 (d); decision 1/CP.16, *supra* note 42, paras. 37 and 50; Paris Agreement, *supra* note 2, art. 14.
[97] UNFCCC, *supra* note 5, art. 2.

measures"[98] would suffice to prevent "dangerous" anthropogenic interference with the climate system, however the latter is defined.

Although no consensus could be reached in 1992 on more specific measures, States agreed to commit to "review the adequacy" of the commitments of Annex I parties at regular intervals "until the objective of the Convention is met."[99] Conducting this review, the COP concluded at its first session, without surprise, that the commitments of developed States were "not adequate" and needed to be strengthened through the adoption of a protocol or another legal instrument.[100] Throughout the subsequent negotiations which would lead to the adoption of the Kyoto Protocol, States agreed on the need "to maintain political momentum."[101]

B. The Kyoto Protocol

The Kyoto Protocol was inspired by a desire to adopt more concrete and specific measures, in particular on climate change mitigation. It centers on the formulation of QELRCs applicable to Annex I parties during a commitment period (2008–2012). It further suggests the collective objective ("a view") of reducing the GhG emissions from Annex I parties by "at least 5 per cent below 1990 levels"[102] during the initial commitment period. This collective objective was simply reflecting the aggregate effect of national commitments. A reduction in GhG emissions within Annex I parties consistent with their commitment would in principle achieve this collective objective.[103] Thus, the Kyoto Protocol did not entail any gap in national commitments. This, however, came at the cost of a limited collective objective, confined to the emissions of Annex I parties within a commitment period. The Kyoto Protocol illustrates the dangerous temptation of "resolving" the gap in national commitments by reducing collective ambition rather than increasing national commitments.

C. The Paris Agreement

In contrast to the Kyoto Protocol, the Paris Agreement initially involved a very considerable gap between national commitments and collective objectives. Because national commitments are nationally determined,[104] there is no guarantee that they will fulfill the objective of "strengthen[ing] the global response to the threat of climate change,"[105] among other things by "holding the increase in the global average temperature to well below 2°C above pre-industrial levels"[106] with "efforts" toward 1.5 degrees Celsius. On the contrary, reviews of the initial national commitments show that they are inconsistent with the objective of holding down the increase in the global average temperature within 2 degrees Celsius above pre-industrial levels, let alone 1.5 degrees

---

[98] *Ibid.*, art. 4.2(a).

[99] *Ibid.*, art. 4.2(d).

[100] Decision 1/CP.1, "The Berlin Mandate" (April 7, 1995), Preamble.

[101] Decision 1/CP.4, "The Buenos Aires Plan of Action" (November 14, 1998).

[102] Kyoto Protocol, *supra* note 7, art. 3.1.

[103] However, a rather theoretical inconsistency results from art. 12 of the Kyoto Protocol, which allows Annex I parties, through the CDM, to report certified emission reductions accruing from such project activities implemented by non-Annex I parties to contribute to compliance with their emission limitation and reduction commitments.

[104] Paris Agreement, *supra* note 2, art. 3.

[105] *Ibid.*, art. 2.1.

[106] *Ibid.*, art. 2.1(a).

Celsius.[107] Existing national commitments on climate change mitigation are most likely to lead to an increase in average global temperature of 3.2 degrees Celsius, with an estimated 5 percent chance of less than 2 degrees Celsius, and a 1 percent chance of less than 1.5 degrees Celsius.[108] Even beyond mitigation commitments, a gap in national commitments was identified in relation to other aspects of international cooperation on climate change, including international support.[109]

The Paris Agreement establishes several mechanisms which seek to reduce this initial gap in national commitments by enhancing the ambition reflected in NDCs. The vague mention that NDCs must represent each party's "highest possible ambition"[110] could feed socio-political or even legal processes and incentivize stronger commitments. The Paris Agreement encourages each party to adjust its NDC at any time "with a view to enhancing its level of ambition."[111] In addition, NDCs represent commitments toward a particular timeframe which then need to be revised, and the Paris Agreement requires that successive NDCs of each party "represent a progression."[112] These provisions clearly indicate a "direction of travel" toward greater ambition: as Lavanya Rajamani argued, "[t]he integrity, rationale and spirit of the Paris Agreement depends on forward movement."[113]

Overall, the Paris Agreement establishes a multilateral "global stocktake" in order "to assess the collective progress towards achieving the purpose of this Agreement and its long-term goals."[114] This review of national commitments is to take place in 2023 and every five years thereafter.[115] It will be informed, in particular, by the overall effect expected from NDCs and the latest reports of the IPCC.[116] The outcome of the global stocktake will "inform Parties in updating and enhancing" their commitments.[117] Most of the modalities of application are to be determined by the Ad Hoc Working Group on the Paris Agreement and should be adopted at the first session of the COP serving as the meeting of the parties to the Paris Agreement.[118] The Talanoa Dialogue, to be conducted in 2018 as an initial review of the ambition of NDCs, will be a test for some of the tools to be included in the global stocktake. The success of the Paris Agreement depends largely on the capacity of the global stocktake, along with other institutional processes within and beyond the UNFCCC regime, to bridge the gap in national commitments.

IV. ADDRESSING THE GAP IN COLLECTIVE OBJECTIVES

The gap in collective objectives is the third component of the compliance gap between the conduct of States and their obligations under general international law. The gap in collective

---

[107] See UNFCCC Secretariat, *supra* note 16; UNEP, *supra* note 10.

[108] Adrian E. Raftery *et al.*, "Less than 2°C warming by 2100 unlikely" (2017) 7 *Nature Climate Change* 637.

[109] See e.g. decision 10/CP.21, "The 2013–2015 review" (December 13, 2015), para. 6.

[110] Paris Agreement, *supra* note 2, art. 4.3.

[111] *Ibid.*, art. 4.11.

[112] *Ibid.*, art. 4.3.

[113] Lavanya Rajamani, "The US and the Paris Agreement: in or out and at what cost?" *EJIL: Talk!* (May 10, 2017), www.ejiltalk.org/the-us-and-the-paris-agreement-in-or-out-and-at-what-cost (accessed January 25, 2018). See also Niklas Höhne, "The Paris Agreement: resolving the inconsistency between global goals and national contributions" (2017) 17:1 *Climate Policy* 16.

[114] Paris Agreement, *supra* note 2, art. 14.1.

[115] *Ibid.*, art. 14.2.

[116] Decision 1/CP.21, *supra* note 2, para. 99.

[117] Paris Agreement, *supra* note 2, art. 14.3.

[118] Decision 1/CP.21, *supra* note 2, para. 101.

objectives relates to possible inconsistencies between the collective objectives defined in international climate agreements and the requirements of general international law. While international climate agreements promote some incremental improvements, the collective objective agreed by States through these agreements falls generally short of fulfilling relevant obligations under general international law, most obviously with regard to the obligation of States responsible for internationally wrongful acts to provide adequate reparation.[119]

Assessing the extent or even the existence of a gap in collective objectives with regard to particular international climate agreements can be challenging. On the one hand, the relevant norms of general international law are often ill-determined for lack of authoritative interpretation.[120] On the other hand, the collective objectives defined through successive international climate agreements are often phrased in ambiguous terms due to hard-fought diplomatic battles. For instance, the objective of the UNFCCC does not define what would constitute "dangerous anthropogenic interference with the climate system."[121] If interpreted as an obligation to prevent changes of the climate system which would cause significant harm to any State, this objective could be consistent with the obligation of States under the no-harm principle, but a gap in collective objectives appears if "dangerous" is to be understood more narrowly. Overall, the objective of the UNFCCC does little to fulfill the obligation of responsible States to provide an adequate reparation to the States affected by excessive GhG emissions.

The collective objective of the Kyoto Protocol is significantly more limited in scope as well as in ambition. Its scope is confined geographically to the emissions of Annex I parties and historically to a commitment period extending from 2008 to 2012. Its ambition is limited to a reduction of 5 percent below 1990 levels.[122] As such, the Kyoto Protocol certainly did not suffice to bring all States, or arguably any States, into compliance with their obligations under the no-harm principle. Excessive GhG emissions would not have stopped even if all Annex I parties had participated and implemented their respective emission limitation and reduction commitments. Here again, there was no provision on climate change reparations.

By contrast, the Copenhagen Accord and the Cancún Agreements endorsed a more ambitious objective: holding the increase in global average temperature to 2 degrees Celsius above pre-industrial temperatures.[123] Going further in this direction, Article 2 of the Paris Agreement detailed a rather demanding interpretation of the "ultimate objective" of the UNFCCC with an explicit emphasis on adaptation and support. Yet, this clarification of the collective objectives pursued by States under the UNFCCC regime reveals more clearly their inconsistency with general international law. The mitigation objective – a firm 2 degrees Celsius target and an aspirational 1.5 degrees Celsius – would certainly not prevent a sea-level rise resulting in the disappearance of some small island developing States, which could easily be considered in itself as an instance of serious harm. Again, nothing in the Paris Agreement could readily be considered as an appropriate form of reparation for the continuing wrongful act constituted by excessive GhG emissions. As such, even though the Paris Agreement formulated ambitious collective objectives, there remains a gap between these objectives and the obligations of States under general international law.

---

[119] See generally Chapter 5, section IV; and Chapter 11, section II.
[120] See generally Chapter 5.
[121] UNFCCC, *supra* note 5, art. 2.
[122] Kyoto Protocol, *supra* note 7, art. 3.1.
[123] See Copenhagen Accord, *supra* note 41, para. 1; decision 1/CP.16, *supra* note 42, para. 4.

When endorsing the 2-degree-Celsius target, the Cancún Agreements called upon the COP to periodically review the adequacy of this objective,[124] including through "consideration of strengthening the long-term global goal ... including in relation to temperature rises of 1.5°C."[125] A first review was completed in 2015;[126] others should be initiated following the adoption of the assessment report of the IPCC or at least every seven years.[127] With regard to climate change mitigation, with an already observed increase in average global temperature of 1.1 degrees Celsius in 2016,[128] it may appear to be too late to pursue a collective objective that is more demanding than that defined in the Paris Agreement. There is, however, a risk that the endorsement of this target could lead to some complacency with regard to the serious harm that it implies for many populations and ecosystems throughout the world. At the very least, much more ambitious collective objectives will need to be defined concerning adaptation, support and the obligation of responsible States to provide adequate reparation.

V. CONCLUSION

National governments are reluctant to undertake extremely expensive measures for the sake of abstract legal or moral principles. Persuading them to comply with their international legal obligations is a daunting task. In contrast to many domestic legal regimes, international law does not rely on strong institutions able to impose sanctions for non-compliance. Rather, it relies on persuasion and advocacy, whether from foreign governments and international organizations or from civil society organizations and voters. The UNFCCC regime can be understood from this perspective as an attempt by States to act together to come gradually into compliance with their obligations under general international law, in particular the no-harm principle, but perhaps with the secondary obligation to pay adequate reparation for injury caused by an internationally wrongful act. As a compliance regime, the UNFCCC regime pushes States to adopt a conduct consistent with procedural and substantive national commitments, to review their commitments in order to fulfill collective objectives and to define collective objectives consistent with general international law.

---

[124] Decision 1/CP.16, *supra* note 42, para. 138.

[125] *Ibid.*, para. 139(a)(iv).

[126] See decision 1/CP.18, "Agreed outcome pursuant to the Bali Action Plan" (December 8, 2012), paras. 86–89; UNFCCC SBSTA and SBI, "Report on the structured expert dialogue on the 2013–2015 review" (May 4, 2015), doc. FCCC/SB/2015/INF.1; decision 10/CP.21, *supra* note 109.

[127] Decision 2/CP.17, *supra* note 39, para. 167.

[128] World Meteorological Organization, *Statement on the State of the Global Climate in 2016* (WMO, 2017) 2. See also L.V. Alexander *et al.*, "Summary for policymakers" in T.F. Stocker *et al.* (eds.), *Climate Change 2013: The Physical Science Basis. Contribution of Working Group I to the Fifth Assessment Report of the Intergovernmental Panel on Climate Change* (Cambridge University Press, 2013) 3, at 5, noting an increase of 0.85 degrees Celsius between 1880 and 2012.

14

Adjudication

The possibility of enforcement by a court of law distinguishes the law from other normative systems such as moral norms. Although compliance rests largely on sociological processes, litigation has come to play a growing role in the development of the international law on climate change in recent years.

Cases regarding the implementation of the international law on climate change could be brought before international and domestic jurisdictions. Yet, access to international jurisdictions is often barred by various procedural or political obstacles, and domestic courts have generally played a more prominent role. Domestic courts have sometimes been able to apply international law directly; in other cases, they relied on domestic law to compel national authorities to adopt a course of conduct consistent with their international obligations. Even when litigation on climate change has faced insurmountable obstacles, it has reinforced socio-political processes through which non-compliance with general international law is publicly named and shamed, thus also, indirectly, promoting compliance.

It is easy to forget, however, that litigation can also hinder climate action. This may be the case when incompatibilities appear between national measures of implementation and the obligations of a State under other fields of international law, for instance, international human rights law or international trade law.[1] Alternatively, these national measures of implementation may be inconsistent with national law, most often with constitutional or administrative law, for instance when a regulation is adopted beyond the authority of its author.[2] Such litigation ensures that responses to climate change do not infringe the rule of law, but it can also be used by economic actors interested in hindering climate action.

This chapter focuses on litigation promoting the implementation of the international law on climate change. It first discusses the prospects of litigation before international courts and tribunals before turning to the role of domestic jurisdictions.

---

[1] See Chapter 16, section I.B.

[2] See, for instance, US Supreme Court, *West Virginia et al.* v. *Environmental Protection Agency*, No. 15A773, order of February 9, 2016 staying the Clean Power Plan.

I. INTERNATIONAL ADJUDICATION

There are multiple courts and tribunals addressing international disputes, although many of them have limited jurisdiction. The International Court of Justice (ICJ) would be the most obvious forum for adjudicating cases regarding the obligations of States in response to climate change.[3] Alternatively, cases could be brought before the International Tribunal for the Law of the Sea (ITLOS), established under the UN Convention on the Law of the Sea (UNCLOS).

Proceedings before international courts or tribunals could be contentious or advisory in nature. A contentious case relates to a dispute between two States: an applicant and a defendant. By contrast, advisory proceedings could be brought by an international institution or through a multilateral agreement on any legal question, whether related to a concrete situation or not. Contentious and advisory proceedings would face a series of legal and political obstacles. If successful, however, any such proceedings could be instrumental to the development of a better understanding of States' obligations under general international law.[4]

A. The Prospects for Contentious Cases before the ICJ

The jurisdiction of the ICJ is established on the basis of consent between the parties to the dispute.[5] *Ad hoc consent* can be given, typically through an agreement concluded between two States, to bring a given dispute before the ICJ. Alternatively, *advance consent* can be established by treaties, often through an optional clause activated by declarations made by States.[6] Lastly, *consent to the compulsory jurisdiction of the ICJ* is granted by declaration of States under Article 36.2 of the Statute of the Court; it binds States to accept the jurisdiction of the Court for any future legal disputes in relation to other States declaring the same.[7]

As a form of *advance consent*, Article 14.2 of the UNFCCC enables any party to recognize the compulsory jurisdiction of the ICJ in relation to any subsequent dispute regarding the interpretation or application of the UNFCCC.[8] Yet, such a declaration has only been issued by Cuba and the Netherlands. A dispute between these two parties could theoretically be introduced by either of them without the agreement of the other. Even then, any claims related to the interpretation or application of the UNFCCC would be hindered by "the crushingly vague nature of the obligations, invariably drafted in such a way as to make it impossible to argue that any particular provision gives rise to a cause of action."[9] The Paris Agreement could

---

[3] Charter of the United Nations, June 26, 1945, 1 *UNTS* XVI, art. 92.

[4] See e.g. Daniel Bodansky, "The role of the International Court of Justice in addressing climate change: some preliminary reflections" (2017) 49 *Arizona State Law Journal* 689.

[5] Statute of the International Court of Justice, June 26, 1945, 3 *Bevans* 1179 (hereinafter ICJ Statute), art. 36.

[6] *Ibid.*, art. 36.

[7] *Ibid.*, art. 36.2.

[8] United Nations Framework Convention on Climate Change, May 9, 1992, 1771 *UNTS* 107 (hereinafter UNFCCC), art. 14.2. See also *ibid.*, art. 14.8; Kyoto Protocol to the United Nations Framework Convention on Climate Change, December 11, 1997, 2303 *UNTS* 162 (hereinafter Kyoto Protocol), art. 19; Paris Agreement, December 12, 2015, in the annex of decision 1/CP.21, "Adoption of the Paris Agreement" (December 12, 2015), art. 24.

[9] Philippe Sands, "Climate change and the rule of law: adjudicating the future in international law" (2016) 28:1 *Journal of Environmental Law* 19, at 28.

provide more specific causes of action, although national commitments remain generally far below the ambit of States' obligations under general international law.[10]

While *advance consent* is by nature limited to the scope of treaty law, *consent to the compulsory jurisdiction of the ICJ* could also allow contentious disputes relating to the rights and obligations of States under general international law. Seventy-two States have issued a declaration under Article 36.2 of the Statute of the ICJ.[11] The compulsory jurisdiction of the Court entailed by such declarations includes not only the interpretation of treaties, but also "any question of international law" and "the nature or extent of the reparation to be made for the breach of an international obligation."[12]

There would be multiple potential obstacles to such disputes. For instance, the existence of a genuine dispute between the parties must be established before the initiation of proceedings. In three parallel cases on the *Obligations Concerning Negotiations relating to Cessation of the Nuclear Arms Race and to Nuclear Disarmament* filed by the Marshall Islands, the ICJ called for evidence "that the respondent was aware, or could not have been unaware, that its views were 'positively opposed' by the applicant."[13] Participation in multilateral negotiations was not considered to be sufficient evidence. In a contentious case related to climate change, prior bilateral consultations could help establish the existence of a dispute.

Another difficulty would relate to the limitation of a contentious dispute to two parties, whereas climate change results from the concurrent conduct of multiple States whose consequences affect all States. Through what came to be called the "Monetary Gold" principle, the ICJ has refused to determine the obligation of a State if, in order to do so, "it would have to rule, as a prerequisite, on the lawfulness" of the conduct of a third State not party to the dispute.[14] On the other hand, in the case between Australia and Japan regarding *Whaling in the Antarctic*, the ICJ found unanimously that it had jurisdiction to entertain the application of a State affected but not directly injured by the violation of a multilateral obligation.[15]

The greatest obstacles to a contentious case before the ICJ might be political rather than legal. States generally have little inclination to engage in proceedings against one another to address global environmental problems, as such proceedings would involve diplomatic costs not justified by the achievement of any specific national interest.[16] Political obstacles

---

[10] On the nature of national commitments on mitigation in the Paris Agreement, see Chapter 7, section II.B.1. On the gap in national commitments, see Chapter 13, section III.

[11] A list is available on the website of the ICJ, "Declarations recognizing the jurisdiction of the Court as compulsory," www.icj-cij.org/en/declarations (accessed January 25, 2018).

[12] ICJ Statute, *supra* note 5, art. 36.2(b) and (d).

[13] ICJ, *Obligations Concerning Negotiations relating to Cessation of the Nuclear Arms Race and to Nuclear Disarmament (Marshall Islands v. United Kingdom)*, judgment of October 5, 2016, para. 41; *ibid. (Marshall Islands v. Pakistan)*, para. 38; *ibid. (Marshall Islands v. India)*, para. 38. This part of the judgment was adopted by eight votes to eight, with the President's casting vote.

[14] ICJ, *East Timor (Portugal v. Australia)*, judgment of June 30, 1995, para. 35. See also ICJ, *Monetary Gold Removed from Rome in 1943 (Italy v. France, United Kingdom and United States)*, judgment of June 15, 1954, at 32; ICJ, *Certain Phosphate Lands in Nauru (Nauru v. Australia)*, judgment of June 26, 1992 on preliminary objections, para. 55. See also, for a radically different approach, ITLOS, *Request for an Advisory Opinion Submitted by the Sub-Regional Fisheries Commission (SRFC)*, Advisory Opinion of April 2, 2015, para. 76 and declaration of Judge Cot, para. 7. See generally Phoebe Okowa, *State Responsibility for Transboundary Air Pollution in International Law* (Oxford University Press, 2000) 202.

[15] ICJ, *Whaling in the Antarctic (Australia v. Japan: New Zealand Intervening)*, judgment of March 31, 2014.

[16] Daniel Bodansky, "The United Nations Framework Convention on Climate Change: a commentary" (1993) 18:2 *Yale Journal of International Law* 451, at 543. In this regard, the case of *Whaling in the Antarctic* is a notable exception.

are greater in circumstances such as climate change where the most affected States are often among the weakest diplomatic powers. The ICJ itself might be reluctant to take a position on a strongly divisive issue. In the *Northern Cameroons* case, the ICJ noted that, even when it has jurisdiction, the Court could refuse to adjudicate on the merits if this "would be inconsistent with its judicial function."[17] A timid Court could use this or other ways to avoid taking a decision in a field which it deems too politically sensitive.

B. The Prospects for Advisory Proceedings before the ICJ

Besides contentious cases, requests for advisory opinions can also be referred to the ICJ. The UN General Assembly and Security Council can request the ICJ to give an advisory opinion "on any legal question." The General Assembly can also authorize other organs of the United Nations and specialized agencies to request advisory opinions on "legal question arising within the scope of their activities."[18] The ICJ "may," but is not under a strict obligation to give an advisory opinion in response to any such request.[19] Nevertheless, it has rarely declined to give an advisory opinion when requested to do so. Requests were entertained despite their far-reaching political implications, provided that the question asked was one of a legal nature.[20] The ICJ refused the World Health Organization's request for an opinion on the legality of nuclear weapons on the ground that the question did not arise "within the scope of the activity" of this specialized agency,[21] but, in a decision taken on the same day, it accepted a similar request from the General Assembly.[22]

An advisory opinion of the ICJ could be an important occasion for an authoritative interpretation of relevant norms of general international law, but it may also be a backlash. Although the ICJ may accept to give an advisory opinion, there is a risk that disproportionate jurisdictional precaution could lead to an excessively conservative interpretation of general international law. The ICJ has already disappointed those who hoped that an advisory opinion would contribute to the development of international law. In its advisory opinion on the *Legality of the Threat or Use of Nuclear Weapons*, it declared itself unable to "conclude definitively whether the threat or use of nuclear weapons would be lawful or unlawful in an extreme circumstance of self-defence, in which the very survival of a State would be at stake."[23] As Philippe Sands justly noted, "[t]hat unhappy conclusion – you might be acting lawfully in extinguishing others if you yourself face extinction – did not seem to bode well for the climate change issue."[24]

However, excessive GhG emissions are much more common among States than whaling, and a State seeking to initiate legal proceedings would face the hostile reaction of more than one counterpart.

[17] ICJ, *Northern Cameroons (Cameroon v. United Kingdom)*, judgment of December 2, 1963, at 37.

[18] UN Charter, *supra* note 3, art. 96.

[19] ICJ Statute, *supra* note 5, art. 65.1.

[20] See e.g. ICJ, *Accordance with International Law of the Unilateral Declaration of Independence in Respect of Kosovo*, Advisory Opinion of July 22, 2010, para. 27. See, however, PCIJ, *Status of Eastern Carelia*, Advisory Opinion of July 23, 1923, PCIJ Ser. B, No. 5.

[21] ICJ, *Legality of the Use by a State of Nuclear Weapons in Armed Conflict*, Advisory Opinion of July 8, 1996, para. 31.

[22] ICJ, *Legality of the Threat or Use of Nuclear Weapons*, Advisory Opinion of July 8, 1996.

[23] *Ibid.*, para. 105.2(E). See also *Advisory Opinion on Accordance with International Law of the Unilateral Declaration of Independence in Respect of Kosovo*, *supra* note 20.

[24] Sands, *supra* note 9, at 20.

The likelihood of a constructive advisory opinion would largely depend on the question(s) asked to the Court. Members of the Court may be reluctant to address questions of historical responsibility, given their great political sensitivity and the lack of direct practical relevance, unless perhaps emphasis were put on international support for climate action. It is likely that the ICJ would be more enthusiastic to venture into discussions regarding the obligations of States on climate change mitigation, including by interpreting relevant norms of general international law such as the no-harm principle and by analyzing the relation of these norms with the rules contained in international climate agreements.

Significant political support would need to be garnered before the UN General Assembly or Security Council decide to request an advisory opinion on questions relating to climate change. For instance, the General Assembly would need a vote, unless a decision could be reached by consensus, to request an advisory opinion from the ICJ. Palau once started a campaign for such a request, but suddenly discontinued it, seemingly after pressure was exercised by the United States, which provides a significant amount of development aid to Palau.[25] Among the UN specialized agencies that the General Assembly authorized to request an advisory opinion,[26] the most likely contenders would be the Food and Agriculture Organization, UNESCO and the WMO. A decision by any of these institutions, however, would need to be made by its plenary body and thus would face similar political difficulties.[27]

C. The Prospects for Adjudication under UNCLOS

UNCLOS provides an alternative venue for contentious or advisory proceedings.[28] The definition of "pollution to the marine environment" under the UNCLOS is sufficiently broad to include impacts of climate change on oceans, such as sea-level rise or ocean acidification.[29] The obligation of the parties to UNCLOS to "protect and preserve the marine environment"[30] could thus be interpreted as involving obligations similar to those arising out of the no-harm principle. Oceans, in other words, would appear as a proxy to bring harms to the climate system under the dispute settlement mechanism established by UNCLOS.

UNCLOS created a robust dispute settlement mechanism. When ratifying UNCLOS, States could indicate their preference for a dispute settlement procedure between ICJ, arbitrage, or ITLOS.[31] In a dispute between parties which have indicated different preferences or have not indicated any preference, arbitrage is the default procedure unless a different agreement is reached by the parties to the dispute.

---

[25] See Stuart Beck and Elizabeth Burleson, "Inside the system, outside the box: Palau's pursuit of climate justice and security at the United Nations" (2014) 3:1 *Transnational Environmental Law* 17, at 26.

[26] See "Jurisdiction of the court and the procedure followed by it" (2014–15) 69 *International Court of Justice Yearbook* 52, at 64–68; "Organs and institutions authorized to request advisory opinions of the court" (Annex 13(A)) (2012–2013) 67 *International Court of Justice Yearbook* 166.

[27] See e.g. *Agreement between the United Nations and the World Meteorological Organization, Approved by the General Assembly of the United Nations*, December 20, 1951, 123 *UNTS* II-415, art. VII.3.

[28] United Nations Convention on the Law of the Sea, December 10, 1982, 1833 *UNTS* 3, part XV. See e.g. Roda Verheyen and Cathrin Zengerling, "International dispute settlement" in John S. Dryzek, Richard B. Norgaard and David Schlosberg (eds.), *The Oxford Handbook of Climate Change and Society* (Cambridge University Press, 2011) 417.

[29] United Nations Convention on the Law of the Sea, *supra* note 28, art. 1.1(4). Pollution of the marine environment is defined as "the introduction by man, directly or indirectly, of substances or energy into the marine environment, including estuaries, which results or is likely to result in such deleterious effects as harm to living resources and marine life, hazards to human health, hindrance to marine activities, including fishing and other legitimate uses of the sea, impairment of quality for use of sea water and reduction of amenities."

[30] *Ibid.*, art. 192.

[31] *Ibid.*, art. 287.

ITLOS is a permanent tribunal seated in Hamburg, Germany.[32] Beyond the dispute settlement mechanism included in UNCLOS, Article 138 of the Rules of ITLOS allows this tribunal to give an advisory opinion on a legal question "if an international agreement related to the purposes of the Convention specifically provides for the submission to the Tribunal of a request for such an opinion."[33] This rather unique provision, added at a late stage of the drafting, creates a jurisdictional avenue through which an authoritative response could be provided to any relevant legal question asked under any related international agreement, whether multilateral or bilateral, provided that this agreement is "related to the purposes" of UNCLOS.[34] Thus, in 2013, ITLOS unanimously recognized its jurisdiction to issue an advisory opinion in response to a request by the Sub-Regional Fisheries Commission (SRFC) that seven Western African States had established through a sub-regional agreement.[35]

Likewise, Article 138 of the Rules of ITLOS could offer an interesting opportunity for an advisory opinion on the application of the obligation of States to protect the marine environment in the context of climate change. It is far from inconceivable that two or several States favorable to an international adjudication on climate change could enter an agreement and request an advisory opinion from ITLOS. Yet, upon finding that it has jurisdiction, ITLOS is not bound to give an advisory opinion[36] and it may be reluctant to do so when an agreement has been formed with the sole purpose of requesting an advisory opinion on the obligations of third parties. Even if the advisory opinion was requested by a large pre-existing organization, ITLOS could refuse to entertain a request for an advisory opinion whose implications would extend far beyond the scope of the agreement from which it arises.[37] In his separate opinion accompanying the advisory opinion requested by the SRFC, Judge Cot highlighted "dangers of abuse and manipulation," suggesting that ITLOS should exercise its discretionary power not to give an advisory opinion in cases where the organization requesting it "seek[s] to gain an advantage over third States and thereby place[s] the Tribunal in an awkward position."[38]

II. DOMESTIC ADJUDICATION

International courts and tribunals do not have a monopoly on international law adjudication. National courts can and do play a role.[39] This role varies between jurisdictions and legal

---

[32] See *ibid.*, Annex VI.

[33] ITLOS, *Rules of the Tribunal* (as of March 17, 2009), doc. ITLOS/8, art. 138.1.

[34] See José Luis Jesus, "Commentary under Article 138" in P. Chandrasekhara Rao and P. Gautier (eds.), *The Rules of the International Tribunal for the Law of the Sea: A Commentary* (Martinus Nijhoff, 2006) 393, at 394; Ki-Jun You, "Advisory opinions of the International Tribunal for the Law of the Sea: Article 138 of the rules of the tribunal, revisited" (2008) 39 *Ocean Development & International Law* 360.

[35] ITLOS, *Request for an Advisory Opinion Submitted by the Sub-Regional Fisheries Commission (SRFC)*, Advisory Opinion of April 2, 2015. See Michael A. Becker, "Request for an advisory opinion submitted by the Sub-Regional Fisheries Commission (SRFC)" (2015) 109:4 *American Journal of International Law* 851.

[36] This is implied by the word "may" in ITLOS, *Rules of the Tribunal, supra* note 33, art. 138.1.

[37] See, by analogy, *Status of Eastern Carelia, supra* note 20, where the PCIJ declined a request of the Council of the League of Nations for an advisory opinion on a dispute between Finland and the USSR on the ground of an objection by the USSR. However, several subsequent cases have been distinguished from *Eastern Carelia*. See generally James Crawford, *Brownlie's Principles of Public International Law*, 8th edn (Oxford University Press, 2012) 730.

[38] ITLOS, *Request for an Advisory Opinion Submitted by the Sub-Regional Fisheries Commission (SRFC)*, Declaration of Judge Cot annex to the Advisory Opinion of April 2, 2015, para. 9.

[39] See generally Brian J. Preston, "The contribution of the courts in tackling climate change" (2016) 28:1 *Journal of Environmental Law* 11. See also Brian J. Preston "Climate change litigation (part 1)" (2011) 5:1 *Carbon & Climate Law Review* 3; Bryan J. Preston "Climate change litigation (part 2)" (2011) 5:2 *Carbon & Climate Law Review* 244.

traditions. In countries with a strong litigation tradition such as the United States and Australia, the many cases that have been brought before domestic courts often focus on common law doctrines such as the tort of nuisance.[40] By contrast, international law could be more central to litigation within monist countries of a civil law tradition.

Here, it is only possible to give a very broad overview of the vast variety of cases occurring before municipal courts. Disputes can be concerned with different aspects of climate laws and policies, including mitigation, adaptation and response measures. They can consist in public law litigation such as judicial review or claims of administrative responsibility, or in private law litigation on tort or "extra-contractual" responsibility. They take place in very diverse legal systems and traditions. Beyond the legal outcome of a case, domestic litigation may be used as a tool for advocacy in attempts to increase public scrutiny on national authorities.

A. Litigation Addressing the Causes of Climate Change

Litigation has repeatedly sought to promote climate change mitigation. Climate change has thus been invoked in a host of disputes on projects ranging from the construction of pipelines to that of power plants or airports.[41] The cases which have had the greatest influence on the conduct of States, however, concerned more generally the national laws and policies on climate change mitigation and their implementation. Three major public law cases are briefly explored in the following: *Massachusetts* v. *Environmental Protection Agency* in the United States, *Urgenda* v. *The Netherlands* in the Netherlands and *Earthlife Africa Johannesburg* v. *Minister of Energy* in South Africa.

Massachusetts v. *Environmental Protection Agency* regarded the petition of several States, local governments and environmental organizations claiming that the US Environmental Protection Agency (EPA) has an obligation to regulate GhG emissions from new motor vehicles.[42] Under the Clean Air Act, the EPA must adopt standards applicable to new motor vehicles on any air pollutants "which in [its] judgment cause, or contribute to, air pollution which may reasonably be anticipated to endanger public health or welfare."[43] In its 2007 judgment in this case, the US Supreme Court concluded that this imposed an obligation on the EPA to regulate GhG emissions from new motor vehicles, unless the EPA could provide "some reasonable explanation" not to regulate such vehicles.[44] While this case did not rely on international law, the Court disregarded the EPA's defense based on the need not to interfere with ongoing negotiations as "impermissible considerations."[45] This case illustrates the existence of multiple provisions in national laws demanding national authorities to act reasonably, consistently with the norms of general international law.

---

[40] See Jacqueline Peel and Hari M. Osofsky, *Climate Change Litigation: Regulatory Pathways to Cleaner Energy* (Cambridge University Press, 2015). See also William C.G. Burns and Hari M. Osofsky, *Adjudicating Climate Change: State, National, and International Approaches* (Cambridge University Press, 2011); Ken Alex, "A period of consequences: global warming as public nuisance" (2007) 26 *Stanford Journal of Environmental Law* 77.

[41] See generally Richard Lord *et al.* (eds.), *Climate Change Liability: Transnational Law and Practice* (Cambridge University Press, 2012).

[42] US Supreme Court, *Massachusetts* v. *Environmental Protection Agency*, judgment of April 2, 2007, 549 *US* 497.

[43] Clean Air Act, 42 U.S.C.A. § 7521(a)(1), para. 202(a)(1).

[44] *Massachusetts* v. *Environmental Protection* Agency, *supra* note 42, at 534.

[45] *Ibid.*, at 501.

By contrast, *Urgenda* v. *The Netherlands*, decided by the District Court of The Hague in June 2015, was directly concerned with the international obligations of the Netherlands.[46] The Urgenda Foundation argued that the national policy target of a 14–17 percent reduction in emissions by 2020 on a 1990 baseline was insufficient for the Netherlands to comply with its international legal obligations. The existence of State obligations under the no-harm principle was not contested by the Dutch government.[47] Relying extensively on the work of the IPCC and on various statements in international negotiations, the District Court considered that the Netherlands was under an international legal obligation to reduce its GhG emissions by at least 25 percent by 2020 on a 1990 baseline.[48] Accordingly, the District Court ordered the State to adopt a national policy target of 25 percent emission by 2020 on a 1990 baseline. The Dutch government appealed against this decision.[49] The *Urgenda* decision inspired similar actions in different countries.[50]

Another case indirectly based on international law, *Earthlife Africa Johannesburg* v. *Minister of Energy*, regarded the procedural obligations of a State to consider the impact of a project on climate change before allowing its implementation. In many countries, projects likely to have a significant environmental impact can only be allowed after undergoing an Environmental Impact Assessment (EIA), a detailed and public process through which such impacts are documented and discussed.[51] The UNFCCC requires all parties to "take climate change considerations into account ... in their relevant social, economic and environmental policies and actions,"[52] but not all EIA legislation clearly includes climate change considerations.[53] When the government of South Africa authorized the construction of a large coal-fired power plant based on an EIA which did not consider its impacts on climate change, Earthlife applied for

[46] District Court of the Hague, *Urgenda Foundation* v. *The State of the Netherlands*, judgment of June 24, 2015. An unofficial translation is available at www.urgenda.nl/documents/VerdictDistrictCourt-UrgendavStaat-24.06.2015.pdf (accessed January 25, 2018). See generally Marjan Peeters, "Case note: *Urgenda Foundation and 886 Individuals* v. *The State of the Netherlands*: The dilemma of more ambitious greenhouse gas reduction action by EU Member States" (2016) 25:1 *Review of European Community & International Environmental Law* 123; K.J. de Graaf and J.H. Jans, "The Urgenda decision: Netherlands liable for role in causing dangerous global climate change" (2015) 27:3 *Journal of Environmental Law* 517; Jolene Lin, "The first successful climate negligence case: a comment on *Urgenda Foundation v. The State of the Netherlands (Ministry of Infrastructure and the Environment)*" (2015) 5:1 *Climate Law* 65; Josephine van Zeben, "Establishing a governmental duty of care for climate change mitigation: will *Urgenda* turn the tide?" (2015) 4:2 *Transnational Environmental Law* 339.

[47] *Urgenda Foundation*, *supra* note 46, para. 4.39.

[48] *Ibid.*, para. 4.29. See also para. 4.42.

[49] Government of the Netherlands, "Cabinet begins implementation of Urgenda ruling but will file appeal" (September 1, 2015), www.government.nl/latest/news/2015/09/01/cabinet-begins-implementation-of-urgenda-ruling-but-will-file-appeal (accessed January 25, 2018).

[50] These include the "Klimaatzaak" in Belgium (see, in French, "Le Procès" on the website of L'Affaire Climat, www.klimaatzaak.eu/fr/le-proces/#klimaatzaak (accessed January 25, 2018)); the Magnolia case in Sweden (see press release of Magnoliamalet on September 15, 2015, www.xn--magnoliamlet-1cb.se/news/press-release-in-english-2016-09-15 (accessed January 25, 2018)); and the case of Rabab Ali in Pakistan (see Naeem Sahoutara, "Seven-year-old girl takes on federal, Sindh governments" *Express Tribune* (July 31, 2017)).

[51] See generally Philippe Sands and Jacqueline Peel, *Principles of International Environmental Law*, 3rd edn (Cambridge University Press, 2012) 601 *et seq.*; Neil Craik, *The International Law of Environmental Impact Assessment* (Cambridge University Press, 2011); Jane Holder, *Environmental Assessment: The Regulation of Decision Making* (Oxford University Press, 2006).

[52] UNFCCC, *supra* note 8, art. 4.1(f).

[53] For an example of a reference to climate change in an EIA legislation, see EU Directive 2011/92/EU on the assessment of the effects of certain public and private projects on the environment (December 13, 2011), doc. 02011L0092, art. 3.1(c).

a judicial review against this decision. The High Court of South Africa settled the case in 2017 in favor of Earthlife, considering that national legislation had to be interpreted consistently with international law, including South Africa's obligations under the UNFCCC.[54] The Court concluded that an assessment of the plant's impacts on the climate was "necessary and relevant to ensuring that the proposed coal-fired power station fits" within South Africa's commitments.[55] A similar reasoning led an Austrian administrative court to strike down the plan for a third runway at Vienna International Airport, although this judgment was later overruled by the Constitutional Court.[56]

Climate change mitigation can also be promoted by litigation brought against private companies. In common law countries, nuisance action has sometimes sought to impose injunctions to reduce GhG emissions, although such cases appear unlikely to succeed when a legislative framework is already in place.[57] Beyond nuisance, the Commission on Human Rights of the Philippines has initiated an investigation regarding the responsibility of carbon majors for human rights violations or threats of violations constituted by the impacts of climate change.[58] In the United States, criminal prosecution could be initiated against large fossil fuel companies who sought to deceive the public or their investors by concealing or denying information.[59]

B. Litigation Addressing the Impacts of Climate Change

Litigation has also occasionally been used as a way to spur responses to the impacts of climate change, for instance, through adaptation action or approaches to address loss and damage.[60] Here too, claimants have alternatively invoked the responsibilities of national authorities, as in *Ashgar Leghari* v. *Pakistan*, and the responsibilities of private actors, as in *Native Village of Kivalina* v. *ExxonMobil Corporation*.

In *Ashgar Leghari* v. *Pakistan*, the High Court of Lahore was approached through public interest litigation claiming that the Pakistani government had failed to implement its National Climate Change Policy and its Framework for Implementation of Climate Change Policy. A Ministry for Climate Change had been established, but its representatives before

---

[54] High Court of South Africa (Western Cape Division), *Earthlife Africa Johannesburg* v. *Minister of Energy*, judgment of April 26, 2017, [2017] *ZAWCHC* 50, para. 83.

[55] *Ibid.*, para. 90.

[56] See Austria Bundesverwaltungsgericht (Administrative Court), case W109 2000179-1/291E, judgment of February 2, 2017, www.bvwg.gv.at/amtstafel/291_ERKENNTNIS_2.2.17_ee.pdf?5spp26 (accessed January 25, 2018); overturned by the Verfassungsgerichtshof (Constitutional Court), case E 875/2017, judgment of June 29, 2017, at www.vfgh.gv.at/downloads/VfGH_E_875-2017_Verkuendungstext_Flughafen.pdf (accessed January 25, 2018).

[57] See in particular US Supreme Court, *American Electric Power* v. *Connecticut*, June 20, 2011, 131 *S.Ct.* 2527. New attempts could lead to new developments. See Michael Burger, "Local governments in California file common law claims against largest fossil fuel companies" *Climate Law Blog* (July 18, 2017), http://blogs.law.columbia.edu/climatechange/2017/07/18/local-governments-in-california-file-common-law-claims-against-largest-fossil-fuel-companies (accessed January 25, 2018).

[58] See e.g. Greenpeace, "Shell to face pressure at AGM for failing to take responsibility for climate-related human rights harms," press release, May 23, 2017, www.greenpeace.org/seasia/ph/press/releases/Shell-to-face-pressure-at-AGM-for-failing-to-take-responsibility-for-climate-related-human-rights-harms (accessed January 25, 2018).

[59] Justin Gillis and Clifford Krauss, "Exxon Mobil investigated for possible climate change lies by New York Attorney General" *New York Times* (November 5, 2015).

[60] Brian Preston, *The Role of the Courts in Facilitating Climate Change Adaptation* (Asia-Pacific Centre for Environmental Law Climate Change Adaptation Platform, 2016) 3–4.

the Court "frankly" acknowledged that little had been implemented and that awareness in other ministries was lacking.[61] In an order on September 4, 2015, Judge Syed Mansoor Ali Shah noted that "the delay and lethargy of the State in implementing" adaptation policies had adverse implications for the constitutional rights of the citizens.[62] The Court directed that concerned ministries appoint focal persons and define action points, while also establishing an inter-ministerial Climate Change Commission to assist the Court in monitoring progress.[63] In a second order ten days later, the High Court of Lahore decided on the composition and terms of reference of the Climate Change Commission.[64] This case illustrates the role that national jurisdictions can play in ensuring that national policies adopted by States in the application of their international commitments do not remain pure window-dressing.[65]

Native Village of Kivalina v. *ExxonMobil Corporation* regarded a coastal Alaskan village which, due to the impacts of climate change on the sea-ice, had become increasingly affected by erosion and storm surge. The population has been considering a collective resettlement plan for 20 years. The village initiated an action in nuisance against 24 major oil, energy and utility companies. Relying on the US political question doctrine, the District Court for the Northern District of California considered that such claims were not for a court to decide.[66] The Court of Appeal for the Ninth Circuit arrived at the same conclusion based on its understanding that the action in nuisance had been displaced by the adoption of specific legislation, namely the Clean Air Act.[67] The Supreme Court denied review of the case.[68] While Kivalina lost its unlikely legal action against the world's largest polluters, it gained considerable media coverage in the process and contributed to raising awareness of the impacts of climate change within developed countries.[69]

Another type of legal action regards the protection of migrants. The impacts of climate change contribute to increased international migration, although this is generally in rather indirect ways. Some so-called "climate refugees" – individuals forced to move abroad due to poverty exacerbated by the adverse impacts of climate change – have claimed asylum in third countries. Such claims have generally been dismissed based on a literal interpretation of the Convention Relating to the Status of Refugees or any related national legislation which

---

[61] High Court of Lahore, *Ashgar Leghari* v. *Federation of Pakistan*, order of September 4, 2015, https://elaw.org/system/files/pk.leghari.090415_0.pdf (accessed January 25, 2018), para. 3.

[62] *Ibid.*, para. 8.

[63] *Ibid.*

[64] *Ibid.*

[65] See e.g. Jolene Lin, "Litigating climate change in Asia" (2014) 4:1–2 *Climate Law* 140.

[66] US District Court, N.D. California, *Native Village of Kivalina* v. *ExxonMobil*, decision of September 30, 2009, 663 F.Supp.2d 863 (N.D.Cal. 2009).

[67] US Court of Appeals for the Ninth Circuit, *Native Village of Kivalina* v. *ExxonMobil*, decision of September 21, 2012, 696 F.3d 849 (9th Cir. 2012). See also *American Electric Power* v. *Connecticut*, *supra* note 57.

[68] US Supreme Court, *Native Village of Kivalina* v. *ExxonMobil*, (2013) 133 S.Ct. 2390. For commentaries on the Kivalina litigation generally, see Qin M. Sorenson, "*Native Village of Kivalina v. ExxonMobil Corp.*: the end of 'climate change' tort litigation?" *Trends* (January–February 2013); Nicole Johnson, "*Native Village of Kivalina v. ExxonMobil Corp*: say goodbye to federal public nuisance claims for greenhouse gas emission" (2013) 40 *Ecology Law Quarterly* 557.

[69] See e.g. Christine Shearer, *Kivalina: A Climate Change Story* (Haymarket Books, 2011).

confines international protection to a narrow category of refugees fleeing persecution rather than misery.[70]

C. Litigation Addressing the Impacts of Response Measures

Litigation can also hinder climate action, including measures which promote climate change mitigation or adaptation to climate change. Response measures often have unintended adverse consequences on populations and the environment as well as economic costs borne by the taxpayer. Large hydroelectric projects, which have been justified for mitigation or sometimes for adaptation purposes, have often given rise to protracted legal disputes between displaced populations and project contenders.[71] In the United States, more than 100 lawsuits were engaged against the EPA when it adopted regulations on GhG emissions after having been urged to do so by the Supreme Court in *Massachusetts* v. *Environmental Protection Agency*.[72] Some disputes on measures taken in response to climate change have no doubt helped to improve the quality of these response measures by ensuring that climate change mitigation and adaptation would not come at a disproportionate price for vulnerable populations, but other disputes might also have simply sought to protect vested interests in the carbon economy by delaying much-needed action.[73]

III. CONCLUSION

Litigation, in particular domestic litigation, is playing a growing role in the development of the international law on climate change and more generally in the promotion of action on climate change. Following the adoption of more ambitious international agreements, some municipal courts have come to feel more confident in recognizing the obligation of national authorities and private actors to comply with relevant obligations. The case of *Urgenda* v. *The Netherlands*, although quite unique, shows that a domestic court could rely on general international law to impose more ambition in the policy of a State on climate change mitigation. This could be the beginning of a trend. When national governments and parliaments fail, by and large, to comply with the norms of general international law, courts could be a last resort for meaningful steps forward.

---

[70] See, for instance, Supreme Court of New Zealand, *Teitota* v. *Chief Executive of the Ministry of Business, Innovation and Employment*, judgment of July 20, 2015, [2015] NZSC 107, www.courtsofnz.govt.nz/cases/ioane-teitiotoa-v-the-chief-executive-of-the-ministry-of-business-innovation-and-employment/at_download/fileDecision (accessed January 25, 2018). See also Convention relating to the Status of Refugees, July 28, 1951, 189 *UNTS* 150, art. 1(A)(2).

[71] See, for instance (although climate change only appeared as a secondary motivation in the projects at issue), Supreme Court of India, *Narmada Bachao Andolan* v. *Union of India*, judgment of March 15, 2005; Tokyo High Court, decision of December 26, 2012 on the Kotopajang dam, discussed in "Statement of protest: we strongly denounce the Tokyo High Court's unfair judgment" (Support Action Center for Kotopanjang Dam Victims, January 10, 2013).

[72] See in particular US Supreme Court, *Utility Air Regulatory Group* v. *EPA*, June 23, 2014, (2014) 573 *US* 573. See generally Lord Carnwath JSC, "Climate change adjudication after Paris: a reflection" (2016) 28:1 *Journal of Environmental Law* 5, at 6.

[73] See e.g. US District Court, N.D. Texas, Fort Worth Division, *ExxonMobil* v. *Healey*, decision of October 13, 2016, Civil Action No. 4:16-CV-469-K.

15

Non-State Actors

International law has long been framed as the law applicable to relations between sovereign nations. This "Westphalian"[1] paradigm continues to prevail, only slightly nuanced by the development of international organizations and the establishment of new fields of international law focusing on more "internal" matters such as the protection of human rights. Accordingly, the international law on climate change generally consists of obligations imposed on States, either based on State consent (treaties) or on the general practice of States accepted as law (customs). Yet, climate change is not just about sovereign rights and obligations; it concerns just about everyone. Making responses to climate change effective requires the fostering of participation and engagement by all relevant non-State actors.

Non-State actors encompass any individuals and any groups, such as businesses, non-profit organizations, subnational administrations or international organizations, the only exception being sovereign States. While some groups are specifically constituted in order to advance particular climate-related action, many other non-State actors are concerned or interested by some impacts of climate change or some response measures, actual or potential, as part of their wider operations. Non-State actors bring rich experience and various ideas to deliberations on climate change, but inevitably they also come with their own vested interests.

Non-State actors may become involved in responses to climate change in two distinct ways. Firstly, they can participate in the making of the law on climate change, including in a formal fashion through a status as an observer organization admitted under the UNFCCC, and thus attempt to inform or influence intergovernmental negotiations. Secondly, non-State actors can also take on voluntary initiatives, for instance, by reducing GhG emissions in their activities, or by taking part in international cooperative initiatives.

I. PARTICIPATION IN THE MAKING OF THE INTERNATIONAL LAW ON CLIMATE CHANGE

Non-State actors play an important role in the making of the international law on climate change. They are central to raising awareness on the obligations of States under general

---

[1] By reference to the 1648 Peace of Westphalia, a series of treaties marking the end of the Thirty Years' War, broadly remembered as recognizing States with exclusive sovereignty within a territory and a population.

international law, for instance, through the discourse on "climate justice."[2] In the absence of an authoritative statement of the relevant general international law, it is mostly down to non-State actors such as scholars and advocates to identify and interpret relevant norms.[3] Although not formally recognized by State-centered processes, non-State actors certainly contribute to the formation of customary international law in many more or less direct ways, most obviously through advocacy and domestic litigation influencing both State practice and its acknowledgment as law.[4] On the other hand, non-State actors have also lobbied against climate action, in particular through the Global Climate Coalition, a now-disbanded coalition of oil companies, car manufacturers and other utility companies opposed to mandatory measures on climate change mitigation.[5]

The role of non-State actors is also particularly important within the UNFCCC regime. To promote climate action, the UNFCCC regime relies on non-State actors to persuade national governments to take steps toward compliance with their obligations under general international law. Without NGOs and media coverage, national governments would surely be far less interested in complying with national commitments. The participation of non-State actors in reviewing ambition, implementation and compliance could be instrumental to the success of the Paris Agreement.[6]

The participation of civil society organizations has been recognized from the outset of the international climate change negotiations, although it has always come with conditions. "Relevant non-governmental organizations" were invited to "make contributions" to the negotiations held toward the adoption of a Framework Convention on Climate Change, "on the understanding that these organizations shall not have any negotiating role during the process."[7] States were willing to involve non-State actors, but they were also concerned to confine them to a purely consultative role. As the COP recalled at its fourth session, "negotiations under the Convention are a matter for the Parties."[8] In addition, some non-democratic States have sometimes appeared reluctant to allow genuinely independent national actors to participate in international negotiations.[9] Progressively, however, the COP placed more emphasis on "the need to engage a broad range of stakeholders at the global, regional, national and local levels, be they government, including subnational and local administrations, private business or civil society, including youth and persons with disability."[10]

---

[2] See e.g. Mary Finley-Brook, "Climate justice advocacy" *Public Diplomacy Magazine* (June 2, 2016); Michael Mintrom and Joannah Luetjens, "Policy entrepreneurs and problem framing: the case of climate change" (2017) 35:5 *Environment and Planning C: Politics and Space* 1362.

[3] See Steven Wheatley, *The Democratic Legitimacy of International Law* (Hart Publishing, 2010) 150.

[4] See ILC, *First Report on the Identification of Customary International Law by Special Rapporteur Michael Wood* (May 17, 2013), doc. A/CN.4/663, para. 98; Hilary Charlesworth, "The unbearable lightness of customary international law" (1998) 92 *Proceedings of the Annual Meeting (American Society of International Law)* 44.

[5] See e.g. Harriet Bulkeley and Peter Newell, *Governing Climate Change* (Routledge, 2015) 115.

[6] See Harro van Asselt, "The role of non-state actors in reviewing ambition, implementation, and compliance under the Paris Agreement" (2016) 6:1–2 *Climate Law* 91; Joseph Szarka, "From climate advocacy to public engagement: an exploration of the roles of environmental non-governmental organisations" (2013) 1:1 *Climate* 12.

[7] UN General Assembly Resolution 45/212, "Protection of global climate for present and future generations of mankind" (December 21, 1990), para. 19.

[8] Decision 18/CP.4, "Attendance of intergovernmental and non-governmental organizations at contact groups" (November 2, 1998), recital 3.

[9] See, for instance, the submission of Egypt in UNFCCC Subsidiary Body for Implementation, reproduced in UNFCCC SBI, *Mechanisms for Consultation with Non-governmental Organizations (NGOs)* (October 10, 1997), doc. FCCC/SBI/1997/MISC.7, 3.

[10] Decision 1/CP.16, "The Cancún Agreements: outcome of the work of the Ad Hoc Working Group on Long-Term Cooperative Action under the Convention" (December 10–11, 2010), para. 7.

To attend negotiations under the UNFCCC regime, an individual needs to be introduced by a civil society organization which, itself, needs to be accredited as an observer organization under the UNFCCC. A unique accreditation system applies to the UNFCCC, the Kyoto Protocol[11] and, in all likelihood, the Paris Agreement.[12] Two categories of observer organizations can be accredited. The first category consists mainly of international organizations and non-States Parties, namely "the United Nations, its specialized agencies and the International Atomic Energy Agency as well as any State member thereof or observers thereto not Party to the Convention."[13] The second, much larger category includes "[a]ny body or agency, whether national or institutional, governmental or non-governmental, which is qualified in matters covered by the Convention."[14] In order to be accredited as observer organizations, these institutions need to "inform ... the secretariat of [their] wish to be represented."[15]

In practice, an organization which wishes to be accredited needs to do much more than simply "inform" the UNFCCC Secretariat. It presently takes at least 18 months for a non-State actor to be admitted as an observer.[16] Criteria to assess whether a particular non-State actor is "qualified in matters covered by the Convention" were established by the Secretariat under the supervision of the COP. Some have proposed allowing individuals, in particular scholars, to be admitted as observers in their individual quality,[17] but the proposal has not been taken further.[18] In the practice of the UNFCCC Secretariat, an applicant cannot be admitted as an observer unless it has an independent juridical personality – independent, in particular, from any national government.[19] While the applicant must be a not-for-profit entity,[20] hundreds of not-for-profit business coalitions were admitted as observers.[21]

As of December 2016, a total of around 2,000 NGOs and 120 international organizations have been admitted as observer organizations to the UNFCCC. These observer organizations encompass a great diversity of horizons and objectives. To facilitate the coordination of admitted NGOs, nine overarching constituencies have been created, gathering observer organizations with similar functions (see Table 15.1).[22]

---

[11] See Kyoto Protocol to the United Nations Framework Convention on Climate Change, December 11, 1997, 2303 *UNTS* 162 (hereinafter Kyoto Protocol), art. 13.8; and decision 36/CMP.1, "Arrangements for the Conference of the Parties serving as the meeting of the parties to the Kyoto Protocol at its first session" (December 9–10, 2005), para. 2(b).

[12] See Paris Agreement, December 12, 2015, in the annex of decision 1/CP.21, "Adoption of the Paris Agreement" (December 12, 2015), art. 16.8.

[13] United Nations Framework Convention on Climate Change, May 9, 1992, 1771 *UNTS* 107 (hereinafter UNFCCC), art. 7.6.

[14] *Ibid.*

[15] *Ibid.* See also UNFCCC, *Draft Rules of Procedure of the Conference of the Parties and its Subsidiary Bodies*, UNFCCC document FCCC/CP/1996/2 (May 22, 1996), rules 6 and 7.

[16] See in particular UNFCCC Secretariat, "Standard admission process for non-governmental organizations NGOs)" (n.d.), http://unfccc.int/files/parties_and_observers/observer_organizations/application/pdf/updated_standard_admission_policy_ngos_english.pdf (accessed January 25, 2018).

[17] UNFCCC SBI, *Mechanisms for Consultations with Non-governmental Organizations. Addendum: The Participation of NGOs in the Convention Process* (June 11, 1997), doc. FCCC/SBI/1997/14/Add.1, para. 14.

[18] This proposal would require a revision of the UNFCCC as it is inconsistent with the UNFCCC, *supra* note 13, art. 7.6.

[19] UNFCCC SBI, *Promoting Effective Participation in the Convention Process: Note by the Secretariat* (April 16, 2004), doc. FCCC/SBI/2004/5, paras. 8–10.

[20] UNFCCC SBI, *supra* note 17, at para. 3.

[21] *Ibid.*, at para. 15.

[22] *Ibid.*, at para. 6.

Table 15.1 Constituency affiliation of observer organizations accredited under the UNFCCC as of December 2016[a]

Constituency	Number of members (*percentage*)
Business and Industry NGOs (BINGOs)	279 (*14%*)
Environmental NGOs (ENGOs)	825 (*41%*)
Farmers NGOs	17 (*1%*)
Indigenous Peoples Organizations (IPOs)	52 (*3%*)
Local Government and Municipal Authorities (LGMAs)	30 (*1%*)
Research and Independent NGOs (RINGOs)	500 (*25%*)
Trade Union NGOs (TUNGOs)	14 (*1%*)
Women and Gender	22 (*1%*)
Youth NGOs (YOUNGOs)	63 (*3%*)
Non-affiliated	199 (*10%*)
Total NGOs	2,001

[a] UNFCCC, "Statistics on observer organizations in the UNFCCC process" (n.d.), http://unfccc.int/parties_and_observers/observer_organizations/items/9545.php#constituency%20affiliation (accessed December 2016).

The extent and modalities of the participation of observer organizations in UNFCCC negotiations constitutes a constant point of contention. Disagreements concern, for instance, the possibility for observer organizations to attend informal negotiating meetings, intervene in meetings, access documentations and submit their views whenever States Parties are invited to provide written submissions.[23] Over time, it seems that the balance has gradually tilted in favor of the greater participation of non-State actors. Contact groups were opened to observer organizations at the discretion of the presiding officers of the Convention bodies unless one-third of the parties present object.[24] New technologies have reduced the cost of providing access to documentation to and accepting written submissions by observer organizations. As these non-State actors were repeatedly called upon to contribute to the realization of the objective of the Convention,[25] their participation gained legitimacy, reinforcing expectations that they would be associated with every process within the UNFCCC regime.[26] On-site side-events and exhibits were organized as formal opportunities for observer organizations to convey their input to negotiators, in addition to many less formal interactions.

Observer organizations certainly contribute to informing international negotiations and raising awareness while making information and analyses readily available to a wider audience, thus contributing to the naming and shaming of non-cooperating governments.[27] Nevertheless, a report of the UNFCCC Secretariat claiming that NGO participation "is both flexible and active, supporting the global trend towards more informed, participatory and

[23] UNFCCC SBI, *supra* note 17.

[24] Decision 18/CP.4, *supra* note 8, para. 1.

[25] See e.g. resolution 1/CP.4, "Solidarity with Central America" (November 14, 1998), para. 3; resolution 1/CP.6, "Solidarity with southern African countries, particularly with Mozambique" (November 25, 2000), paras. 2 and 3.

[26] See Sébastien Duyck, "MRV in the 2015 Climate Agreement: promoting compliance through transparency and the participation of NGOs" (2014) 9:3 *Carbon & Climate Law Review* 175.

[27] See, for instance, the website of Climate Action Tracker, http://climateactiontracker.org (accessed January 25, 2018); the website of Climate Analytics, http://climateanalytics.org (accessed January 25, 2018); and the website of the WRI's Paris Agreement tracker, www.wri.org/resources/maps/paris-agreement-tracker (accessed January 25, 2018). See also, generally, Kentaro Tamura, Takeshi Kuramochi and Jusen Asuka, "A process for making nationally-determined mitigation contributions more ambitious" (2013) 7:4 *Carbon & Climate Law Review* 231.

responsible societies"[28] appears as a slight overstatement. The greatest limitation on the participation of non-State actors under the UNFCCC remains their blatant lack of representativeness. About three-quarters of the observer organizations are based in developed countries. In particular, just three countries – the United States, the United Kingdom and Germany – host more than one-third of observer organizations. Among developed as well as developing countries, English-speaking countries appear significantly better represented than others. While many NGOs from developing countries apply for accreditation when a session of the COP is held in a developing State, their representatives are less likely to join subsequent sessions in other venues.[29] Financial difficulties explain some of this gap. In addition, the accreditation process might be more challenging in a developing State with weaker institutions, where the required documentation needs to be translated into an official UN language or where few NGOs are genuinely and demonstrably independent of the government.

II. VOLUNTARY COMMITMENTS

The role of non-State actors in responses to climate change is not limited to influencing State-led processes. In addition, non-State actors participate directly in the implementation of climate actions, especially with regard to climate change mitigation. Often, such actions are mandated by relevant national laws, but non-State actors can and do also take extra steps on a voluntary basis and, often, seek an international acknowledgement of their contribution. In recent years, a broader understanding of the limitations on States' efforts led to an expansion of voluntary commitments by non-State actors, including businesses and financial institutions as well as subnational authorities and charities. Various institutional arrangements were established to galvanize effective climate action by non-State actors through what some have called a "hybrid multilateralism."[30] Standard-setting initiatives and broader international cooperative initiatives developed rules to measure, report and verify the integrity and the effective implementation of voluntary commitments made by non-State actors. Overarching databases were established to try and provide a comprehensive picture of these burgeoning developments.

A. Overview

Voluntary commitments have been undertaken by various non-State actors, including subnational administrations as well as businesses and financial institutions.

Subnational administrations often play an important role in promoting adaptation to climate change at the local scale, whether through dedicated action or simply by mainstreaming adaptation into other plans.[31] Likewise, they have also come to play a role with regard to

---

[28] UNFCCC Subsidiary Body for Implementation, *supra* note 9, at para. 12.

[29] Statistics developed by the author based on the list of NGOs available on the UNFCCC website, consulted in November 2016.

[30] Karin Bäckstrand *et al.*, "Non-state actors in global climate governance: from Copenhagen to Paris and beyond" (2017) 26:4 *Environmental Politics* 1.

[31] See e.g. Jonathan Struggles, "Climate disasters and cities: the role of local government in increasing urban resilience" (2016) 18 *Asia Pacific Journal of Environmental Law* 91; Hannah Reid and Saleemul Huq, "Mainstreaming community-based adaptation into national and local planning" (2014) 6:4 *Climate & Development* 291; Susanne

climate change mitigation, in particular when national governments appeared reluctant to take such action.[32] Thus, as it became clear that the United States would not participate in the Kyoto Protocol, California announced its own legislation on climate change mitigation[33] and many municipal governments undertook similar commitments.[34] Likewise, as the federal government of Canada showed little interest in complying with its QELRC, provincial governments developed their own initiatives to promote climate change mitigation in British Columbia, Ontario and Quebec.[35]

Thus, US President Trump's announcement that the United States would pull out of the Paris Agreement led many American cities, States and companies to turn to similar pledges again.[36] Only a few days after this announcement, Michael Bloomberg, the UN Secretary General's special envoy for Cities and Climate Change, communicated the intention of more than 1,000 US governors, mayors, businesses, universities and others to continue their efforts to meet the goals of the Paris Agreement.[37]

The nature and ambit of voluntary actions by subnational administrations largely depend on their resources and power, which vary from country to country. Cities do not always have the authority or the means to take impactful actions in key economic sectors.[38] Some, however, do; the Tokyo Metropolitan Government's Emissions Trading Scheme, for instance, was reported to have significantly contributed to reductions in GhG emissions in Japan after a project for a nationwide carbon market was rejected.[39]

Besides subnational administrations, businesses and financial institutions have also played a growing role in promoting climate change mitigation. Such business initiatives include public commitments to save energy (e.g. Sony), to purchase only electricity made from renewable energy (Google) and to promote sustainable commodities (IKEA), to reduce emissions from transportation (FedEx) or from industrial activities (Gazprom), and to turn to more efficient

Lorenz et al., "Adaptation planning and the use of climate change projections in local government in England and Germany" (2017) 17:2 Regional Environmental Change 425; Heleen-Kydeke P. Mees and Peter P.J. Driessen, "Adaptation to climate change in urban areas: climate-greening London, Rotterdam, and Toronto" (2011) 2:2 Climate Law 251; and generally Liliana B. Andonova, Michele M. Betsill and Harriet Bulkeley, "Transnational climate governance" (2009) 9:2 Global Environmental Politics 52.

[32] See e.g. Sheridan Bartless and David Satterthwaite (eds.), Cities on a Finite Planet: Towards Transformative Responses to Climate Change (Routledge, 2016); Benjamin J. Richardson (ed.), Local Climate Change Law: Environmental Regulation in Cities and Other Localities (Edward Elgar, 2012); Anatole Boute, "Renewable energy federalism in Russia: regions as new actors for the promotion of clean energy" (2013) 25:2 Journal of Environmental Law 261.

[33] California, Global Warming Solutions Act, 2006, Assembly Bill No. 32. See generally Ann E. Carlson, "Regulatory capacity and state environmental leadership: California's climate policy" (2012–2013) 24:1 Fordham Environmental Law Review 63.

[34] New York Department of Environmental Protection, Assessment and Action Plan: Report 1 (May 2008).

[35] See British Columbia, Climate Action Plan, 2008; Québec Action Plan 2006–2012, Québec and Climate Change: A Change for the Future, June 2008; Ontario, Action Plan on Climate Change (August 2007). Each of these policies has been extended and intensified in subsequent periods.

[36] Hiroko Tabuchi and Henry Fountain, "Bucking Trump, these cities, States and companies commit to Paris accord" New York Times (June 1, 2017).

[37] Press release, "Leaders in U.S. economy say 'we are still in' on Paris Climate Agreement" (June 5, 2017), www.wearestillin.com (accessed January 25, 2018).

[38] See Magali Dreyfus, "Are cities a relevant scale of action to tackle climate change?" (2013) 7:4 Carbon & Climate Law Review 283.

[39] Sven Rudolph and Toru Morotomi, "Acting local! An evaluation of the first compliance period of Tokyo's carbon market" (2016) 10:1 Carbon & Climate Law Review 75.

production processes in general (Lego).[40] In public companies whose managers were reluctant to commit to such action, shareholders have sometimes passed resolutions requiring the disclosure of GhG emissions and measures to reduce them.[41] Likewise, financial institutions have increasingly considered GhG emissions and reliance on fossil fuels in their investment strategies.[42] The private sector as a whole has massively invested in renewable energies.[43]

The voluntary commitments of private actors initially were part of a larger trend toward corporate social responsibility.[44] In addition, investing in sustainable operations has increasingly appeared to make business sense. Today, some voluntary commitments are certainly supported by a strong economic justification. Large economic gains can be made by the businesses which move to new sectors of a greener economy.[45] The commitment of some financial institutions to "divestment" from fossil fuel sectors relates in no small part to a belief that the current levels of investment in these sectors are disproportionately high and hence risky.[46] Beyond such economic interests, however, businesses and financial institutions – or their managers or shareholders – are sometimes genuinely sensitive to the need to avoid a catastrophic climate crisis. Some businesses are certainly able to look beyond short-term profit maximization without losing the game. Just like national governments, some large corporations or financial actors have come to an understanding that short-term economic interests cannot be pursued in isolation from long-term environmental concerns.

Besides any genuine economic interests or sincere concerns for the preservation of the climate system, the efforts made by non-State actors to publicly communicate their voluntary commitments indicate the desire to ensure some form of public acknowledgment. Just like States, non-State actors appear interested in how displaying their commitment to tackling climate change could improve their reputation as a form of symbolic capital. Galvanizing voluntary climate action by non-State actors requires the development of standards and institutional arrangements which would ensure that public commitments are implemented or, else, that violation come at a reputational cost. Yet, whereas national commitments are verified through complex review mechanisms under the UNFCCC regime, no centralized institutional arrangement provides measurement, reporting and verification (MRV) of the implementation of non-State actors' voluntary commitments.[47]

B. Standard-Setting

Measuring mitigation action by non-State actors raises some difficulties due to the absence of any central authority in charge of determining the relevant standards and methodologies. Although

---

[40] Christopher Wright and Daniel Nyberg, *Climate Change, Capitalism, and Corporations: Processes of Creative Self-Destruction* (Cambridge University Press, 2015) 18–22.

[41] See, for instance, Megan Darby, "Exxon shareholder rebellion gains momentum ahead of climate vote" *Climate Home* (May 29, 2017).

[42] See Harriet Bulkeley and Peter Newell, *Governing Climate Change* (Routledge, 2015) 124; *Equator Principles* (June 2013).

[43] See Barbara K. Buchner *et al.*, *Global Landscape of Climate Finance 2015* (CPI, November 2015).

[44] See e.g. Thomas P. Lyon and John W. Maxwell, "Corporate social responsibility and the environment: a theoretical perspective" (2008) 2:2 *Review of Environmental Economics Policy* 240.

[45] Bulkeley and Newell, *supra* note 43, at 118.

[46] See e.g. Atif Ansar, Ben Caldecott and James Tilbury, *Stranded Assets and the Fossil Fuel Divestment Campaign: What Does Divestment Mean for the Valuation of Fossil Fuel Assets?* (Stranded Asset Programme, October 2013).

[47] Regarding the MRV system applicable to national commitments, see Chapter 13, section II.B.

the IPCC guidelines for national GhG inventories can be used, these guidelines were developed to account for GhG emissions within a territory and cannot always readily be applied to measure the GhG emissions attributable to a given city, a given project or a given business.[48] The Greenhouse Gas Protocol and the International Standardization Organization (ISO), an international association of national standards organizations founded in 1947, have led two parallel efforts to fill this gap by developing standards specifically applicable to mitigation commitments by non–State actors.[49]

The Greenhouse Gas Protocol was launched in 1998 by the World Resources Institute, a research organization, and the World Business Council for Sustainable Development, a CEO-led organization of over 200 companies. In addition to standards for corporate and project accounting,[50] the Greenhouse Gas Protocol has developed standards for cities[51] and for assessing emissions from a product life cycle.[52] These standards are regularly updated and completed by supplements, for instance, for accounting for emissions from the value chain[53] or for grid-connected electricity projects.[54]

A comparable initiative was carried out by the ISO. New standards have been developed on GhG management within the "family" of ISO 14000 standards concerning environmental management. Some of these standards overlap with standards developed by the Greenhouse Gas Protocol, for instance, with regard to the quantification, monitoring and reporting of GhG emissions at the project or organizational level[55] or through a product life cycle.[56] Others touch on more specific aspects of accreditation of verification bodies.[57] However, the impact of these standards is limited by the ISO's strict copyright policies and hefty fees that small or even medium-sized non–State actors could probably not afford. By contrast, the Greenhouse Gas Protocol allows free online access to all its standards.

C. International Cooperative Initiatives

Since the late 1990s, non–State actors have coalesced in hundreds of international cooperative initiatives through which they pledge, together, to implement actions on climate change

---

[48] Simon Eggleston *et al.*, *2006 IPCC Guidelines for National Greenhouse Gas Inventories* (IGES, 2006).

[49] Other, sector-specific standards have also been developed, for instance, for international shipping. See generally Joanne Scott *et al.*, "The promise and limits of private standards in reducing greenhouse gas emissions from shipping" (2017) 29:2 *Journal of Environmental Law* 231.

[50] The Greenhouse Gas Protocol, *A Corporate Accounting and Reporting Standard* (revised edn, 2004); and *The Greenhouse Gas Protocol for Project Accounting* (2005).

[51] The Greenhouse Gas Protocol, *Global Protocol for Community-Scale Greenhouse Gas Emission Inventories: An Accounting and Reporting Standard for Cities* (2014).

[52] The Greenhouse Gas Protocol, *Product Life Cycle Accounting and Reporting Standard* (2011).

[53] The Greenhouse Gas Protocol, *Corporate Value Chain (Scope 3) Accounting and Reporting Standard, Supplement to the Greenhouse Gas Protocol Corporate Accounting and Reporting Standard* (2011).

[54] The Greenhouse Gas Protocol, *Guidelines for Quantifying GHG Reductions from Grid-Connected Electricity Projects* (2007).

[55] See ISO 14064-1:2006, *Greenhouse Gases – Part 1: Specification with Guidance at the Organization Level for Quantification and Reporting of Greenhouse Gas Emissions and Removals*; ISO 14064-2:2006, *Greenhouse Gases – Part 2: Specification with Guidance at the Project Level for Quantification, Monitoring and Reporting of Greenhouse Gas Emission Reductions or Removal Enhancements*; ISO 14064-3:2006, *Greenhouse Gases – Part 3: Specification with Guidance for the Validation and Verification of Greenhouse Gas Assertions*.

[56] See ISO/CD 14067, *Carbon Footprint of Products*.

[57] See ISO 14065:2007, *Greenhouse Gases – Requirements for Greenhouse Gas Validation and Verification Bodies for Use in Accreditation or Other Forms of Recognition*.

mitigation and, occasionally, support for climate change adaptation. International cooperation initiatives were sometimes created by pre-existing associations of non-State actors, while others are created by ad hoc consortia. Some of these initiatives are general; others are limited to specific actors, sectors, geographic areas or types of commitments.[58] While some international cooperative initiatives essentially consist in a template commitment that signatories implement subject to some adjustments, others only consist in a network where partners can exchange good practice and reflect on their experience. Only a few of these initiatives define solid quantitative commitments whose implementation is to be regularly reported, while many more define rather blurred qualitative pledges. An exhaustive review of all international cooperative initiatives would go well beyond the scope of this book. Instead, four illustrations – selected among the most prominent and successful initiatives – are briefly discussed in the following.

Firstly, the *Compact of Mayors* was launched in 2014 by the ICLEI-Local Governments for Sustainability, a network of more than 1,500 cities throughout the world. About 500 signatories, municipalities of all sizes, aim to grab "the opportunity to be recognized as leaders in local climate change."[59] They promised to implement specific commitments on climate change mitigation and adaptation, including measuring GhG emissions within their territory, the adoption of a quantitative target and the development of an action plan. They also committed to reporting on their action every year, using the standards developed by the Greenhouse Gas Protocol. In June 2016, the Compact of Mayors and another international cooperative initiative for climate action in cities, the EU Covenant of Mayors, merged into a new, unique initiative, the *Global Covenant of Mayors for Climate and Energy*. This new coalition covers 7,447 cities that, with 675 million people, represent almost 10 percent of the world's population.[60]

Secondly, the *Declaration on Climate Leadership* was adopted in 2014 by the UITP, an international association of public transport companies and administrations.[61] The Declaration itself only contains a very general commitment of its signatories "to be Climate Leaders," in particular "through actions that embrace clean energy, boost efficiency, and limit GhG emissions."[62] Under the umbrella of this Declaration, over 110 member organizations from 80 cities made about 350 concomitant pledges to climate action ranging from improved efficiency to sectorial extension.[63] It is likely that some of these pledges overlap with national commitments or that their outcomes will be reported by national governments in the fulfillment of national commitments made under their NDCs. But even if this initiative does not directly enhance global mitigation action, it may at least contribute to maintaining an international momentum for climate action by promoting public transport throughout the world and by circulating ideas and experience among interested stakeholders.

[58] See Harriet Bulkeley *et al.*, *Transnational Climate Change Governance* (Cambridge University Press, 2014) Chapter 2.

[59] ICLEI website, "Compact of Mayors: definition of compliance," http://e-lib.iclei.org/wp-content/uploads/2015/08/Compact-of-Mayors_Definition-of-compliance-EN.pdf (accessed January 25, 2018). See generally ICLEI, "The Compact of Mayors: goals, objectives and commitments" (September 25, 2014).

[60] See Global Covenant of Mayors for Climate & Energy, "About" (n.d.), www.globalcovenantofmayors.org/about (accessed January 25, 2018). See generally Veerle Heyvaert, "What's in a name? The Covenant of Mayors as transnational environmental regulation" (2013) 22:1 *Review of European Community & International Environmental Law* 78.

[61] UITP stands for the French "Union Internationale des Transports Publics."

[62] UITP, "International Association of Public Transport – UITP Declaration on Climate Leadership" (2014), paras. 8 and 9.

[63] See UITP, *Climate Action and Public Transport: Analysis of Planned Actions* (2014), www.uitp.org/sites/default/files/documents/Advocacy/Climate%20action%20and%20PT.pdf (accessed January 25, 2018).

Thirdly, the *Science-Based Targets initiative* was launched in 2014 by a partnership between the Carbon Disclosure Project (CDP), the World Resource Institute, the World Wide Fund for Nature (WWF) and the UN Global Compact. Under this initiative, companies and financial institutions are invited to submit a commitment letter before they engage in the process of developing an emission reduction target fulfilling particular criteria. A close oversight of the target is imposed to ensure that it involves a genuine contribution to climate change mitigation.[64] This process of bringing committed companies into a forum where further negotiations could be held on their individual targets reflects the modus operandi of the UNFCCC regime, where it was only after years of negotiations that States agreed to specific mitigation commitments. As of early 2018, about 350 companies had committed to participate, but targets had been approved for fewer than 100.[65]

Lastly, the *Montréal Carbon Pledge* was launched by ten investors in 2015 to promote the public disclosure of corporate GhG emissions. Its signatories pledge to measure the GhG emissions implied by their investment portfolios and to report publicly on an annual basis. Unlike other voluntary commitments, the Montréal Carbon Pledge does not require its signatories to divest from any particular companies. Instead, it merely contributes to making information widely available and thus, potentially, to naming and shaming those financial actors which invest heavily and irresponsibly in the fossil fuel industry. Because the reporting commitment itself is not particularly costly, this international cooperative initiative has been rapidly joined by key actors – as of early 2018, its 120 signatories based around 20 different countries represented more than USD 10 trillion of assets under management [66] – and most complied with the terms of their voluntary commitments.[67] The website of the initiative highlights the availability of complementary international cooperative initiatives which involve portfolio decarbonization.[68] Some of the signatories of the Montréal Carbon Pledge are already among the 27 members of the Portfolio Decarbonization Coalition, a divestment initiative which imposes more substantive commitments, covering over USD 600 billion.[69]

D. Taking Stock of Non-State Actors' Contributions

Some databases have been developed as attempts to take stock of this plethora of voluntary commitments and international cooperative initiatives by non-State actors. In particular, the Non-State Actor Zone for Climate Action (NAZCA) was launched by the Peruvian Presidency of COP20 and anchored the following year in the COP decision on the Adoption of the Paris Agreement.[70] NAZCA is essentially an online platform that seeks to increase the visibility of multiple international cooperative initiatives and self-standing voluntary commitments. As of early 2018, it listed more than 70 international cooperative initiatives involving more

---

[64] See Science Based Target, *SBTi Call to Action Guidelines* (April 5, 2017); Science Based Target, *SBTi Criteria and Recommendations* (February 24, 2017).

[65] See Science Based Target, "Meet the companies already setting their emissions reduction targets in line with climate science," http://sciencebasedtargets.org/companies-taking-action (accessed January 25, 2018).

[66] See the website of the Montréal Carbon Pledge (n.d.), http://montrealpledge.org (accessed January 25, 2018).

[67] Novethic, *Montréal Carbon Pledge, Accelerating Investor Climate Sisclosure* (PRI Initiative, September 2016).

[68] See the website of the Montréal Carbon Pledge, *supra* note 67.

[69] See the website of the Portfolio Decarbonization Coalition, http://unepfi.org/pdc (accessed January 25, 2018).

[70] Decision 1/CP.21, *supra* note 12, para. 117.

than 12,500 commitments from 2,500 cities, 200 regions, 2,000 companies and close to 500 investors from 180 countries.[71] Under the aegis of the UNFCCC, NAZCA has succeeded in providing "a much higher level of recognition than before to the role of non-state actors"[72] in response to climate change.

Other general databases have been developed by actors outside of the UNFCCC regime, including non-State actors. In 2014, Ecofys, the University of Cambridge Institute for Sustainability Leadership and the World Resource Institute launched the Climate Initiatives Platform. This platform was then taken over by the UN Environment and the UNEP DTU Partnership, a research and advisory institution, with funding from the Dutch Ministry of Infrastructure and the Environment. As of mid-2017, this platform listed 221 international cooperative initiatives. Other, more specialized databases include the "Investor Platform for Climate Action," launched in 2015 by a consortium of associations representing financial institutions.[73]

Despite the development of such databases, it remains nearly impossible to quantify the overall contribution of voluntary commitments by non-State actors to climate change mitigation.[74] Firstly, information on the commitments of non-State actors is not readily available. Hundreds of international cooperative initiatives and many more unilateral initiatives implement commitments of different sorts based on various assumptions, standards and methodologies. Few of these initiatives provide a quantitative estimate of a quantity of GhG emission reduction; even fewer verify the effective implementation of the commitments. Some studies suggest that, once reputational benefits are achieved through communicating voluntary commitments, non-State actors make little effort to implement these commitments.[75] Absent any legal obligation, even good faith efforts toward compliance can be interrupted by a change in the management of a business or in the administration of a subnational government.

Secondly, aggregating mitigation efforts announced by different non-State actors would raise the risks of double-counting, where a single reduction in GhG emissions is reported by two different actuators. Double-counting may appear amongst non-State actors when their voluntary commitments overlap. When an electricity utility shifts to more efficient operations, for instance, mitigation outcomes could be reported by the electricity utility as well as by its clients (depending on the methodology they use).[76] Coordination to reduce double-counting is hindered by the multiplicity of voluntary commitments and international cooperative initiatives, as well as a lack of harmonized methodology.

---

[71] UNFCCC, "Global climate action: NAZCA" (n.d.), http://climateaction.unfccc.int (accessed January 25, 2018).

[72] UNEP, *The Emissions Gap Report 2016* (November 2016) 24. See also David Wei, "Linking non-State action with the UN Framework Convention on Climate Change" (C2ES, October 2016).

[73] See UNFCCC website, Investor Platform for Climate Actions, http://investorsonclimatechange.org (accessed January 25, 2018).

[74] See e.g. Jakob Graichen *et al.*, *International Climate Initiatives – A Way Forward to Close the Emissions Gap? Initiatives' Potential and Role under the Paris Agreement* (Federal Environment Agency, 2016); Fatemeh Bakhtiari, "International cooperative initiatives and the United Nations Framework Convention on Climate Change" *Climate Policy* (forthcoming).

[75] A study of 330 voluntary initiatives made by non-State actors in 2002 on diverse aspects of sustainability showed that, a decade later, "38 per cent showed no measurable activity at all, while 26 per cent pursued activities not directly related to their stated goals." Sander Chan *et al.*, "Reinvigorating international climate policy: a comprehensive framework for effective nonstate action" (2015) 6:4 *Global Policy* 466, at 469.

[76] Indirect GhG emissions from electricity consumption are commonly considered as "scope 2" emissions; they are within the scope of required reporting according to most standards. See, for instance, *A Corporate Accounting and Reporting Standard*, *supra* note 51, at 25.

Thirdly, there is also a risk of double-counting between the voluntary commitments of non-State actors and national commitments. If a State's mitigation efforts are limited to compulsory technical standards, it is relatively straightforward to verify that the action reported by a non-State actor is genuinely additional to its municipal legal obligations. Yet, many States seek to promote climate change mitigation through other measures, including through price-based incentives, such as carbon taxes or market-based mechanisms, or through subsidies. In such cases, it is difficult to determine whether a non-State actor's mitigation commitment is additional to the mitigation efforts it would have been incentivized to achieve anyway. Within a market-based mechanism, the voluntary effort of one actor to reduce its GhG emissions frees up its emission allowances, which then permits another actor to emit more GhGs, thus resulting in possibly no overall benefit.[77] Such commitments may "give credibility to the national pledges"[78] and generally contribute to raising political support for international cooperation. They may also persuade national governments to adopt more ambitious national commitments on climate change mitigation.

III. CONCLUSION

Non-State actors play a role in the development of the international law on climate change which is often overlooked. NGOs, businesses or subnational governments participate in international negotiations as observers and often encourage State negotiators to conclude fair and ambitious agreements. Voluntary commitments play a role too, especially where States fail most blatantly to implement their obligations under general international law. However, non-State actors have yet to develop a rigorous MRV system allowing them to take stock of their additional contribution to global climate change mitigation in a consistent and reliable way.

[77] Nevertheless, even rigorous international cooperative initiatives approve voluntary commitments taking place within an ETS. The Science Based Targets initiative, for instance, approved the emission reduction target of Enel, an electricity utility, for its operations in Italy, which are subjected to the EU ETS.

[78] Thomas Hale, "'All hands on deck': the Paris Agreement and nonstate climate action" (2016) 16:3 *Global Environmental Politics* 12.

16

International Law in Times of Climate Change

Law often evolves in reaction to crises. The history of international law is replete with violent crises which spurred fundamental changes. The Peace of Westphalia, in 1648, promoted a modern conception of sovereignty as a way to rebuild a Europe ravaged by the Thirty Years' War. The end of the First World War saw the establishment of the League of Nations; the Second World War was followed by the establishment of the United Nations and a new system of international institutions. International law was described as "a discipline of crisis"[1] because it is often unable to respond to structural issues before they precipitated acute crises. Sometimes, however, less violent or less abrupt events could also precipitate important changes. Slow-onset "crises" such as decolonization or peaceful transitions such as the end of the Cold War have led to substantial changes in international law.

It would be naïve to think that, while efforts to address climate change through international law are gaining momentum, the tool itself – international law – will remain unaltered.[2] Climate change prompts new reflections on the nature and content of general international law. It calls for the development of new rules through specific agreements. The international law on climate change has its own characteristics which differ from other fields of international law. For example, collective objectives and national commitments contributing to their realization formed a compliance regime – a system of transitory rules through which States agreed to take steps to come back to compliance with their obligations under general international law. Beyond this, climate change, as a creeping crisis, is bringing to light a number of pre-existing shortcomings in international law and institutions, for instance with regard to the lack of international solidarity when a State is unable to provide effective protection to its population. New questioning of existing laws and institutions could lead to substantial reforms. New circumstances could lead to new applications of general international law.

This chapter develops a reflection on the evolution of international law in times of climate change. Thus, beyond the rules adopted to address climate change, it seeks to understand how international law as a whole is adapting to this creeping crisis. The first section discusses the links between climate change and other aspects of the international law on sustainable development. The second section turns to a reflection on changes in general international law

---

[1] See Hilary Charlesworth, "International law: a discipline of crisis" (2002) 65:3 *Modern Law Review* 377.
[2] See generally Rosemary Rayfuse and Shirley V. Scott (eds.), *International Law in the Era of Climate Change* (Edward Elgar, 2012).

whereby the no-harm principle and the law of State responsibility, for example, could be the object of new debates and new authoritative interpretations. Lastly, the third section explores the potential impact of climate change on the axioms of international law, in particular its sources, its implementation and the relevant actors, as well as the scope of international cooperation more generally.

I. LINKS WITH OTHER ASPECTS OF THE INTERNATIONAL LAW ON SUSTAINABLE DEVELOPMENT

Beyond the UNFCCC regime and other relevant treaty regimes where climate action has been orchestrated, efforts to address climate change have also had far-reaching implications on a range of other treaty regimes promoting sustainable development.[3] Synergies and tensions appeared between climate change and other aspects of the international law on sustainable development, including the protection of human rights, environmental protection and international trade and investments.

A. Synergies

Responses to climate change often contribute more generally to the advancement of sustainable development. For instance, synergies have appeared between climate change and the objectives pursued in international human rights law, with regard to poverty eradication, and in the protection of the environment more generally. Because the impacts of climate change hinder the enjoyment of human rights and efforts to eradicate poverty, climate action can generally be justified as human rights protection.[4] Likewise, because GhG emissions are often associated with emissions in other pollutants affecting public health and the local environment, there is often a synergy between efforts to tackle local air pollution and action on climate change mitigation.[5] Lastly, biodiversity and the conservation of diverse ecosystems such as wetlands and marine ecosystems also call for global action on climate change, given the impact of climate change on the environment.[6]

These synergies have increasingly been recognized in various international legal regimes as the impacts of climate change become more palpable. Since 2008, the UN Human Rights Council passed a series of resolutions highlighting the current and predictable impacts of climate change on the effective enjoyment of human rights throughout the world.[7] The impacts of climate change have also been extensively discussed in the reports of special procedures under the UN Human Rights Council, for instance, by the Special Rapporteur on the human rights

---

[3] See generally Harro van Asselt, *The Fragmentation of Global Climate Governance: Consequences and Management of Regime Interactions* (Edward Elgar, 2014) 52–58. See also Margaret A. Young, "Climate change law and regime interaction" (2011) 5:2 *Carbon & Climate Law Review* 147.

[4] See Chapter 5, section III.

[5] See Lydia McMullen-Laird *et al.*, "Air pollution governance as a driver of recent climate policies in China" (2015) 9:3 *Carbon & Climate Law Review* 243.

[6] Chris Hilson, "It's all about climate change, stupid! Exploring the relationship between environmental law and climate law" (2013) 25:3 *Journal of Environmental Law* 359.

[7] HRC, 35/20, "Human rights and climate change" (June 22, 2017); 32/33, "Human rights and climate change" (July 1, 2016); 29/15, "Human rights and climate change" (July 2, 2015); 18/22, "Climate change and human rights" (September 30, 2011); 7/23, "Human rights and climate change" (March 28, 2008); 10/4, "Human rights and climate change" (March 25, 2009); 7/23, "Human rights and climate change" (March 28, 2008).

of migrants,[8] the Special Rapporteur on the human rights of internally displaced persons,[9] and the Special Rapporteur on adequate housing as a component of the right to an adequate standard of living and on the right to non-discrimination in this context.[10] In 2017, the CESCR recommended to Australia that it "revise its climate change and energy policy" as a way of protecting the enjoyment of economic, social and cultural rights, in particular by Indigenous peoples.[11] The need to "take urgent action to combat climate change and its impacts" was also acknowledged by the UN General Assembly, in 2015, as one of the Sustainable Development Goals.[12] Accordingly, States were called upon to "[i]ntegrate climate change measures into national policies, strategies and planning."[13] Multilateral development agencies are striving to mainstream climate change mitigation and adaptation within their operations.[14] Likewise, the COP to the Convention on Biological Diversity has recognized the impact of climate change on biodiversity.[15]

This, in turn, led to discussions within the UNFCCC regime about the need to recognize such impacts of climate change on other regimes. Following an intense civil society advocacy, the Cancún Agreements noted the existence of Resolution 10/4 of the UN Human Rights Council and recognized, for the first time in a COP decision, "that the adverse effects of climate change have a range of direct and indirect implications for the effective enjoyment of human rights."[16] Article 2 of the Paris Agreement, which defines its objectives, includes a reference to "the context of sustainable development and efforts to eradicate poverty."[17] The Preamble to the Paris Agreement further recognizes "the intrinsic relationship that climate change actions, responses and impacts have with equitable access to sustainable development and eradication of poverty."[18] It also notes the "importance of ensuring the integrity of all ecosystems, including oceans, and the protection of biodiversity."[19]

Beyond such recognition of the synergies between climate change and other aspects of sustainable development, some advocates have sought to use the climate regime as a vehicle for promoting their agenda. In the run-up to the Paris Agreement, the Climate Action Network,

---

[8] HRC, *Report of the Special Rapporteur on the Human Rights of Migrants, François Crépeau* (August 13, 2012), doc. A/67/299.

[9] HRC, *Report of the Special Rapporteur on the Human Rights of Internally Displaced Persons, Chaloka Beyani* (August 9, 2011), doc. A/66/285.

[10] HRC, *Report of the Special Rapporteur on Adequate Housing as a Component of the Right to an Adequate Standard of Living, and on the Right to Non-discrimination in this Context, Raquel Rolnik* (August 6, 2009), doc. A/64/255.

[11] CESCR, "Concluding observations on the fifth periodic report of Australia" (June 23, 2017), doc. E/C.12/AUS/CO/5, para. 12. See generally Edward Cameron and Marc Limon, "Restoring the climate by realizing rights: the role of the international human rights system" (2012) 21:3 *Review of European Community & International Environmental Law* 204.

[12] UN General Assembly Resolution 70/1, "Transforming our world: the 2030 Agenda for Sustainable Development" (September 25, 2015), goal 13.

[13] *Ibid.*, goal 13.2.

[14] See, for instance, Kira Vinke *et al.*, *A Region at Risk: The Human Dimensions of Climate Change in Asia and the Pacific* (Asian Development Bank, 2017).

[15] See e.g. CBD decision X/33, "Biodiversity and climate change" (Nagoya, October 18–29, 2010).

[16] Decision 1/CP.16, "The Cancún Agreements: outcome of the work of the Ad Hoc Working Group on Long-Term Cooperative Action under the Convention" (December 10–11, 2010), recital 8.

[17] Paris Agreement, December 12, 2015, in the annex of decision 1/CP.21, "Adoption of the Paris Agreement" (December 12, 2015), art. 2.1.

[18] *Ibid.*, recital 9.

[19] United Nations Framework Convention on Climate Change, May 9, 1992, 1771 *UNTS* 107 (hereinafter UNFCCC), recital 14.

a global network of 1,100 civil society organizations, called for an inclusion of human rights among the objectives defined in Article 2 of the Paris Agreement "as a means to protect the rights of the people and communities that are most vulnerable but least responsible for climate change."[20] States showed little support for this idea. Norway, for instance, contended that, despite obvious links between climate change and human rights objectives, an international climate agreement should have a clear "climate goal"[21] rather than a human rights goal. Other attempts have been more successful. Successive COP decisions have sought to "improv[e] the participation of women in the representation of Parties in bodies established under the [UNFCCC] and the Kyoto Protocol."[22] The Cancún Agreements emphasized the need to engage a "broad range of stakeholders," including "youth and persons with disability," and the importance of "gender equality and the effective participation of women and indigenous peoples."[23] These goals are important, but they are not critical to climate action, and there is a risk that they may distract from the many challenging tasks that the climate regime is seeking to achieve.

B. Tensions

Climate action may also be in tension with other aspects of sustainable development. For instance, despite the synergies between climate action and the protection of human rights at a systemic level, specific climate-related projects, programs or policies could infringe human rights or impede poverty eradication. This tension is not one between contradicting obligations: the obligations of a State under the international law on climate change do not exclude their obligation under international human rights law.[24] As John Knox, the UN Special Rapporteur on human rights and the environment, put it: "Governments do not check their human rights obligations at the door when they respond to climate change."[25] Rather, this is a tension between competing priorities. If not carefully checked, attempts to increase the efficiency of climate action could come at the cost of an effective protection of human rights – all the more so with regard to transnational projects. The proponents of a hydroelectricity project supported by a flexibility mechanism, for instance, may be more interested in cutting costs than in offering proper compensation to the populations resettled by the project.[26]

---

[20] Climate Action Network, Closing intervention at ADP 2–11 (October 2015), www.climatenetwork.org/sites/default/files/climate-action-network-closing-intervention-adp-2-11-october-2015.pdf (accessed January 25, 2018), at 3.

[21] Website of the Government of Norway, "COP 21: Indigenous peoples, human rights and climate changes" (December 7, 2015), www.regjeringen.no/no/aktuelt/cop21-indigenous-peoples-human-rights-and-climat-changes/id2466047 (accessed January 25, 2018).

[22] Decision 36/CP.7 "Improving the participation of women in the representation of Parties in bodies established under the United Nations Framework Convention on Climate Change and the Kyoto Protocol" (November 9, 2001). See also decision 23/CP.18 "Promoting gender balance and improving the participation of women in UNFCCC negotiations and in the representation of parties in bodies established pursuant to the Convention or the Kyoto Protocol" (December 7, 2012); decision 18/CP.20, "Lima work programme on gender" (December 12, 2014); and decision 21/CP.22, "Gender and climate change" (November 17, 2016).

[23] Decision 1/CP.16, *supra* note 16, para. 7.

[24] See ILC, *Fragmentation of International Law: Difficulties Arising from the Diversification and Expansion of International Law* (April 13, 2006), doc. A/CN.4/L.682, para. 4 (principle of harmonization).

[25] See OHCHR, "States' human rights obligations encompass climate change" (December 3, 1995). See generally Damilola S. Olawuyi, *The Human Rights-Based Approach to Carbon Finance* (Cambridge University Press, 2016).

[26] See e.g. WCED, *Dams and Development: A Framework for Decision Making* (Earthscan, November 2000).

Similar concerns arise in relation to environmental protection. The alternatives to fossil fuels are not always "clean" energies. Nuclear energy creates extraordinarily dangerous and long-lasting wastes. Hydroelectric dams may be highly disturbing for natural ecosystems, for instance, by hindering the migration of species of fish. The massive deployment of monoculture (the cultivation of a single crop in a particular area) for biofuel production in particular regions of the world may also have an impact on the local environment, most obviously when forests are being burned to leave room for such plantations.[27] Even solar or wind power generation can affect ecosystems if it is not appropriately managed.[28]

Safeguard policies can help to ensure that climate action does not come at the expense of other aspects of sustainable development. Safeguards have been discussed in the context of some cooperation programs[29] and imposed by some financial institutions,[30] but a more systematic approach is surely necessary. The Cancún Agreements recognized that "Parties should, in all climate change related actions, fully respect human rights."[31] The Paris Agreement followed suit with a recital acknowledging that:

> Parties should, when taking action to address climate change, respect, promote and consider their respective obligations on human rights, the right to health, the rights of indigenous peoples, local communities, migrants, children, persons with disabilities and people in vulnerable situations and the right to development, as well as gender equality, empowerment of women and intergenerational equity.[32]

Beyond the Paris Agreement, there remains a need for more specific provisions on the minimum standards of human rights protection in all climate actions, including those conducted with international support.

Besides human rights and environmental protection, climate action could also clash with provisions of international economic law. While international economic law promotes unimpeded economic interactions between States, a State could be tempted to impose restrictions to trade or investment in order to reduce GhG emissions. To try and avoid such inconsistencies with international economic law, the UNFCCC stated that "[m]easures taken to combat climate change, including unilateral ones, should not constitute a means of arbitrary or unjustifiable discrimination or a disguised restriction on international trade."[33] On the other hand, Article XX of the General Agreement on Tariffs and Trade (GATT) contains general exceptions for measures "necessary to protect human, animal or plant life or health" or "relating to the conservation of exhaustible natural resources," on the condition that such measures do not

---

[27] See J.D. Neidel *et al.* (eds.), *Biofuels: The Impact of Oil Palm on Forests and Climate* (Yale University and Smithsonian Tropical Research Institute, 2012).

[28] See John Upton, "Solar farms threaten birds: certain avian species seem to crash into large solar power arrays or get burned by the concentrated rays" *Scientific American* (August 27, 2014).

[29] See Chapter 8, section II (clean development mechanism) and Chapter 12, section III (REDD+).

[30] See e.g. GEF, *Agency Minimum Standards on Environmental and Social Safeguards* (Policy SD/PL03, last updated on February 19, 2015); *The World Bank Environmental and Social Framework* (World Bank, 2017).

[31] Decision 1/CP.16, *supra* note 16, para. 8.

[32] Paris Agreement, *supra* note 17, recital 8.

[33] UNFCCC, *supra* note 19, art. 3.5. See also Kyoto Protocol to the United Nations Framework Convention on Climate Change, December 11, 1997, 2303 *UNTS* 162 (hereinafter Kyoto Protocol), art. 2.3.

"constitute a means of arbitrary or unjustifiable discrimination between countries where the same conditions prevail, or a disguised restriction on international trade."[34] Measures adopted through multilateral environmental agreements are likely to be considered as falling squarely within the scope of a general exception.[35] Unilateral policies are more likely to attract scrutiny, although they may still be considered as justified under Article XX of the GATT if they do not constitute disguised protectionist measures.[36] In the other parts of international economic law, bilateral investment treaties have increasingly been designed at least to encourage consistency with international climate agreements and sometimes – rarely – to actively promote investments in renewable energy.[37]

A particularly sensitive question regards the lawfulness of Border-Adjustment Tax (BAT). Measures on climate change mitigation often impose additional costs on domestic producers, for instance, through a carbon tax or a market-based mechanism. Consequently, a State which refuses to cooperate on climate change mitigation could benefit from lower production costs as a result. This State would thus attract economic activity in GhG-intensive sectors, a phenomenon called "carbon leakage." BAT refers to a tax which would seek to prevent such carbon leakage by imposing an additional cost on goods and services from a State without equivalent mitigation measures. BAT could also be imposed as a means of economic pressure on a country systematically disregarding its international obligations under the international law on climate change. The decision of US President Trump to pull the United States out of the Paris Agreement gave new relevance to the question of a BAT. A non-punitive BAT imposed on a State not participating in quasi-universal agreements and not taking any equivalent action, aimed at the protection of the climate system, would arguably not constitute a "means of arbitrary or unjustifiable discrimination between countries where the same conditions prevail"[38] and thus could be justified under Article XX of the GATT.[39]

II. CHANGES IN GENERAL INTERNATIONAL LAW

Climate change and responses to it may contribute to new developments in general international law. The norms of general international law emerged when they were first applied to given

---

[34] General Agreement on Tariffs and Trade, October 30, 1947, 55 *UNTS* 194 (hereinafter GATT), art. XX, paras. (b) and (g). See also Agreement on Technical Barriers to Trade, April 12, 1979, 1868 *UNTS* 120, art. 2.2; General Agreement on Trade-Related Aspects of Intellectual Property, April 15, 1994, 1869 *UNTS* 299, art. 27.2.

[35] See van Asselt, *supra* note 3, at 162.

[36] See e.g. WTO DSB, DS58: *United States – Import Prohibition of Certain Shrimp and Shrimp Products*; DS135: *European Communities – Measures Affecting Asbestos and Asbestos Containing Products*.

[37] See J. Anthony VanDuzer, "The complex relationship between international investment law and climate change initiatives: exploring the tension" in Panagiotis Delimatsis (ed.), *Research Handbook on Climate Change and Trade Law* (Edward Elgar, 2016) 434, at 446–456.

[38] GATT, *supra* note 34, art. XX.

[39] For more information, see e.g. Deok-Young Park (ed.), *Legal Issues on Climate Change and International Trade Law* (Springer, 2016); James W. Coleman, "Unilateral climate regulation" (2014) 38:1 *Harvard Environmental Law Review* 87; Margaret A. Young, "Trade measures to address environmental concerns in faraway places: jurisdictional issues" (2014) 23:3 *Review of European, Comparative & International Environmental Law* 302; Rafael Leal-Arcas, "Climate change mitigation from the bottom up: using preferential trade agreements to promote climate change mitigation" (2013) 88:2 *Carbon & Climate Law Review* 34; Navraj Singh Ghaleigh & David Rossati, "The spectre of carbon border-adjustment measures" (2011) 2:1 *Climate Law* 63; Christina Voigt, *Sustainable Development as a Principle of International Law: Resolving Conflicts between Climate Measures and WTO Law* (Martinus Nijhoff, 2009).

circumstances; new circumstances may lead to refinements. Already, some yet ill-determined aspects of general international law are being clarified. With regard to the no-harm principle, new reflections are emerging in relation to modalities of application to global environmental harm. With regard to the law of State responsibility, reconsideration of the notion that a responsible State must pay "full reparation" is needed: a quantum of reparations determined solely on the basis of the injury appears particularly unlikely in relation to climate change.

A. Clarifying the Modalities of the No-Harm Principle

Although the no-harm principle is generally recognized as an integral part of general international law, it has never been applied in relation to environmental damage of a global nature. There is little doubt that this principle, as a corollary of the principle of sovereign equality, applies to climate change. The opposite conclusion would be absurd: how could relatively minor transboundary environmental harms be prohibited, but not cause harm to entire planetary systems, causing the disappearance of the entire territory of some States and, in the worst-case scenario, threatening human civilization as a whole? The Preamble to the UNFCCC recognizes the no-harm principle as one of the "pertinent principles" of the Stockholm Declaration,[40] and several small island States declared upon adoption of successive international climate agreements that none of their provisions could "be interpreted as derogating from the principles of general international law."[41] Yet, important questions remain regarding the modalities of application of the no-harm principle, where ongoing debates and future developments could lead to a better understanding of the contents of general international law.

1. Geographic Scope

An ill-defined modality of the no-harm principle regards the geographic scope of the obligation of a State not to permit activities causing transboundary environmental harm. Whereas the *Trail Smelter* arbitral award related to the "use of a *territory*"[42] for activities causing transboundary harm, the Stockholm Declaration on the Human Environment referred to "activities within [States'] *jurisdiction* or *control*."[43] Following the terminology of the Stockholm Declaration, the ICJ alluded to "activities within [States'] *jurisdiction* or *control*"[44] in its advisory opinion on the *Legality of the Threat or Use of Nuclear Weapons*. In the case of the *Pulp Mills on the River Uruguay*, the same Court considered that a State's obligation to regulate applies to "activities which take place in its *territory*, or in any area under its *jurisdiction*,"[45] without mentioning

---

[40] See UNFCCC, *supra* note 19, recitals 8 and 9. See also UNCHE, Stockholm Declaration on the Human Environment, available in (1972) 11 *ILM* 1416 (June 5–16, 1972) (hereinafter Stockholm Declaration).

[41] See, for instance, the declarations of Kiribati, Fiji, Nauru and Tuvalu upon signature of the UNFCCC (1992), 1771 *UNTS* 317–318.

[42] *Trail Smelter (U.S.* v. *Canada)*, Arbitral Award of March 11, 1941, (1949) III *UNRIAA* 1938, at 1965. See also ICJ, *Corfu Channel (United Kingdom* v. *Albania)*, judgment of April 9, 1949, at 22, where the ICJ recognized "every State's obligation not to allow knowingly *its territory* to be used for acts contrary to the rights of other States" (emphasis added).

[43] Stockholm Declaration, *supra* note 40, Principle 21 (emphasis added).

[44] ICJ, *Legality of the Threat or Use of Nuclear Weapons*, Advisory Opinion of July 8, 1996, para. 29. See also UNCED, Rio Declaration on Environment and Development (June 3–14, 1992), available in (1992) 31 *ILM* 874 (hereinafter Rio Declaration), Principle 2.

[45] ICJ, *Pulp Mills on the River Uruguay (Argentina* v. *Uruguay)*, judgment of April 20, 2010, para. 101 (emphasis added).

control as an alternative basis. However, it remains likely that control could be a basis for responsibility. In that sense, the ICJ affirmed in its advisory on the *Legal Consequences for States of the Continue Presence of South Africa in Namibia* that "[p]hysical control of a territory, and not sovereignty or legitimacy of title, is the basis of State liability for acts affecting other States."[46]

In classical cases regarding transboundary air pollution, the geographic scope of the no-harm principle remained undebated because activities that produce great environmental harm are usually permanent facilities within a State's own territory, over which a State has jurisdiction *and* exercises effective control. The *Iron Rhine* case between Belgium and the Netherlands was an exception, as it involved "the exercise of a treaty-guaranteed right of one state in the territory of another state and a possible impact of such exercise on the territory of the latter state."[47] In this case, the arbitral tribunal accepted, "by analogy" with the no-harm principle, that "*where a state exercises a right under international law within the territory of another state,* considerations of environmental protection also apply."[48] While the Tribunal thus recognized the exercise of a sovereign right beyond a State's territory as a ground for responsibility, it avoided taking a position on the broader question of (effective) control, whether with or without a sovereign right.[49]

By contrast, the emerging practice with regard to climate change has placed more emphasis on territory than on jurisdiction or control. Despite some arguments about consumption-based or profit-based responsibility,[50] States' mitigation obligations were generally understood to apply to production processes taking place within their own territory. This approach was implicitly accepted under the Kyoto Protocol through the endorsement of the technical guidelines adopted by the IPCC.[51] A brief attempt was made by the European Union to apply its emissions trading scheme to international flights departing from or bound to a Member State, including for GhG emissions occurring beyond its airspace.[52] Despite a favorable judgment by the European Court of Justice,[53] the step remained controversial in its international context, as reflected

---

[46] ICJ, *Legal Consequences for States of the Continued Presence of South Africa in Namibia (South West Africa) Notwithstanding Security Council Resolution 276*, Advisory Opinion of June 21, 1971, para. 118.

[47] Arbitral award of May 24, 2005 in the case of the *Iron Rhine Railway (Belgium* v. *Netherlands)*, (2005) XXVII *UNRIAA* 35, para. 223.

[48] *Ibid.* (emphasis added).

[49] The question has been the object of greater attention with regard to the responsibility of a State for human rights abuses. See in particular ECtHR Grand Chamber, *Al Skeini* v. *United Kingdom*, judgment of July 7, 2011; and generally Marko Milanovic, *Extraterritorial Application of Human Rights Treaties: Law, Principles, and Policy* (Oxford University Press, 2011); Marko Milanovic, "European court decides Al-Skeini and Al-Jedda" *EJIL: Talk!* (July 7, 2011), www.ejiltalk.org/european-court-decides-al-skeini-and-al-jedda (accessed January 25, 2018).

[50] See e.g. Alexandra Marques *et al.*, "Income-based environmental responsibility" (2012) 84 *Ecological Economics* 57.

[51] See decision 24/CP.19, "Revision of the UNFCCC reporting guidelines on annual inventories for parties included in Annex I to the Convention" (November 22, 2013); decision 17/CP.8 "Guidelines for the preparation of national communications from parties not included in Annex I to the Convention" (November 1, 2002); and decision 2/CP.17 (December 11, 2011). See also Simon Eggleston *et al.*, *2006 IPCC Guidelines for National Greenhouse Gas Inventories* (IGES, 2006).

[52] EU Directive 2008/101/EC of 19 November 2008 amending Directive 2003/87/EC so as to include aviation activities in the scheme for greenhouse gas emission allowance trading within the Community (January 13, 2009), doc. 32008L0101.

[53] ECJ judgment (Grand Chamber), Case C-366/10, *ATA* v. *Secretary of State for Energy*, judgment of December 21, 2011, doc. 62010CJ0366.

in an intense academic debate about whether it was permissible – let alone compulsory – for a State to seek to limit GhG emissions beyond its own territory.[54]

2. The Standard of Due Diligence

Another yet ill-defined aspect of the no-harm principle regards the nature of the obligation of a State under this principle. The obligation of a State to prevent activities within its territory or under its jurisdiction is generally understood as a due diligence obligation. Yet, no authority has yet defined how much effort a State must make to discharge this due diligence obligation. Here again, this modality was generally of little relevance to cases where the no-harm principle was previously applied, as such cases were decided by international courts or tribunals after intensive diplomatic consultations where the unwillingness of the responsible State to take any action was typically well established. The ICJ mentioned due diligence in the *Pulp Mills* case, for instance, only to briefly characterize it as an obligation of the State "to use all the means at its disposal."[55] The International Tribunal for the Law of the Sea (ITLOS) stated very cautiously that "[t]he content of 'due diligence' obligations may not easily be described in precise terms," as "[i]t may change over time as measures considered sufficiently diligent at a certain moment may become not diligent enough in light, for instance, of new scientific or technologic knowledge."[56]

The content of the due diligence obligation stemming from the no-harm principle is of particular relevance with regard to climate change. Determining the standard of due diligence would be essential to defining the application of the no-harm principle to historical GhG emissions. Despite some relatively vague allusions to the historical responsibility of developed States, the question remains largely unsettled for the time being.[57] More generally, arguments on different grounds for differentiation relate to different understandings of this due diligence obligation. International negotiations on climate change have recognized the relevance of a State's capacity and other "national circumstances" to determining what this State could be expected to contribute to global efforts on climate change mitigation.[58] This may be taken to suggest that the same elements should be taken into consideration in determining States' due diligence obligations. The principle of common but differentiated responsibilities has been treated as evidence that, in general international law, "lack of economic and technological capacity may mitigate the attendant obligations for developing countries."[59] Yet, the actual

---

[54] See Glan Plant's analysis in (2013) 107:1 *American Journal of International Law* 183; Snaja Bogojević, "Legalising environmental leadership: a comment on the CJEU's ruling in C-366/10 on the inclusion of aviation in the EU Emissions Trading Scheme" (2012) 24:2 *Journal of Environmental Law* 345; Andrea Gattini, "Between splendid isolation and tentative imperialism: the EU's extension of its Emission Trading Scheme to international aviation and the ECJ's judgment in the ATA case" (2012) 61:4 *International & Comparative Law Quarterly* 977; Joanne Scott, "The geographical scope of the EU's climate responsibilities" (2015) 17 *Cambridge Yearbook of European Legal Studies* 92; Joanne Scott and Lavanya Rajamani, "EU climate change unilateralism" (2012) 23:2 *European Journal of International Law* 469; An Hertogen, "Sovereignty as decisional independence over domestic affairs: the dispute over aviation in the EU emissions trading system" (2012) 1:2 *Transnational Environmental Law* 281.

[55] *Pulp Mills on the River Uruguay*, *supra* note 45, para. 101.

[56] ITLOS Seabed Dispute Chamber, *Responsibilities and Obligations of States Sponsoring Persons and Entities with Respect to Activities in the Area*, Advisory Opinion of February 1, 2011, para. 117.

[57] See e.g. UNFCCC, *supra* note 19, recital 4; decision 1/CP.16, *supra* note 16, second recital before para. 36; decision 1/CP.21, *supra* note 17, para. 52.

[58] See UNFCCC, *supra* note 19, art. 3.1; Paris Agreement, *supra* note 17, art. 2.2; and generally Chapter 6.

[59] Timo Koivurova, "Due diligence" in Rüdiger Wolfrum *et al.* (ed.), *Max Planck Encyclopaedia of Public International Law* (Oxford University Press, 2010), para. 19.

significance of the principle of common but differentiated responsibilities remains disputed. Although the Paris Agreement formally leaves it to States to determine their own contribution to climate action, further debates will unavoidably take place during global stocktake exercises.[60]

B. Reconsideration of Certain Aspects of the Law of State Responsibility

Climate change could also trigger some refinements in the general international law on State responsibility. While every State has accepted that developed States bear some form of political responsibility in combating climate change and its impacts,[61] developed States remain generally reluctant to acknowledge their legal responsibility for their failure to prevent excessive GhG emissions under their jurisdiction. Some of the most vulnerable States, however, have maintained claims for such reparations.[62] These States could, one day, seek to bring their claims before an international court or tribunal.[63] Alternatively, clarification could come from the work program of the International Law Commission (ILC) on the protection of the atmosphere, despite the narrow definition imposed on its syllabus.[64]

An authoritative interpretation of the law of State responsibility would likely start with a reference to the Articles on the Responsibility of States for Internationally Wrongful Acts that the ILC adopted in 2001. These articles affirm that a State responsible for an international wrongful act is "under an obligation to make full reparation for the injury caused by the internationally wrongful act."[65] It would be extremely difficult to determine what constitutes the injury caused by excessive GhG emissions, given the complex causal relation between excessive GhG emissions and actual damages. Valuation could also be challenging when harm is caused to non-economic values or unfolds over a long period of time. However, it remains the case that, by any possible account, the quantum of damages would likely be much greater than what any court or tribunal has attributed in any case so far.

Rather than full reparation assessed exclusively on the basis of the injury, it is far more likely that an international court or tribunal would incline toward a more moderate interpretation of the general international law on State responsibility. This interpretation would reflect the general practice of States of asking for less than full reparation, reflected in the decisions of some international courts and tribunals, in cases of colossal damages.[66] Rather than a vain exercise in assessing the value of the injury, an international court or tribunal could call on States to negotiate and agree on a reasonable lump-sum payment, as States have often done

---

[60] Paris Agreement, *supra* note 17, art. 14.

[61] See e.g. UNFCCC, *supra* note 19, art. 3.1.

[62] See *supra* note 41.

[63] See Philippe Sands, "Climate change and the rule of law: adjudicating the future in international law" (2016) 28:1 *Journal of Environmental Law* 19. See also Chapter 14, section I.

[64] See Peter H. Sand and Jonathan B. Wiener, "Towards a new international law of the atmosphere?" (2016) 7:2 *Göttingen Journal of International Law* 195; and Chapter 5, section VI.

[65] ILC, *Draft Articles on Responsibility of States for Internationally Wrongful Acts with Commentaries*, in (2001) *Yearbook of the International Law Commission*, vol. II, part two (hereinafter *Articles on State Responsibility*), art. 31.1.

[66] See, for instance, Eritrea-Ethiopia Claims Commission (EECC), decision of August 17, 2009, Final Award on Ethiopia's Damages Claims, in (2009) XXVI *UNRIAA* 631, paras. 18–22; and generally Benoit Mayer, "Climate change reparations and the law and practice of state responsibility" (2016) 7:1 *Asian Journal of International Law* 185; Benoit Mayer, "Less-than-full reparation in international law" (2016) 56:3–4 *Indian Journal of International Law* 465.

in relation to large-scale injuries.[67] Instead of compensation,[68] developing States could possibly consent, as part of a broader agreement, to the provision of reparation through financial or technical assistance. These developments would affect our understanding of general international law not just for a one-off application to climate change, but also, potentially, for further applications to any analogous issues.

III. CHANGES TO OUR UNDERSTANDING OF INTERNATIONAL LAW

Global crises have repeatedly transformed international law. The aftermaths of the First and Second World Wars saw not just subtle refinements to existing rules, but rather a complete overhaul of the foundations of international cooperation. Although climate change does not unfold in the same way as a global armed conflict, it may also bring about profound changes in our expectations of international institutions and in our conception of international law. Structural changes could have already started to take place regarding the sources, the implementation and the actors of international law, as well as, more fundamentally, the very scope of international cooperation.

A. Sources

The international law on climate change may contribute to an evolution of the sources of international law. The Statute of the ICJ recognizes three main sources of international law: international conventions "establishing rules expressly recognized" by States, international customs "as evidence of a general practice accepted as law" and "the general principles of law recognized by civilized nations."[69] In addition, the UNFCCC regime has given a prominent role to secondary instruments.[70] The UNFCCC empowered its COP to "make, within its mandate, the decisions necessary to promote the effective implementation of the Convention."[71] Over a quarter of a century of negotiations, the COP to the UNFCCC and the Meeting of the Parties to the Kyoto Protocol adopted about 600 decisions, some of them with far-reaching normative implications. In particular, the Cancún Agreements established a comprehensive system of national commitments by developed and developing States going much further than the UNFCCC itself.[72] Two years later, the decision adopting the Doha Amendment to the Kyoto Protocol encouraged parties to apply it provisionally, pending its entry into force, and to notify the UNFCCC Secretariat of their intent to do so.[73] These two decisions remained, until

---

[67] See generally Richard B. Lillich and H. Burns, Weston, *International Claims: Their Settlement by Lump Sum Agreements* (University Press of Virginia, 1975); Burns H. Weston, David J. Bederman and Richard B. Lillich, *International Claims: Their Settlement by Lump Sum Agreements, 1975–1995* (Martinus Nijhoff, 1999). See also *Responsabilité de l'Allemagne à raison des dommages causés dans les colonies portugaises du sud de l'Afrique (Portugal* v. *Germany)*, award of July 31, 1928, II *UNRIAA* 1011, 1032–1033.

[68] See *Articles on State Responsibility*, *supra* note 65, art. 34, providing that reparation "shall take the form of restitution, compensation and satisfaction." In practice, compensation remains the most common form of reparation for damages that are financially assessable.

[69] Statute of the International Court of Justice, June 26, 1945, 3 *Bevans* 1179 (hereinafter ICJ Statute), art. 38.1.

[70] Duncan French and Lavanya Rajamani, "Climate change and international environmental law: musings on a journey to somewhere" (2013) 25:3 *Journal of Environmental Law* 437, at 443–448.

[71] UNFCCC, *supra* note 19, art. 7.2.

[72] Decision 1/CP.16, *supra* note 16. See also Chapter 3, section III.A and Chapter 7, section II.B.1.

[73] Decision 1/CMP.8, "Amendment to the Kyoto Protocol pursuant to its Article 3, paragraph 9 (Doha Amendment)" (December 8, 2012), para. 5.

the entry into force of the Paris Agreement, the only source of specific national commitments on climate change.

Besides COP decisions, the UNFCCC regime has also increasingly relied on unilateral declarations to define national commitments. Like the Cancún pledges, NDCs are documents communicated by States to international institutions through which States unilaterally determine the ambit of their obligations.[74] This approach marks a shift away from the more classical approach, attempted through the Kyoto Protocol, of defining respective national obligations in a treaty. The legal status of NDCs will remain a subject of academic debate for years to come. These documents derive some legal value from the terms of the Paris Agreement;[75] they may also be considered binding in their own right, as unilateral declarations of States capable of creating legal obligations.[76] Although NDCs may not fall squarely within any unique category of legal instruments, many NDCs and contemporary statements indicate the intention of States to be bound under international law.[77]

B. Implementation

The difficulty of ensuring cooperation among multiple States has spurred an innovative approach to implementation in the international law on climate change, in particular within the UNFCCC regime. The latter regime has arguably gone further than any previous multilateral environmental regime in seeking to facilitate implementation rather than trying to impose it. Significant support – including finance as well as transfer of technology and capacity building – is provided to countries which lack the resources to address climate change effectively. Compliance is promoted through public deliberations rather than through blunt condemnation. Education and the participation of civil societies in the negotiations help to ensure that effective naming-and-shaming processes take place within States. The UNFCCC regime as a whole can be approached as a means to slowly leverage compliance with pre-existing norms of general international law.[78] Litigation, in this context, is just one of many political tools available to trigger stronger ambition and stricter compliance.

C. Actors

This quest for an extended political support has come along with an increased recognition of non-State actors in the international law on climate change. Non-State actors are essential to the naming-and-shaming processes which drive national governments to

---

[74] See decision 1/CP.16, *supra* note 16, paras. 36, 37, 49 and 50; Paris Agreement, *supra* note 17, art. 3. See also Copenhagen Accord, in the annex of decision 2/CP.15 (December 18–19, 2009), paras. 4 and 5.

[75] For instance, the parties to the Paris Agreement have committed to "pursue domestic mitigation measures, with the aim of achieving the objective" of their contribution. See Paris Agreement, *supra* note 17, art. 4.2.

[76] See generally Benoit Mayer, "International law obligations arising in relation to Nationally Determined Contributions" *Transnational Environmental Law* (forthcoming). See also ILC, *Guiding Principles Applicable to Unilateral Declarations of States Capable of Creating Legal Obligations, with Commentaries Thereto*, in (2006) *Yearbook of the International Law Commission*, vol. II part two, principle 1, recognizing that "[d]eclarations publicly made and manifesting the will to be bound may have the effect of creating legal obligations."

[77] This is not to suggest that everything contained in NDCs is binding upon States. Rather, NDCs are communicated with the understanding that this commits them to achieve a specified objective, but not to implement a particular list of measures. See Paris Agreement, *supra* note 17, art.4.2; and the discussion in Chapter 7, section II.B.1.

[78] See Chapter 13.

act. Beyond, the international law on climate change has increasingly recognized the role that corporations, local governments and not-for-profit organizations can and do play in addressing climate change. Non-State actors have made and implemented voluntary commitments on climate change mitigation and adaptation. They have established parallel normative systems where commitments to reduce GhG emissions are reported and, increasingly, negotiated and verified, based on standards that non-State actors have developed.[79] Following US President Trump's announcement that the United States would pull out of the Paris Agreement, some non-State actors have been contemplating ways to implement the commitments contained in the INDC of the United States, even without the participation of the federal government.[80]

While these initiatives reinforce the effectiveness of international action against climate change, they may also result in greater complexity, possibly creating some confusion in the respective role of multiple actors. Thus, it may be difficult to determine the actual impact of the voluntary commitments pledged by non-State actors, particularly when their operations are taking place under financial incentives such as a market-based mechanism. While non-State actors could come to play a more pronounced role in transnational cooperation, this role will generally remain subjected to the monopolistic territorial authority of the State.

D. The Scope of International Cooperation

Lastly, some more fundamental changes may be occurring with regard to the limits of general international law. Despite a formal affirmation of "the principle of sovereignty of States in international cooperation to address climate change"[81] in the Preamble to the UNFCCC, international cooperation on climate change has extended deep into what was formerly considered to be internal matters. Calls for the "widest possible cooperation by all countries"[82] have extended beyond what is strictly necessary to address a collective action issue, climate change, through efforts on climate change mitigation. For example, States have also agreed that they should cooperate in addressing the local impacts of climate change.[83] Scholars, too, have increasingly expressed support for and hope in "a flourishing ecological governance paradigm that respects all life on Earth now and in the future."[84]

Thus, climate change has contributed to a gradual extension of the sphere of international cooperation, reinforcing a long-lasting evolution taking place, among others, in international human rights law, international cooperation on development or through the promotion of so-called "good governance" standards by financial institutions. Responses to climate change further reinforce faith in international cooperation, which at times may appear somewhat naïve,

---

[79] See Chapter 15.

[80] See "Leaders in U.S. economy say 'we are still in' on Paris Climate Agreement" (June 5, 2017), www.wearestillin. com (accessed January 25, 2018).

[81] UNFCCC, *supra* note 19, recital 10.

[82] *Ibid.*, recital 7.

[83] See, for instance, UN General Assembly Resolution 70/1, *supra* note 12, para. 31, calling for the "widest possible cooperation" aimed at "accelerating the reduction of global greenhouse gas emissions and addressing adaptation to the adverse impacts of climate change."

[84] Burns H. Weston and David Bollier, *Green Governance: Ecological Survival, Human Rights, and the Law of the Commons* (Cambridge University Press, 2014), at 3.

as if international cooperation were the panacea for most of the world's problems.[85] There is a risk that this naïve faith in international cooperation may lead to the imposition of a unique model of development.

IV. CONCLUSION

While efforts are made to address climate change through international law, the latter is also being transformed. The international law on climate change interacts with other aspects of the international law on sustainable development by exploiting synergies and addressing potential conflicts. More fundamentally, general international law continues to evolve as new questions are raised, new State practice is emerging and new ideas are accepted as law. Overall, our very understanding of international law – its sources, its implementation mechanisms and its central actors, as well as its scope – is changing.

This, however, is only the start of the story. After a quarter of a century, the international law on climate change is still in its infancy. Much, much more is needed to ensure that a cataclysmic climatic crisis will be avoided – a crisis which could lead to our end, as civilizations or perhaps even as a species. This is the endeavor of a generation. Much more effort will be made; this may or may not be sufficient. Most of the story remains to be written.

---

[85] Olanrewaju Fagbohun, "Cultural legitimacy of mitigation and adaptation to climate change: an analytical framework" (2011) 6:3 *Carbon & Climate Law Review* 308. See also Thoko Kaime, *International Climate Change Law and Policy: Cultural Legitimacy in Adaptation and Mitigation* (Routledge, 2014); Siri H, Eriksen, Andrea J. Nightingale and Hallie Eakin, "Reframing adaptation: the political nature of climate change adaptation" (2015) 35 *Global Environmental Change* 523.

References*

BOOKS

Ahmad, Tanveer, *Climate Change Governance in International Civil Aviation* (Eleven, 2016).

Andersen, Stephen O. and K. Madhava Sarma, *Protecting the Ozone Layer: The United Nations History* (UNEP and Earthscan, 2002).

Andersen, Stephen O., K. Madhava Sarma and Kristen N. Taddonio (eds.), *Technology Transfer for the Ozone Layer: Lessons for Climate Change* (Earthscan, 2007).

Arnold, Denis (ed.), *The Ethics of Global Climate Change* (Cambridge University Press, 2011).

Asselt, Harro van, *The Fragmentation of Global Climate Governance: Consequences and Management of Regime Interactions* (Edward Elgar, 2014).

Bartless, Sheridan and David Satterthwaite (eds.), *Cities on a Finite Planet: Towards Transformative Responses to Climate Change* (Routledge, 2016).

Baumert, Kevin A. *et al.* (eds.), *Building on the Kyoto Protocol: Options for Protecting the Climate* (World Resources Institute, 2002).

Bernthal, Frederick M. *et al.*, *Climate Change: The IPCC Response Strategies. Report Prepared for Intergovernmental Panel on Climate Change by Working Group III* (Island Press, 1990).

Betts, Alexander, *Survival Migration: Failed Governance and the Crisis of Displacement* (Cornell University Press, 2013).

Bowman, Micahel and Alan Boyle (eds.), *Environmental Damage in International and Comparative Law: Problems of Definition and Valuation* (Oxford University Press, 2002).

Bryner, Gary with Robert J. Duffy, *Integrating Climate, Energy, and Air Pollution Policies* (MIT Press, 2012).

Bulkeley, Harriet and Peter Newell, *Governing Climate Change* (Routledge, 2015).

Bulkeley, Harriet *et al.*, *Transnational Climate Change Governance* (Cambridge University Press, 2014).

Burns, William C.G. and Andrew L. Strauss (eds.), *Climate Change Geoengineering: Philosophical Perspectives, Legal Issues, and Governance Frameworks* (Cambridge University Press, 2015).

Burns, William C.G. and Hari M. Osofsky, *Adjudicating Climate Change: State, National, and International Approaches* (Cambridge University Press, 2011).

Carlarne, Cinnamon P., Kevin R. Gray and Richard Tarasofsky (eds.), *The Oxford Handbook of International Climate Change Law* (Oxford University Press, 2016).

Chatterjee, Deen K. (ed.), *Encyclopedia of Global Justice* (Springer, 2011).

Chayes, Abram and Antonia Handler Chayes, *The New Sovereignty Compliance with International Regulatory Agreements* (Harvard University Press, 1995).

Cohen, Roberta and Francis M. Deng, *Masses in Flight: The Global Crisis of Internal Displacement* (Brookings Institution Press, 1998).

* Institutional reports, negotiation documents, working papers and other unpublished papers are not listed.

Condon, Bradly J. and Tapen Sinha, *The Role of Climate Change in Global Economic Governance* (Oxford University Press, 2013).

Craik, Neil, *The International Law of Environmental Impact Assessment* (Cambridge University Press, 2011).

Crawford, James, *Brownlie's Principles of Public International Law*, 8th edn (Oxford University Press, 2012).

Crawford, James, *State Responsibility: The General Part* (Cambridge University Press, 2013).

Crawford, James, Alain Pellet and Simon Olleson (eds.), *The Law of International Responsibility* (Oxford University Press, 2010).

Cullet, Philippe, *Differential Treatment in International Environmental Law* (Routledge, 2003).

Delimatsis, Panagiotis (ed.), *Research Handbook on Climate Change and Trade Law* (Edward Elgar, 2016).

Deng, Francis *et al.*, *Sovereignty as Responsibility: Conflict Management in Africa* (Brookings Institution Press, 1996).

Diamond, Jared M., *Collapse: How Societies Choose to Fail or Succeed* (Penguin, 2011).

Dryzek, John S., Richard B. Norgaard and David Schlosberg (eds.), *The Oxford Handbook of Climate Change and Society* (Oxford University Press, 2011).

Edenhofer, Ottmar *et al.* (eds.), *IPCC Special Report on Renewable Energy Sources and Climate Change Mitigation* (Cambridge University Press, 2011).

Eggleston, Simon *et al.*, *2006 IPCC Guidelines for National Greenhouse Gas Inventories* (Institute for Global Environmental Strategies, 2006).

Ehlers, Eckart and Thomas Krafft (eds.), *Earth System Science in the Anthropocene* (Springer, 2006).

Faure, Michael G. and Roy A. Partain, *Carbon Capture and Storage: Efficient Legal Policies for Risk Governance and Compensation* (MIT Press, 2017).

Field, Christopher B. *et al.* (eds.), *Climate Change 2014: Impacts, Adaptation, and Vulnerability. Working Group II Contribution to the Fifth Assessment Report of the Intergovernmental Panel on Climate Change*, two vols. (Cambridge University Press, 2014).

Franck, Thomas M., *The Power of Legitimacy among Nations* (Oxford University Press, 1990).

Gardiner, Stephen, *A Perfect Moral Storm: The Ethical Tragedy of Climate Change* (Oxford University Press, 2011).

Gardiner, Stephen and David A. Weisbach, *Debating Climate Ethics* (Oxford University Press, 2016).

Gardiner, Stephen *et al.* (eds.), *Climate Ethics: Essential Readings* (Oxford University Press, 2010).

Gerrard, Michael B. and Tracy Hester (eds.), *Climate Engineering and the Law: Regulation and Liability for Solar Radiation Management and Carbon Dioxide Removal* (Cambridge University Press, 2018).

Guo, Rongxing, *How the Chinese Economy Works* (Palgrave Macmillan, 2017).

Guzman, Andrew T., *Overheated: The Human Cost of Climate Change* (Oxford University Press, 2013).

Haites, Erik (ed.), *International Climate Finance* (Routledge, 2013).

Hamilton, Clive, *Earthmasters: The Dawn of the Age of Climate Engineering* (Yale University Press, 2013).

Hampson, Fen Osler and Judith Reppy (eds.), *Earthly Goods: Environmental Change and Social Justice* (Cornell University Press, 1996).

Hartman, Chester and Gregory D. Squires (eds.), *There Is No Such Thing as a Natural Disaster: Race, Class, and Hurricane Katrina* (Routledge, 2006).

Hecht, Susanna and Alexander Cockburn, *The Fate of the Forest: Developers, Destroyers, and Defenders of the Amazon*, updated edn (University of Chicago Press, 2010).

Helm, Dieter and Cameron Hepburn (eds.), *The Economics and Politics of Climate Change* (Oxford University Press, 2011).

Henkin, Louis, *How Nations Behave*, 2nd edn (Columbia University Press, 1979).

Holder, Jane, *Environmental Assessment: The Regulation of Decision Making* (Oxford University Press, 2006).

Hollo, Erkki J., Kati Kulovesi and Michael Mehling (eds.), *Climate Change and the Law* (Springer, 2013).

Houghton, John T., G.J. Jenkins and J.J. Ephaums (eds.), *Climate Change: The IPCC Scientific Assessment. Report Prepared for Intergovernmental Panel on Climate Change by Working Group I* (Cambridge University Press, 1990).

Howarth, David, *The Shadow of the Dam* (Macmillan, 1961).

Hulme, Mike, *Why We Disagree about Climate Change* (Cambridge University Press, 2009).

Hume, David, *A Treatise of Human Nature* (Clarendon Press, 1896).

Humphreys, Stephen (ed.), *Human Rights and Climate Change* (Cambridge University Press, 2010).

Jayakumar, S. *et al.* (eds.), *Transboundary Pollution: Evolving Issues of International Law and Policy* (Edward Elgar, 2015).

Jiang, Xiaoyi, *Legal Issues for Implementing the Clean Development Mechanism in China* (Springer, 2013).

Kaime, Thoko, *International Climate Change Law and Policy: Cultural Legitimacy in Adaptation and Mitigation* (Routledge, 2014).

Karim, Saiful Md., *Prevention of Pollution of the Marine Environment from Vessels: The Potential and Limits of the International Maritime Organisation* (Springer, 2015).

Keith, David, *A Case for Climate Engineering* (MIT Press, 2013).

Keukeleire, Stephan and Tom Delreux, *The Foreign Policy of the European Union*, 2nd edn (Palgrave Macmillan, 2014).

Klein, Daniel *et al.* (eds.), *The Paris Climate Agreement: Analysis and Commentary* (Cambridge University Press, 2017).

Koh, Kheng-Lian *et al.* (eds.), *Adaptation to Climate Change* (World Scientific, 2015).

Kosolapova, Elena, *Interstate Liability for Climate Change-Related Damage* (Eleven, 2013).

Kreiser, Larry *et al.* (eds.), *Carbon Pricing: Design, Experiences and Issues* (Edward Elgar, 2015).

Kuokkanen, Thomas *et al.* (eds.), *International Environmental Law-Making and Diplomacy: Insights and Overviews* (Routledge, 2016).

Lillich, Richard B. and Burns H. Weston, *International Claims: Their Settlement by Lump Sum Agreements* (University Press of Virginia, 1975).

Lord, Richard *et al.* (eds.), *Climate Change Liability: Transnational Law and Practice* (Cambridge University Press, 2012).

Mayer, Benoit, *The Concept of Climate Migration: Advocacies and its Prospects* (Edward Elgar, 2016).

Metz, Bert *et al.* (eds.), *Methodological and Technological Issues in Technology Transfer: A Special Report of IPCC Working Group III* (Cambridge University Press, 2000).

Metz, Bert *et al.*, *IPCC Special Report on Carbon Dioxide Capture and Storage. Prepared by Working Group III of the Intergovernmental Panel on Climate Change* (Cambridge University Press, 2005).

Meyer, Lukas H. and Pranay Sanklecha (eds.), *Climate Justice and Historical Emissions* (Cambridge University Press, 2017).

Milanovic, Marko, *Extraterritorial Application of Human Rights Treaties: Law, Principles, and Policy* (Oxford University Press, 2011).

Miles, Edward L. *et al.*, *Environmental Regime Effectiveness: Confronting Theory with Evidence* (MIT Press, 2002).

Mintzer, Irving M. and J. Amber Leonard (eds.), *Negotiating Climate Change: The Inside Story of the Rio Convention* (Cambridge University Press, 1994).

Newhall, Christopher G. and Raymundo S. Punongbayan (eds.), *Fire and Mud: The Eruptions and Lahars of Mount Pinatubo, Philippines* (University of Washington Press, 1996).

Oberthür, Sebastian and Hermann E. Ott, *The Kyoto Protocol: International Climate Policy for the 21st Century* (Springer, 1999).

Okowa, Phoebe, *State Responsibility for Transboundary Air Pollution in International Law* (Oxford University Press, 2000).

Olawuyi, Damilola S., *The Human Rights-Based Approach to Carbon Finance* (Cambridge University Press, 2016).

Olson, Mancur, *The Logic of Collective Action* (Harvard University Press, 1965).

Pachauri, Rajendra K. *et al.*, *Climate Change 2014: Synthesis Report. Contribution of Working Groups I, II and III to the Fifth Assessment Report of the Intergovernmental Panel on Climate Change* (IPCC, 2015).

Park, Deok-Young (ed.), *Legal Issues on Climate Change and International Trade Law* (Springer, 2016).

Parson, Edward A., *Protecting the Ozone Layer: Science and Strategy* (Oxford University Press, 2003).

Peel, Jacqueline and Hari M. Osofsky, *Climate Change Litigation: Regulatory Pathways to Cleaner Energy* (Cambridge University Press, 2015).

Pichs-Madruga, Ramon *et al.* (eds.), *Climate Change 2014: Mitigation of Climate Change. Contribution of Working Group III to the Fifth Assessment Report of the Intergovernmental Panel on Climate Change* (Cambridge University Press, 2014).

Pogge, Thomas, *World Poverty and Human Rights: Cosmopolitan Responsibilities and Reforms* (Blackwell, 2002).

Posner, Eric A. and David A. Weisbach, *Climate Change Justice* (Princeton University Press, 2010).

Preston, Christopher (ed.), *Engineering the Climate: The Ethics of Solar Radiation Management* (Lexington, 2012).

Rajamani, Lavanya, *Differential Treatment in International Environmental Law* (Oxford University Press, 2006).

Rao, P. Chandrasekhara and P. Gautier (eds.), *The Rules of the International Tribunal for the Law of the Sea: A Commentary* (Martinus Nijhoff, 2006).

Rayfuse, Rosemary and Shirley V. Scott (eds.), *International Law in the Era of Climate Change* (Edward Elgar, 2012).

Richardson, Benjamin J. (ed.), *Local Climate Change Law: Environmental Regulation in Cities and Other Localities* (Edward Elgar, 2012).

Richardson, Benjamin J. *et al.* (eds.), *Climate Law and Developing Countries: Legal and Policy Challenges for the World Economy* (Edward Elgar, 2009).

Romera, Beatriz Martinez, *Regime Interaction and Climate Change: The Case of International Aviation and Maritime Transport* (Routledge, 2018).

Sabbioni, Cristina, Peter Brimblecombe and May Cassar, *The Atlas of Climate Change Impact on European Cultural Heritage: Scientific Analysis and Management Strategies* (Anthem Press, 2010).

Sands, Philippe and Jacqueline Peel, *Principles of International Environmental Law*, 3rd edn (Cambridge University Press, 2012).

Shearer, Christine, *Kivalina: A Climate Change Story* (Haymarket Books, 2011).

Shelton, Dinah (ed.), *The Oxford Handbook of International Human Rights Law* (Oxford University Press, 2013).

Solorio, Israel and Helge Jörgens (eds.), *A Guide to EU Renewable Energy Policy: Comparing Europeanization and Domestic Policy Change in EU Member States* (Edward Elgar, 2017).

Sornarajah, Muthucumaraswamy, *The International Law on Foreign Investment*, 3rd edn (Cambridge University Press, 2010).

Stern, Nicholas, *The Economics of Climate Change* (Cambridge University Press, 2007).

Stewart, Richard B. and Jonathan B. Wiener, *Reconstructing Climate Policy: Beyond Kyoto* (AEI Press, 2003).

Stocker, Thomas F. *et al.* (eds.), *Climate Change 2013: The Physical Science Basis. Contribution of Working Group I to the Fifth Assessment Report of the Intergovernmental Panel on Climate Change* (Cambridge University Press, 2013).

Tegart, W.J. McG., G.W. Sheldon and D.C. Griffiths (eds.), *Climate Change: The IPCC Impacts Assessment. Report Prepared for Intergovernmental Panel on Climate Change by Working Group II* (Australian Government, 1990).

Thorp, Teresa M., *Climate Justice: A Voice for the Future* (Palgrave Macmillan, 2014).

Tol, Richard S.J., *Climate Change: Economics Analysis of Climate, Climate Change and Climate Policy* (Edward Elgar, 2014).

Torney, Diarmuid, *European Climate Leadership in Question* (MIT Press, 2015).

Vachani, Sushil and Jawed Usmani (eds.), *Adaptation to Climate Change in Asia* (Edward Elgar, 2014).

Valdés, Alejandro Piera, *Greenhouse Gas Emissions from International Aviation: Legal and Policy Challenges* (Eleven, 2015).

Vattel, Emer de, *The Law of Nations*, Joseph Chitty trans. (Sweet, 1758).

Verheyen, Roda, *Climate Change Damage and International Law: Prevention Duties and State Responsibility* (Brill, 2005).

Verschuuren, Jonathan (ed.), *Research Handbook on Climate Change Adaptation Law* (Edward Elgar, 2013).

Vinke, Kira *et al.*, *A Region at Risk: The Human Dimensions of Climate Change in Asia and the Pacific* (Asian Development Bank, 2017).

Viñuales, Jorge E. (ed.), *The Rio Declaration on Environment and Development: A Commentary* (Oxford University Press, 2015).

Voigt, Christina, *Sustainable Development as a Principle of International Law: Resolving Conflicts between Climate Measures and WTO Law* (Martinus Nijhoff, 2009).

Voigt, Christina (ed.), *Research Handbook on REDD-Plus and International Law* (Edward Elgar, 2016).

Waton, Robert T. *et al.*, *Land Use, Land-Use Change and Forestry: A Special Report of the Intergovernmental Panel on Climate Change* (Cambridge University Press, 2000).

Weart, Spencer R., *The Discovery of Global Warming*, revised and expanded edn (Harvard University Press, 2008).

Weishaar, Stefan E., *Emissions Trading Design: A Critical Overview* (Edward Elgar, 2014).

Weston, Burns H. and David Bollier, *Green Governance: Ecological Survival, Human Rights, and the Law of the Commons* (Cambridge University Press, 2014).

Weston, Burns H., David J. Bederman and Richard B. Lillich, *International Claims: Their Settlement by Lump Sum Agreements, 1975–1995* (Martinus Nijhoff, 1999).

Wheatley, Steven, *The Democratic Legitimacy of International Law* (Hart Publishing, 2010).

Wolfrum, Rüdiger *et al.* (ed.), *Max Planck Encyclopaedia of Public International Law* (Oxford University Press, 2010).

World Commission on Dams, *Dams and Development: A Framework for Decision Making* (Earthscan, November 2000).

World Commission on Environment and Development, *Our Common Future* (Oxford University Press, 1987).

Wright, Christopher and Daniel Nyberg, *Climate Change, Capitalism, and Corporations: Processes of Creative Self-Destruction* (Cambridge University Press, 2015).

Wurzel, Rüdiger K.W. and James Connelly (eds.), *The European Union as a Leader in Climate Change Politics* (Routledge, 2011).

Wyatt, Tanya (ed.), *Detecting and Preventing Green Crimes* (Springer, 2016).

Yamin, Farhana and Joanna Depledge, *The International Climate Change Regime: A Guide to Rules, Institutions and Procedures* (Cambridge University Press, 2004).

Yan, Jinyue *et al.* (eds.), *Handbook of Clean Energy Systems* (Wiley, 2015).

Zahar, Alexander, *Climate Change Finance and International Law* (Routledge, 2017).

Zahar, Alexander, *International Climate Change Law and State Compliance* (Routledge, 2015).

JOURNAL ARTICLES

Abdel-Latif, Ahmed, "Intellectual property rights and the transfer of climate change technologies: issues, challenges, and way forward" (2015) 15:1 *Climate Policy* 103.

Adenle, Ademola A., Hossein Azadi and Joseph Arbiol, "Global assessment of technological innovation for climate change adaptation and mitigation in developing world" (2015) 161 *Journal of Environmental Management* 261.

Adger, Neil *et al.*, "Cultural dimensions of climate change impacts and adaptation" (2013) 3:2 *Nature Climate Change* 112.

Alex, Ken, "A period of consequences: global warming as public nuisance" (2007) 26 *Stanford Journal of Environmental Law* 77.

Alston, Lee J. and Krister Andersson, "Reducing greenhouse gas emissions by forest protection: the transaction costs of implementing REDD" (2011) 2:2 *Climate Law* 281.

Ames, James Barr, "Law and morals" (1908) 22 *Harvard Law Review* 97.

Amon, Robert, "Bioenergy carbon capture and storage in global climate policy: examining the issues" (2016) 10:4 *Carbon & Climate Law Review* 187.

Anderson, Kevin and Glen Peters, "The trouble with negative emissions" (2016) 354:6309 *Science* 182.

Andonova, Liliana B., Michele M. Betsill and Harriet Bulkeley, "Transnational climate governance" (2009) 9:2 *Global Environmental Politics* 52.

Asselt, Harro van, "The role of non-state actors in reviewing ambition, implementation, and compliance under the Paris Agreement" (2016) 6:1–2 *Climate Law* 91.

Babiker, Mustafa H., "Climate change policy, market structure, and carbon leakage" (2005) 65:2 *Journal of International Economics* 421.

Bäckstrand, Karin *et al.*, "Non-State actors in global climate governance: from Copenhagen to Paris and beyond" (2017) 26:4 *Environmental Politics* 1.

Bakhtiari, Fatemeh, "International cooperative initiatives and the United Nations Framework Convention on Climate Change" *Climate Policy* (forthcoming).

Bakker, Christine, "The Paris Agreement on climate change: balancing 'legal force' and 'geographical scope'" (2016) 25:1 *Italian Yearbook of International Law* 299.

Bauer, Anja, Judith Feichtinger and Reinhard Steurer, "The governance of climate change adaptation in 10 OECD countries: challenges and approaches" (2012) 14:3 *Journal of Environmental Policy & Planning* 279.

Beck, Stuart and Elizabeth Burleson, "Inside the system, outside the box: Palau's pursuit of climate justice and security at the United Nations" (2014) 3:1 *Transnational Environmental Law* 17.

Becker, Michael A., "Request for an advisory opinion submitted by the Sub-Regional Fisheries Commission (SRFC)" (2015) 109:4 *American Journal of International Law* 851.

Blaxekjær, Lau Øfjord and Tobias Dan Nielsen, "Mapping the narrative position of new political groups under the UNFCCC" (2015) 15:6 *Climate Policy* 751.

Bodansky, Daniel, "The legal character of the Paris Agreement" (2016) 25:2 *Review of European Comparative & International Environmental Law* 142.

Bodansky, Daniel, "The Paris Climate Change Agreement: a new hope?" (2016) 110:2 *American Journal of International Law* 288.

Bodansky, Daniel, "The role of the International Court of Justice in addressing climate change: some preliminary reflections" (2017) 49 *Arizona State Law Journal 689*.

Bodansky, Daniel, "The United Nations Framework Convention on Climate Change: a commentary" (1993) 18:2 *Yale Journal of International Law* 451.

Bodansky, Daniel *et al.*, "Facilitating linkage of climate policies through the Paris outcome" (2016) 16:8 *Climate Policy* 956.

Bodle, Ralph, "Geoengineering and international law: the search for common legal ground" (2010) 46:2 *Tulsa Law Review* 305.

Bogojević, Snaja, "Legalising environmental leadership: a comment on the CJEU's ruling in C-366/10 on the inclusion of aviation in the EU Emissions Trading Scheme" (2012) 24:2 *Journal of Environmental Law* 345.

Boisson de Chazournes, Laurence, "The Global Environment Facility (GEF): a unique and crucial institution" (2005) 14:3 *Review of European, Comparative & International Environmental Law* 193.

Boisson de Chazournes, Laurence, "One swallow does not a summer make, but might the Paris Agreement on Climate Change a better future create?" (2016) 27:2 *European Journal of International Law* 253.

Boot-Handford, Matthew E., "Carbon capture and storage update" (2014) 7 *Energy & Environmental Sciences* 130.

Bos, Astrid B. *et al.*, "Comparing methods for assessing the effectiveness of subnational REDD+ initiatives" (2017) 12:7 *Environmental Research Letters* 1.

Boute, Anatole, "The impossible transplant of the EU Emissions Trading Scheme: the challenge of energy market regulation" (2017) 6:1 *Transnational Environmental Law* 59.

Boute, Anatole, "Renewable energy federalism in Russia: regions as new actors for the promotion of clean energy" (2013) 25:2 *Journal of Environmental Law* 261.

Boyd, Emily, "Governing the Clean Development Mechanism: global rhetoric versus local realities in carbon sequestration projects" (2009) 41:10 *Environment & Planning A* 2380.

Boyle, Alan, "State responsibility and international liability for injurious consequences of acts not prohibited by international law: a necessary distinction?" (1990) 31:1 *International & Comparative Law Quarterly* 1.

Bremer, Nicolas, "Post-Environmental Impact Assessment monitoring of measures of activities with significant transboundary impact: an assessment of customary international law" (2017) 26:1 *Review of European, Comparative & International Environmental Law* 80.

Brent, Kerryn, Jeffrey McGee and Amy Maguire, "Does the 'no-harm' rule have a role in preventing transboundary harm and harm to the global atmospheric commons from geoengineering?" (2015) 5:1 *Climate Law* 35.

Briffa, K.R. *et al.*, "Influence of volcanic eruptions on Northern Hemisphere summer temperature over the past 600 years" (1998) 393:6684 *Nature* 450.

Brunnée, Jutta, "Of sense and sensibility: reflections on international liability regimes as tools for environmental protection" (2004) 53:2 *International & Comparative Law Quarterly* 351.

Brunnée, Jutta and Charlotte Streck, "The UNFCCC as a negotiation forum: towards common but more differentiated responsibilities" (2013) 13:5 *Climate Policy* 589.

Burkett, Maxine, "Loss and damage" (2014) 4:1–2 *Climate Law* 119.

Burkett, Maxine, "Reading between the red lines: loss and damage and the Paris outcome" (2016) 6:1–2 *Climate Law* 118.

Burns, William C.G., "Potential causes of action for climate change damages in international fora: the Law of the Sea Convention" (2006) 2:1 *McGill International Journal of Sustainable Development Law & Policy* 27.

Busby, Josha W., "Who cares about the weather? Climate change and U.S. national security" (2008) 17:3 *Security Studies* 468.

Cameron, Edward and Marc Limon, "Restoring the climate by realizing rights: the role of the international human rights system" (2012) 21:3 *Review of European Community & International Environmental Law* 204.

Caney, Simon, "Cosmopolitan justice, responsibility, and global climate change" (2005) 18:4 *Leiden Journal of International Law* 747.

Carlson, Ann E., "Regulatory capacity and state environmental leadership: California's climate policy" (2012–2013) 24:1 *Fordham Environmental Law Review* 63.

Carnwath, Lord Robert John Anderson, "Climate change adjudication after Paris: a reflection" (2016) 28:1 *Journal of Environmental Law* 5.

Chan, Sander *et al.*, "Reinvigorating international climate policy: a comprehensive framework for effective nonstate action" (2015) 6:4 *Global Policy* 466.

Charlesworth, Hilary, "International law: a discipline of crisis" (2002) 65:3 *Modern Law Review* 377.

Charlesworth, Hilary, "The unbearable lightness of customary international law" (1998) 92 *Proceedings of the Annual Meeting (American Society of International Law)* 44.

Chaturvedi, Ipshita, "The 'Carbon Tax Package': an appraisal of its efficiency in India's clean energy future" (2016) 10:4 *Carbon & Climate Law Review* 194.

Clark, Peter U. *et al.*, "Consequences of twenty-first-century policy for multi-millennial climate and sea-level change" (2016) 6:4 *Nature Climate Change* 360.

Coleman, James W., "Unilateral climate regulation" (2014) 38:1 *Harvard Environmental Law Review* 87.

Cooper, Charles F. and William C. Jolly, "Ecological effects of silver iodide and other weather modification agents: a review" (1970) 6:1 *Water Resources Research* 88.

Craik, Neil, "International EIA law and geoengineering: do emerging technologies require special rules?" (2015) 5:2–4 *Climate Law* 111.

Crawford, Neta, "Homo politicus and argument (nearly) all the way down: persuasion in politics" (2009) 7:1 *Perspectives on Politics* 103.

Crutzen, Paul J., "Albedo enhancement by stratospheric sulphur injections: a contribution to resolve a policy dilemma?" (2006) 77:221 *Climatic Change* 211.

Cullet, Philippe, "Differential treatment in environmental law: addressing critiques and conceptualizing the next steps" (2016) 5:2 *Transnational Environmental Law* 305.

Cullet, Philippe, "Differential treatment in international law: towards a new paradigm of inter-state relations" (1999) 10:3 *European Journal of International Law* 549.

Deleuil, Thomas and Tuula Honkonen, "Vertical, horizontal, concentric: the mechanics of differential treatment in the climate regime" (2015) 5:1 *Climate Law* 82.

Dernbach, John C. and Federico Cheever, "Sustainable development and its discontents" (2015) 4:2 *Transnational Environmental Law* 247.

Dixon, Tim *et al.*, "CCS projects as Kyoto Protocol CDM activities" (2013) 37 *Energy Procedia* 7596.

Doelle, Meinhard, "Climate change and the use of the dispute settlement regime of the Law of the Sea Convention" (2006) 37:3–4 *Ocean Development & International Law* 319.

Doelle, Meinhard, "Early experience with the Kyoto System: possible lessons from MEA compliance system design" (2010) 1:2 *Climate Law* 237.

Doelle, Meinhard, "The birth of the Warsaw Loss & Damage Mechanism: planting a seed to grow ambition?" (2014) 8:1 *Carbon & Climate Law Review* 35.

Doelle, Meinhard and Emily Lukaweski, "Carbon capture and storage in the CDM: finding its place among climate mitigation options?" (2012) 3:1 *Climate Law* 49.

Doelle, Meinhard, Steven Evans and Tony George Puthucherril, "The role of the UNFCCC regime in ensuring effective adaptation in developing countries: lessons from Bangladesh" (2014) 4:3–4 *Climate Law* 327.

Dreyfus, Magali, "Are cities a relevant scale of action to tackle climate change?" (2013) 7:4 *Carbon & Climate Law Review* 283.

Driesen, David M., "The limits of pricing carbon" (2014) 4:1–2 *Climate Law* 107.

Duyck, Sébastien, "MRV in the 2015 Climate Agreement: promoting compliance through transparency and the participation of NGOs" (2014) 9:3 *Carbon & Climate Law Review* 175.

Etzioni, Amitai, "Sovereignty as responsibility" (2006) 50:1 *Orbis* 71.

Fagbohun, Olanrewaju, "Cultural legitimacy of mitigation and adaptation to climate change: an analytical framework" (2011) 6:3 *Carbon & Climate Law Review* 308.

Farber, Daniel A., "Beyond the North-South dichotomy in international climate law: the distinctive adaptation responsibilities of the emerging economies" (2013) 22:1 *Review of European Community & International Environmental Law* 42.

Farber, Daniel A., "The challenge of climate change adaptation: learning from national planning efforts in Britain, China, and the USA" (2011) 23:3 *Journal of Environmental Law* 359.

Faure, Michael G. and André Nollkaemper, "International liability as an instrument to prevent and compensate for climate change" (2007) 43 *Stanford Journal of International Law* 123.

Fearnside, Philip M., "Tropical hydropower in the Clean Development Mechanism: Brazil's Santo Antônio Dam as an example of the need for change" (August 2015) 131:4 *Climatic Change* 575.

Figueres, Christiana, "Sectoral CDM: opening the CDM to the yet unrealized goal of sustainable development" (2006) 2:1 *McGill International Journal of Sustainable Development Law & Policy* 5.

Finnemore, Martha and Kathryn Sikkink, "International norm dynamics and political change" (1998) 52:4 *International Organization* 887.

Fitzmaurice, Sir Gerald, "The law and procedure of the International Court of Justice 1951–4: treaty interpretation and other treaty points" (1957) 33 *British Yearbook of International Law* 203.

Fleming, James R., "The pathological history of weather and climate modification: three cycles of promise and hype" (2006) 37:1 *Historical Studies in the Physical & Biological Sciences* 3.

Ford, J.D. *et al.*, "Adaptation tracking for a post-2015 climate agreement" (2015) 5:11 *Nature Climate Change* 967.

French, Duncan and Lavanya Rajamani, "Climate change and international environmental law: musings on a journey to somewhere" (2013) 25:3 *Journal of Environmental Law* 437.

Frenzen, Donald, "Weather modification: law and policy" (1971) 12:4 *Boston College Law Review* 503.

Gattini, Andrea, "Between splendid isolation and tentative imperialism: the EU's extension of its Emission Trading Scheme to international aviation and the ECJ's judgment in the ATA case" (2012) 61:4 *International & Comparative Law Quarterly* 977.

Geest, Kees van der and Koko Warner, "Loss and damage from climate change: emerging perspectives" (2015) 8:2 *International Journal of Global Warming* 133.

Ghaleigh, Navraj Singh and David Rossati, "The spectre of carbon border-adjustment measures" (2011) 2:1 *Climate Law* 63.

Ginzky, Harald and Robyn Frost, "Marine geo-engineering: legally binding regulation under the London Protocol" (2014) 9:2 *Carbon & Climate Law Review* 82.

Gomez-Echeverri, Luis, "The changing geopolitics of climate change finance" (2013) 13:5 *Climate Policy* 632.

Goodin, Robert E., "Selling environmental indulgences" (1994) 47:4 *KYKLOS International Review for Social Sciences* 573.

Graaf, K.J. de and J.H. Jans, "The Urgenda Decision: Netherlands liable for role in causing dangerous global climate change" (2015) 27:3 *Journal of Environmental Law* 517.

Grubb, M.J., D.G. Victor and C.W. Hope, "Pragmatics in the greenhouse" (1991) 354:6352 *Nature* 348.

Gruber, Stefan, "The impact of climate change on cultural heritage sites: environmental law and adaptation" (2011) 5:2 *Carbon & Climate Law Review* 209.

Gruber, Stefan, "Protecting China's cultural heritage sites in times of rapid change: current developments, practice and law" (2010) 10:3–4 *Asia Pacific Journal of Environmental Law* 253.

Gupta, Joyeeta, "The global environmental facility in its North-South context" (1995) 4:1 *Environmental Politics* 19.

Hale, Thomas, "'All hands on deck': the Paris Agreement and nonstate climate action" (2016) 16:3 *Global Environmental Politics* 12.

Hao, Zhang, "Designing the regulatory framework of an emissions trading programme in China: lessons from Tianjin" (2012) 6:2 *Carbon & Climate Law Review* 329.

Hardin, Garrett, "The tragedy of the commons" (1968) 162.3859 *Science* 1243.

Harris, Paul G. and Jonathan Symons, "Norm conflict in climate governance: greenhouse gas accounting and the problem of consumption" (2013) 13:1 *Global Environmental Politics* 9.

Harris, Paul G. and Taedong Lee, "Compliance with climate change agreements: the constraints of consumption" (2017) 17:6 *International Environmental Agreements* 779.

Hathaway, James, "A reconsideration of the underlying premise of refugee law" (1990) 31:1 *Harvard International Law Journal* 129.

He, Xiangbai, "Setting the legal enabling environment for adaptation mainstreaming into environmental management in China: applying key environmental law principles" (2014) 17 *Asia Pacific Journal of Environmental Law* 23.

Hedges, Andrew, "Carbon units as property: guidance from analogous common law cases" (2016) 13:3 *Carbon & Climate Law Review* 190.

Heindl, Peter, "The impact of administrative transaction costs in the EU emissions trading system" (2017) 17:3 *Climate Policy* 314.

Hertogen, An, "Sovereignty as decisional independence over domestic affairs: the dispute over aviation in the EU emissions trading system" (2012) 1:2 *Transnational Environmental Law* 281.

Heyvaert, Veerle, "What's in a name? The Covenant of Mayors as transnational environmental regulation" (2013) 22:1 *Review of European Community & International Environmental Law* 78.

Heyward, Clare, "A growing problem? Dealing with population increases in climate justice" (2012) 19:4 *Ethical Perspectives* 703.

Hilson, Chris, "It's all about climate change, stupid! Exploring the relationship between environmental law and climate law" (2013) 25:3 *Journal of Environmental Law* 359.

Hintermann, Beat, "Allowance price drivers in the first phase of the EU ETS" (2010) 59:1 *Journal of Environmental Economics & Management* 43.

Höhne, Niklas, "The Paris Agreement: resolving the inconsistency between global goals and national contributions" (2017) 17:1 *Climate Policy* 16.

Honadle, Beth Walter, "A capacity-building framework: a search for concept and purpose" (1981) 41:5 *Public Administration Review* 575.

Horstmann, Britta and Achala Chandani Abeysinghe, "The Adaptation Fund of the Kyoto Protocol: a model for financing adaptation to climate change?" (2011) 2:3 *Climate Law* 415.

Hulme, Mike, "Attributing weather extremes to 'climate change': a review" (2014) 38:4 *Progress in Physical Geography* 499.

Huq, Saleemul, Erin Roberts and Adrian Fenton, "Loss and damage" (2013) 3:11 *Nature Climate Change* 947.

Hurwitz, Margaret M. *et al.*, "Early action on HFCs mitigates future atmospheric change" (2016) 11:11 *Environment Research Letters* 114019.

Jalan, Jyotsna and Martin Ravallion, "Are the poor less well insured? Evidence on vulnerability to income risk in rural China" (1999) 58:1 *Journal of Development Economics* 61.

Jamieson, Dale, "Ethics and intentional climate change" (1996) 33:3 *Climatic Change* 323.

Johnson, Nicole, "*Native Village of Kivalina v. ExxonMobil Corp*: say goodbye to federal public nuisance claims for greenhouse gas emission" (2013) 40 *Ecology Law Quarterly* 557.

Johnson, Hope *et al.*, "Towards an international emissions trading scheme: legal specification of tradeable emissions entitlements" (2017) 34:1 *Environment & Planning Law Journal* 3.

Jones, Andy *et al.*, "The impact of abrupt suspension of solar radiation management (termination effect) in experiment G2 of the Geoengineering Model Intercomparison Project (GeoMIP)" (2013) 118:17 *Atmospheres: Journal of Geophysical Research* 9743.

Karlsson-Vinkhuyzen, Sylvia I. *et al.*, "Entry into force and then? The Paris Agreement and state accountability" *Climate Policy* (forthcoming).

Keohane, Robert O. and David G. Victor, "The regime complex for climate change" (2011) 9:1 *Perspectives on Politics* 7.

Kirschke, Stefanie *et al.*, "Three decades of global methane sources and sinks" (2013) 6:10 *Nature Geoscience* 813.

Klepper, Gernot and Sonja Peterson, "Trading hot-air: the influence of permit allocation rules, market power and the US withdrawal from the Kyoto Protocol" (2015) 32:2 *Environmental & Resource Economics* 205.

Kling, George W. *et al.*, "The 1986 Lake Nyos gas disaster in Cameroon, West Africa" (1987) 236:4798 *Science* 169.

Knox, John, "The myth and reality of transboundary environmental impact assessment" (2002) 96:2 *American Journal of International Law* 291.

Knutti, Reto *et al.*, "A scientific critique of the two-degree climate change target" (2016) 9 *Nature Geoscience* 13.

Koh, Harold Hongju, "Why do nations obey international law?" (1997) 106:8 *Yale Law Journal* 2599.

Kulovesi, Kati, "'Make your own special song, even if nobody else sings along': international aviation emissions and the EU Emissions Trading Scheme" (2011) 2:4 *Climate Law* 535.

Kysar, Douglas A., "Global environmental constitutionalism: getting there from here" (2012) 1:1 *Transnational Environmental Law* 83.

Lal, Rattan, "Carbon sequestration" (2008) 363:1492 *Philosophical Transactions of the Royal Society B: Biological Sciences* 815.

Leal-Arcas, Rafael, "Climate change mitigation from the bottom up: using preferential trade agreements to promote climate change mitigation" (2013) 88:2 *Carbon & Climate Law Review* 34.

Lees, Emma, "Responsibility and liability for climate loss and damage after Paris" (2017) 17:1 *Climate Policy* 59.

Lehman, Glen, "Environmental accounting: pollution permits or selling the environment" (1996) 7:6 *Critical Perspective on Accounting* 667.

Leiserowitz, Anthony, "Climate change risk perception and policy preferences: the role of affect, imagery, and values" (2006) 77:1–2 *Climatic Change* 45.

Lesnikowski, Alexandra *et al.*, "What does the Paris Agreement mean for adaptation?" (2017) 17:7 *Climate Policy* 825.

Li, Lei *et al.*, "A review of research progress on CO2 capture, storage, and utilization in Chinese Academy of Sciences" (2013) 108 *Fuel* 112.

Lin, Boqiang and Chuanwang Sun, "Evaluating carbon dioxide emissions in international trade of China" (2010) 38:1 *Energy Policy* 613.

Lin, Jolene, "Litigating climate change in Asia" (2014) 4:1–2 *Climate Law* 140.

Lin, Jolene, "The first successful climate negligence case: a comment on *Urgenda Foundation v. The State of the Netherlands (Ministry of Infrastructure and the Environment)*" (2015) 5:1 *Climate Law* 65.

Linnerooth-Bayer, Joanne, Reinhard Mechler and Stefan Hochrainer-Stigler, "Insurance against losses from natural disasters in developing countries: evidence, gaps and the way forward" (2014) 7 *International Journal of Disaster Risk Reduction* 154.

Lorenz, Susanne *et al.*, "Adaptation planning and the use of climate change projections in local government in England and Germany" (2017) 17:2 *Regional Environmental Change* 425.

Low, Kelvin F.K. and Jolene Lin, "Carbon credits as EU like it: property, immunity, tragiCO2medy? (2015) 27:3 *Journal of Environmental Law* 377.

Luterbacher, J. and C. Pfister, "The year without a summer" (2015) 0:4 *Nature Geoscience* 246.

Lyon, Thomas P. and John W. Maxwell, "Corporate social responsibility and the environment: a theoretical perspective" (2008) 2:2 *Review of Environmental Economics Policy* 240.

Lyster, Rosemary, "A fossil fuel-funded climate disaster response fund under the Warsaw International Mechanism for Loss and Damage Associated with Climate Change Impacts" (2015) 4:1 *Transnational Environmental Law* 125.

Mace, M.J., "Mitigation commitments under the Paris Agreement and the way forward" (2016) 6:1–2 *Climate Law* 21.

Magnan, Alexandre K. and Teresa Ribera, "Global adaptation after Paris" (2016) 352:6291 *Science* 1280.

Manea, Sabina, "Defining emissions entitlements in the Constitution of the EU Emissions Trading System" (2012) 1:2 *Transnational Environmental Law* 303.

Maosheng, Duan, "From carbon emissions trading pilots to national system: the road map for China" (2015) 9:3 *Carbon & Climate Law Review* 231.

Marques, Alexandra *et al.*, "Income-based environmental responsibility" (2012) 84 *Ecological Economics* 57.

Marzeion, Ben and Anders Levermann, "Loss of cultural world heritage and currently inhabited places to sea-level rise" (2014) 9 *Environmental Research Letters* 034001.

Mayer, Benoit, "The applicability of the principle of prevention to climate change: a response to Zahar" (2015) 5:1 *Climate Law* 1.

Mayer, Benoit, "*ATA v. Secretary of State for Energy*" (2012) 49:3 *Common Market Law Review* 1113.

Mayer, Benoit, "Climate change reparations and the law and practice of state responsibility" (2016) 7:1 *Asian Journal of International Law* 185.

Mayer, Benoit, "Construing international climate change law as a compliance regime" (2018) 7:1 *Transnational Environmental Law* 115.

Mayer, Benoit, "Human rights in the Paris Agreement" (2016) 6:1–2 *Climate Law* 109.

Mayer, Benoit, "International law obligations arising in relation to Nationally Determined Contributions" *Transnational Environmental Law* (forthcoming).

Mayer, Benoit, "Less-than-full reparation in international law" (2016) 56:3–4 *Indian Journal of International Law* 465.

Mayer, Benoit, "Migration in the UNFCCC Workstream on Loss and Damage: an assessment of alternative framings and conceivable responses" (2017) 6:1 *Transnational Environmental Law* 107.

Mayer, Benoit, "Obligations of conduct in the international law on climate change: a defence" *Review of European, Comparative and International Environmental Law* (forthcoming).

Mayer, Benoit, "State responsibility and climate change governance: a light through the storm" (2014) 13:3 *Chinese Journal of International Law* 539.

Mayer, Benoit, "Whose 'loss and damage'? Promoting the agency of beneficiary states" (2014) 4:3–4 *Climate Law* 267.

Mayer, Benoit and Mikko Rajavuori, "National fossil fuel companies and climate change mitigation under international law" (2016) 44:1 *Syracuse Journal of International Law & Commerce* 55.

Mayer, Benoit, Mikko Rajavuori and Fang Meng, "The contribution of state-owned enterprises to climate change mitigation in China" (2017) 7:2–3 *Climate Law* 97.

Mbatu, Richard S., "Domestic and international forest regime nexus in Cameroon: an assessment of the effectiveness of REDD + policy design strategy in the context of the climate change regime" (2015) 52 *Forest Policy & Economics* 46.

McDonald, Jan, "The role of law in adapting to climate change" (2011) 2:2 *Wiley Interdisciplinary Reviews: Climate Change* 283.

McMullen-Laird, Lydia *et al.*, "Air pollution governance as a driver of recent climate policies in China" (2015) 9:3 *Carbon & Climate Law Review* 243.

Mees, Heleen-Kydeke P. and Peter P.J. Driessen, "Adaptation to climate change in urban areas: climate-greening London, Rotterdam, and Toronto" (2011) 2:2 *Climate Law* 251.

Meyer, Lukas H., "Climate justice and historical emissions" (2010) 13:1 *Critical Review of International Social & Political Philosophy* 229.

Mintrom, Michael and Joannah Luetjens, "Policy entrepreneurs and problem framing: the case of climate change" *Environment and Planning C: Politics and Space* (2017) 35:5 *Environment and Planning C: Politics and Space* 1362.

Mooney, Erin, "The concept of internal displacement and the case for internally displaced persons as a category of concern" (2005) 24:3 *Refugee Survey Quarterly* 9.

Morrisette, Peter M., "The evolution of policy responses to stratospheric ozone depletion" (1989) 29:3 *Natural Resources Journal* 793.

Morseletto, Piero, Frank Biermann and Philipp Pattberg, "Governing by targets: Reduction ad unum and evolution of the two-degree climate target" *International Environmental Agreements: Politics, Law & Economics* (2017) 17:5 *International Environmental Agreements: Politics, Law & Economics* 655.

Naeem, S. *et al.*, "Get the science right when paying for nature's services" (2015) 347:6227 *Science* 1206.

Naess, Arne, "The shallow and the deep, long-range ecology movement: a summary" (1973) 16:1–4 *Inquiry (Oslo)* 95.

Nordhaus, William D., "A review of the Stern Review on the Economics of Climate Change" (2007) 45:3 *Journal of Economic Literature* 686.

Oberthür, Sebastian and Claire Roche Kelly, "EU leadership in international climate policy: achievements and challenges" (2008) 43:3 *International Spectator* 35.

Oberthür, Sebastian, "Global climate governance after Cancún: options for EU leadership" (2011) 46:1 *International Spectator* 5.

Ochieng, R.M. *et al.*, "Institutional effectiveness of REDD+ MRV: countries progress in implementing technical guidelines and good governance requirements" (2016) 61 *Environmental Science & Policy* 42.

Ockwell, David G. *et al.*, "Intellectual property rights and low carbon technology transfer: conflicting discourses of diffusion and development" (2010) 20:4 *Global Environmental Change* 729.

Okereke, Chukwumerije, "Climate justice and the international regime" (2010) 1:3 *Wires Climate Change* 462.

Oppenheimer, Michael and Annie Petsonk, "Article 2 of the UNFCCC: historical origins, recent interpretations" (2005) 73:3 *Climatic Change* 195.

Ostrom, Elinor *et al.*, "Revisiting the commons: local lessons, global challenges" (1999) 284:5412 *Science* 178.

Ott, Hermann E., Volfgang Sterk and Rie Watanabe, "The Bali roadmap: new horizons for global climate policy" (2008) 8:1 *Climate Policy* 91.

Page, Edward A., "Cashing in on climate change: political theory and global emissions trading" (2011) 14:2 *Critical Review of International Social & Political Philosophy* 259.

Pall, Pardeep *et al.*, "Anthropogenic greenhouse gas contribution to flood risk in England and Wales in autumn 2000" (2011) 470:7334 *Nature* 382.

Parmesan, Camille *et al.*, "Beyond climate change attribution in conservation and ecological research" (2013) 16:1 *Ecology Letters* 58.

Peeters, Marjan, "Case note: *Urgenda Foundation and 886 Individuals v. The State of the Netherlands*: the dilemma of more ambitious greenhouse gas reduction action by EU Member States" (2016) 25:1 *Review of European Community & International Environmental Law* 123.

Peeters, Marjan and Thomas Schomerus, "Modifying our society with law: the case of EU renewable energy law" (2014) 4:1–2 *Climate Law* 131.

Perry, Jim, "World heritage hot spots: a global model identifies the 16 natural heritage properties on the World Heritage List most at risk from climate change" (2011) 17:5 *International Journal of Heritage Studies* 426.

Peters, Glen P. *et al.*, "Growth in emission transfers via international trade from 1990 to 2008" (2011) 108:21 *Proceedings of the National Academy of Sciences* 8903.

Plant, Glan, "*ATA v. Secretary of State for Energy*" (2013) 107:1 *American Journal of International Law* 183.

Popp, David, "International technology transfer, climate change, and the Clean Development Mechanism" (2011) 5:1 *Review of Environmental Economics & Policy* 131.

Preston, Brian J. "Climate change litigation (part 1)" (2011) 5:1 *Carbon & Climate Law Review* 3.

Preston, Brian J. "Climate change litigation (part 2)" (2011) 5:2 *Carbon & Climate Law Review* 244.

Preston, Brian J., "The contribution of the courts in tackling climate change" (2016) 28:1 *Journal of Environmental Law* 11.

Psaraftis, Harilaos N., "Market-based measures for greenhouse gas emissions from ships: a review" (2012) 11:2 *WMU Journal of Maritime Affairs* 211.

Raftery, Adrian E. *et al.*, "Less than 2°C warming by 2100 unlikely" (2017) 7 *Nature Climate Change* 637.

Rai, Varun, Kanye Schultz and Erik Funkhouser, "International low carbon technology transfer: do intellectual property regimes matter?" (2014) 24 *Global Environmental Change* 60.

Rajamani, Lavanya, "The 2015 Paris Agreement: interplay between hard, soft and non-obligations" (2016) 28:2 *Journal of Environmental Law* 337.

Rajamani, Lavanya, "Ambition and differentiation in the 2015 Paris Agreement: interpretative possibilities and underlying politics" (2016) 65:2 *International & Comparative Law Quarterly* 493.

Rajamani, Lavanya, "Differentiation in the emerging climate regime" (2013) 14:1 *Theoretical Inquiries in Law* 151.

Reichwein, David *et al.*, "State responsibility for environmental harm from climate engineering" (2015) 5:2–4 *Climate Law* 142.

Reid, Hannah and Saleemul Huq, "Mainstreaming community-based adaptation into national and local planning" (2014) 6:4 *Climate & Development* 291.

Reins, Leonie *et al.*, "China's climate strategy and evolving energy mix: policies, strategies and challenges" (2015) 9:3 *Carbon & Climate Law Review* 256.

Ricke, Katharine L., M. Granger Morgan and Myles R. Allen, "Regional climate response to solar-radiation management" (2010) 3:8 *Nature Geoscience* 537.

Ringius, Lasse, Asbjørn Torvanger and Arild Underdal, "Burden sharing and fairness principles in international climate policy" (2002) 2:1 *International Environmental Agreements* 1.

Risse, Thomas, "'Let's argue!': communicative action in world politics" (2000) 54:1 *International Organization* 1.

Rogelj, Joeri *et al.*, "Paris Agreement climate proposals need a boost to keep warming well below 2°C" (2016) 534:7609 *Nature* 631.

Romera, Beatriz Martinez and Harro van Asselt, "The international regulation of aviation emissions: putting differential treatment into practice" (2015) 27:2 *Journal of Environmental Law* 259.

Rudolph, Sven and Toru Morotomi, "Acting local! An evaluation of the first compliance period of Tokyo's carbon market" (2016) 10:1 *Carbon & Climate Law Review* 75.

Ruhl, J.B., "Climate change adaptation and the structural transformation of environmental law" (2010) 40:2 *Environmental Law* 363.

Sand, Peter H. and Jonathan B. Wiener, "Towards a new international law of the atmosphere?" (2016) 7:2 *Göttingen Journal of International Law* 195.

Sand, Peter H., "Protecting the ozone layer: the Vienna Convention is adopted" (1985) 27:5 *Environment: Science & Policy for Sustainable Development* 18.

Sands, Philippe, "Climate change and the rule of law: adjudicating the future in international law" (2016) 28:1 *Journal of Environmental Law* 19.

Sands, Philippe, "The United Nations Framework Convention on Climate Change" (1992) 1:3 *Review of European Community & International Environmental Law* 270.

Sanna, A. *et al.*, "A review of mineral carbonation technologies to sequester CO2" (2014) 43 *Chemical Society Reviews* 8049.

Scaccia, Brian, "California's Renewable Energy Transmission Initiative as a model for state renewable resource development and transmission planning" (2012) 3:1 *Climate Law* 25.

Schipper, Lisa, "Conceptual history of adaptation in the UNFCCC process" (2006) 15:1 *Review of European, Comparative & International Environmental Law* 82.

Schneider, Lambert and Anja Kollmuss, "Perverse effects of carbon markets on HFC-23 and SF6 abatement projects in Russia" (2015) 5 *Nature Climate Change* 1061.

Scott, Joanne, "The geographical scope of the EU's climate responsibilities" (2015) 17 *Cambridge Yearbook of European Legal Studies* 92.

Scott, Joanne and Lavanya Rajamani, "EU climate change unilateralism" (2012) 23:2 *European Journal of International Law* 469.

Scott, Joanne et al., "The promise and limits of private standards in reducing greenhouse gas emissions from shipping" (2017) 29:2 *Journal of Environmental Law* 231.

Scott, Karen N., "International law in the Anthropocene: responding to the geoengineering challenge" (2013) 34:2 *Michigan Journal of International Law* 309.

Sépibus, Joëlle de, "Green Climate Fund: how attractive is it to donor countries?" (2015) 9:4 *Carbon & Climate Law Review* 298.

Seymour, F., "REDD reckoning: a review of research on a rapidly moving target" (2012) *Plant Sciences Reviews* 147.

Sharma, Anju, "Precaution and post-caution in the Paris Agreement: adaptation, loss and damage and finance" (2017) 17:1 *Climate Policy* 33.

Shen, Ying, "Crossing the river by groping for stones: China's pilot emissions trading schemes and the challenges for a national scheme" (2016) 18 *Asia Pacific Journal of Environmental Law* 1.

Shindell, D. et al., "A climate policy pathway for near- and long-term benefits" (2017) 356:6337 *Science* 493.

Shishlov, Igor, Romain Morel and Valentin Bellassen, "Compliance of the parties to the Kyoto Protocol in the first commitment period" (2016) 16:6 *Climate Policy* 770.

Shue, Henry, "Global environment and international inequality" (1999) 75:3 *International Affairs* 531.

Shue, Henry, "Subsistence emissions and luxury emissions" (1993) 15:1 *Law & Policy* 39.

Siri H, Eriksen, Andrea J. Nightingale and Hallie Eakin, "Reframing adaptation: the political nature of climate change adaptation" (2015) 35 *Global Environmental Change* 523.

Solomon, Susan, "Stratospheric ozone depletion: a review of concepts and history" (1999) 37:3 *Review of Geophysics* 275.

Solomon, Susan et al., "Emergence of healing in the Antarctic ozone layer" (2015) 353:6296 *Science* 269.

Spash, Clive L., "The brave new world of carbon trading" (2010) 15:2 *New Political Economy* 169.

Steffen, Will et al., "The Anthropocene: conceptual and historical perspectives" (2011) 369:1938 *Philosophical Transactions of the Royal Society, A: Mathematical, Physical & Engineering Sciences* 842.

Stephens, Jennie C., "Carbon capture and storage: a controversial climate mitigation approach" (2015) 50:1 *International Spectator: Italian Journal of International Affairs* 74.

Stone, Christopher D., "Common but differentiated responsibilities in international law" (2004) 98:2 *American Journal of International Law* 276.

Stott, Peter A. et al., "Attribution of extreme weather and climate-related events" (2016) 7:1 *WIREs Climate Change* 23.

Streck, Charlotte, "Ensuring new finance and real emission reduction: a critical review of additionality concept" (2011) 5:2 *Carbon & Climate Law Review* 158.

Streck, Charlotte and Moritz von Unger, "Creating, regulating and allocating rights to offset and pollute: carbon rights in practice" (2016) 10:3 *Carbon & Climate Law Review* 178.

Struggles, Jonathan, "Climate disasters and cities: the role of local government in increasing urban resilience" (2016) 18 *Asia Pacific Journal of Environmental Law* 91.

Sunstein, Cass R., "On the divergent American reactions to terrorism and climate change" (2007) 107:2 *Columbia Law Review* 503.

Surminski, Swenja and Dalioma Oramas-Dorta, "Flood insurance schemes and climate adaptation in developing countries" (2014) 7 *International Journal of Disaster Risk Reduction* 154.

Sykes, Alan O., "Economic 'necessity' in international law" (2015) 109:2 *American Journal of International Law* 296.

Szarka, Joseph, "From climate advocacy to public engagement: an exploration of the roles of environmental non-governmental organisations" (2013) 1:1 *Climate* 12.

Tamura, Kentaro, Takeshi Kuramochi and Jusen Asuka, "A process for making nationally-determined mitigation contributions more ambitious" (2013) 7:4 *Carbon & Climate Law Review* 231.

Tanaka, Yoshifumi, "Regulation of greenhouse gas emissions from international shipping and jurisdiction of states" (2016) 25:3 *Review of European Community & International Environmental Law* 333.

Tingzhen, Ming *et al.*, "Fighting global warming by climate engineering: is the Earth radiation management and the solar radiation management any option for fighting climate change?" (2014) 31 *Renewable & Sustainable Energy Reviews* 792.

Tollefson, Jeff, "Fossil-fuel divestment campaign hits resistance" (2015) 521:7550 *Nature* 16.

Tomuschat, Christian, "International law: ensuring the survival of mankind on the eve of a new century: general course on public international law" (1999) 281 *Collected Courses of the Hague Academy of International Law* 1.

Trip, James T.B., "The UNEP Montreal Protocol: industrialized and developing countries sharing the responsibility for protecting the stratospheric ozone layer" (1987–1988) 20:3 *New York University Journal of International Law & Politics* 733.

Tvinnereim, Endre and Michael Mehling, "Carbon pricing and the 1.5°C target: near-term decarbonisation and the importance of an instrument mix *Carbon and Climate Law Review* (forthcoming).

Urban, Mark C., "Accelerating extinction risk from climate change" (2015) 348:6234 *Science* 571.

Vaughan, Naomi E. and Timothy M. Lenton, "A review of climate geoengineering proposals" (2011) 109:3–4 *Climatic Change* 745.

Velders, Guus J.M. *et al.*, "The importance of the Montreal Protocol in protecting climate" (2007) 104:12 *Proceedings of the National Academy of Sciences of the United States of America* 4814.

Velders, Guus J.M. *et al.*, "Preserving Montreal Protocol climate benefits by limiting HFCs" (2012) 335:6071 *Science* 922.

Venuti, Stephanie, "REDD+ in Papua New Guinea and the protection of the REDD+ safeguard to ensure the full and effective participation of indigenous peoples and local communities" (2015) 17 *Asia Pacific Journal of Environmental Law* 131.

Verschuuren, Jonathan, "Towards a regulatory design for reducing emissions from agriculture: lessons from Australia's carbon farming initiative" (2014) 7:1 *Climate Law* 1.

Vihma, Antto, "Climate of consensus: managing decision making in the UN climate change negotiations" (2015) 24:1 *Review of European Community & International Environmental Law* 58.

Viñuales, Jorge E., "The rise and fall of sustainable development" (2013) 22:1 *Review of European Community & International Environmental Law* 3.

Voigt, Christina, "Equity in the 2015 Climate Agreement: lessons from differential treatment in multilateral environmental agreements" (2014) 4:1–2 *Climate Law* 50.

Voigt, Christina, "State responsibility for climate change damages" (2008) 77:1 *Nordic Journal of International Law* 1.

Voigt, Christina and Felipe Ferreira, "'Dynamic differentiation': the principles of CBDR-RC, progression and highest possible ambition in the Paris Agreement" (2016) 5:2 *Transnational Environmental Law* 285.

Wallimann-Helmer, Ivo, "Justice for climate loss and damage" (2015) 133:3 *Climatic Change* 469.

Wara, Michael, "Is the global carbon market working?" (2007) 445:7128 *Nature* 595.

Warren, R. *et al.*, "Quantifying the benefit of early climate change mitigation in avoiding biodiversity loss" (2013) 3:7 *Nature Climate Change* 678.

Weitzman, Martin L., "On modeling and interpreting the economics of catastrophic climate change" (2009) 91:1 *The Review of Economics & Statistics* 1.

Welch, Graham Donnelly, "HFC smuggling: preventing the illicit (and lucrative) sale of greenhouse gases" (2017) 44:2 *Boston College Environmental Affairs Law Review* 525.

Winkler, Harald and Lavanya Rajamani, "CBDR&RC in a regime applicable to all" (2014) 14:1 *Climate Policy* 102.

Winter, Gerd, "The climate is no commodity: taking stock of the Emissions Trading System" (2010) 22:1 *Journal of Environmental Law* 1.

Woolf, Dominic *et al.*, "Sustainable biochar to mitigate global climate change" (2010) 1:1 *Nature Communications* 56.

You, Ki-Jun, "Advisory opinions of the International Tribunal for the Law of the Sea: Article 138 of the rules of the tribunal, revisited" (2008) 39 *Ocean Development & International Law* 360.

Young, Margaret A., "Climate change law and regime interaction" (2011) 5:2 *Carbon & Climate Law Review* 147.

Young, Margaret A., "Trade measures to address environmental concerns in faraway places: jurisdictional issues" (2014) 23:3 *Review of European, Comparative & International Environmental Law* 302.

Zahar, Alexander, "A bottom-up compliance mechanism for the Paris Agreement" (2017) 1:1 *Chinese Journal of Environmental Law* 69.

Zahar, Alexander, "Mediated versus cumulative environmental damage and the International Law Association's legal principles on climate change" (2014) 4:3–4 *Climate Law* 217.

Zahar, Alexander, "Methodological issues in climate law" (2015) 5:1 *Climate Law* 25.

Zeben, Josephine van, "Establishing a governmental duty of care for climate change mitigation: will *Urgenda* turn the tide?" (2015) 4:2 *Transnational Environmental Law* 339.

Index

293